Nature Lost?

NATURE LOST?

Natural Science and the German Theological Traditions of the Nineteenth Century

Frederick Gregory

Harvard University Press
Cambridge, Massachusetts
London, England   1992

BL
245
.G74
1992

Copyright © 1992 by the President and Fellows of Harvard College
All rights reserved
Printed in the United States of America
10 9 8 7 6 5 4 3 2 1

This book is printed on acid-free paper, and its binding materials have been chosen for strength and durability.

Library of Congress Cataloging-in-Publication Data
Gregory, Frederick, 1942–
  Nature lost? : natural science and the German theological traditions of the nineteenth century / Frederick Gregory.
      p.   cm.
  Includes bibliographical references and index.
  ISBN 0-674-60483-0 (alk. paper)
  1. Religion and science—Germany—History—19th century.
2. Theology, Doctrinal—Germany—History—19th century. I. Title.
BL245.G74   1992
261.5'5'094309034—dc20

91-24804
     CIP

for Tricia,
who taught me how to dance

Acknowledgments

I SHOULD LIKE TO THANK numerous individuals who have played a part in the completion of this book. Portions of the manuscript were read by David Hollinger, David Lashmet, Ronald Numbers, and Claude Welch. I have benefited from their helpful criticisms and have been encouraged by their enthusiasm. To my colleague Eldon Turner at the University of Florida I owe a special debt of gratitude. Not only did he read the entire manuscript, but through many enjoyable conversations he also made clear to me the importance of viewing the various nineteenth century theological treatments of natural science as competing systems in the larger society of knowledge.

Elke Hahn of Humboldt University in Berlin went out of her way to procure biographical materials about Otto Zöckler for me. I am likewise indebted to the staff of the Württembergische Landesbibliothek in Stuttgart for their assistance in locating and obtaining elusive literary remains of Rudolf Schmid. I should also like to record my gratitude to William Woodward, whom I first met when our families both found themselves in Heidelberg during a research year abroad, and whose grasp of the history of nineteenth century German thought has repeatedly been an invaluable resource to me. Bill's instinctively generous nature has meant that he has always stood ready to help whenever I have turned to him.

The investigation of the life and work of Jakob Fries was made possible by a grant from the Alexander von Humboldt-Stiftung, whose support I am pleased to acknowledge. Portions of "Darwin and the German Theologians," pp. 269–278 in *World Views and Scientific Discipline Formation,* ed. W. R. Woodward and R. S. Cohen,

© 1991 Kluwer Academic Publishers, are contained in Chapters 3 and 4; reprinted by permission of Kluwer Academic Publishers. Through the Department of History at the University of Florida I was able to arrange for time to complete the final stage of writing. Without this arrangement the project would still be unfinished.

Finally, there are others whose assistance has been less direct but no less appreciated. However difficult it may be to assess precisely the stimulation and learning that comes through interaction with graduate students, the contribution of the future historians of science in my seminars has been undeniable. The same can be said about the forced involvement in this endeavor of my wife, Patricia, and my daughters, Kalée and Laura. They patiently tolerated my repeatedly drawing them into academic discussions about a subject they had not previously realized was so crucial.

Contents

Part I. The Historical Challenge of Religion and Science

1. Historiographical Approaches to German Religion and Science *3*
2. The Shape of German Protestant Theology in the Nineteenth Century *24*

Part II. Nature Retained

3. The New Hegelian Faith of David Friedrich Strauss *67*
4. Otto Zöckler, the Orthodox School, and the Problem of Creation *112*
5. Rudolf Schmid and the Reconciliation of Science and Religion *160*

Part III. Nature Lost

6. Wilhelm Herrmann's Encounter with the Theology of Albrecht Ritschl *201*
7. The Existential Critique of Science and Theology *231*

Epilogue: The Future Challenge of Religion and Science *261*

Bibliography *267*
Notes *281*
Index *333*

PART I

The Historical Challenge of Religion and Science

Where reason is concerned, the basis for the external world's susceptibility to the ideal stamp of our will lies in the purely transcendental identity of the ideal and the real.

Friedrich Schleiermacher

The truth is, we cannot get to the Eternal by raising natural knowledge to the Absolute, for therein we overextend ourselves and lose ourselves in absolute emptiness.

Jakob Friedrich Fries

1

Historiographical Approaches to German Religion and Science

IN A FORUM on university teaching at the University of Florida a distinguished physicist was asked a question by one of the nonscientists in his audience. "Why are physicists not viewed by most people as ordinary individuals like the rest of us?" Few who were present could have anticipated the reply. "Physics contains bad news," observed my scientific colleague. "The universe is not a very nice place for human beings because it is completely indifferent to their existence and their concerns. We physicists are the bearers of this bad news and we are held responsible for it."

The bad news about the physical world has in fact been circulating in Western society since at least the middle of the nineteenth century. If the above assessment of the present social perception of physicists has any merit at all, then one must conclude that at least a substantial portion of Western society is still very much perturbed by the bad news, so much so that it even colors the way in which science and scientists are regarded. Not everyone, apparently, is willing simply to concede that nature is so alien to humankind that one can no longer reasonably insist on uncovering the meaning residing *in nature* while articulating the meaning of human existence. In this sense, then, one can say that nature has retained its place as a vital source of purpose and value for some, while to others the indifference of nature forces humans to abandon all attempts to discover transcendent significance in the cosmos itself.

Ultimate questions about the meaning of nature and human existence are the crux of the relationship between natural science and religion. Occasionally we are reminded that this age-old question is still with us. Except for religiously conservative groups, however, the relationship between science and religion seems to have lost its urgency throughout most of the twentieth century. This was certainly not true in the nineteenth. Someone has said that the nineteenth century was the last one in which human beings *knew* what they were doing, the implication being that since then the status of knowledge itself has become problematical. Nowhere did this issue emerge with more force than in the fields of theology and natural science, both of whose practitioners saw themselves as possessing unique insights where matters of knowledge and truth were concerned.

This book attempts to describe how nineteenth century German theologians struggled with the prospect of losing nature from the theological enterprise. It chronicles among other things the variety of their responses to the new and forceful possibility of unbelief, which has been so eloquently described by James Turner.[1] It traces their differing attempts to retain confidence in the knowledge which they possessed in the face of factors gnawing away at its roots. It is an account of our fellow human beings' sometimes desperate attempts to articulate and thereby to justify to themselves the greater meaning of their lives in an age which, because it *knew* what it was doing, required such justification from its theologians.

The world of nineteenth century German theology, however, is a realm lost to all but a very few in our day. Some names are still recognizable—Schleiermacher's above all others. But for the most part German theology, like theology in general, represents an unexplored source of cultural history. While many historians have come to realize that an important vehicle for understanding past culture is available in the natural science of former times, including so-called wrong science, the "irrelevant" musings of professional theologians have for the most part been left to a handful of church historians.

But a great deal is missed if one gives in to the assumption that individuals whose lives were devoted to theological ideas can be of little use as vehicles for gaining insight into past societies. The theologians discussed in this study were spokesmen for large groups of people. Although it is certainly true that many people did not understand or did not care about the details of the theological or for that matter the scientific argumentation, the same cannot be said about the overall results defended by the theological representatives examined below.

We shall follow the various attempts of German theologians in the previous century to determine the relevance of nature and of natural science to their task as theologians. Our purpose is to make understandable the varying impact exerted on the theologian's craft by the growing authority of natural science over the course of the nineteenth century. How did those theologians who refused to acknowledge a loss of nature from their theological systems respond to the challenge facing them from the "bad news" of physics in the second half of the nineteenth century? And what did it mean that for some theologians nature was lost by the century's close?

A satisfactory account of the relationship between natural science and religion has never been a simple matter. How is one to integrate into a coherent account the very different claims made by natural scientists and theologians regarding the knowledge of nature? In the nineteenth century this was not a problem which concerned elites exclusively. Virtually anyone could appreciate that theologians in the late nineteenth century were having greater and greater difficulty retaining their authority in general, to say nothing of their ability to speak normatively about nature.

Much of twentieth century theology stands in marked contrast to earlier attempts by virtue of the uneasy truce it has declared between natural science and theology. Since roughly the turn of the century many theologians (and scientists) have gone about their business convinced that science and religion are radically separate enterprises. From this perspective the sometimes acrimonious charges of the post-Darwinian controversies represent hasty oversimplifications that were due to the emotional pitch of an unfortunate climate of warfare. In spite of occasional spectacles such as the Scopes Trial in the United States, there arose among numerous professional scientists the conviction that if one really understood the complexities of modern science, one realized that there was no genuine conflict with religion. Likewise, many theologians in this century frequently implied that if one really understood the motivation of religion, one knew why there could be no real clash with natural science.

Among those from earlier in the century who helped set this

tone are Germany's leading theologians. Rudolf Bultmann, Karl Barth, Paul Tillich, and others have proceeded in accordance with what Ueli Hasler has described as an atmosphere of the "amicable juxtaposition" *(das schiedlich-friedliche Nebeneinander)* of science and theology.[2] For them the significance of the truce meant that they could understand the subject matter of theology and natural science to be so qualitatively different that conflict between the two just could not occur. Put simply, scientific investigation, which was confined to the How, did not intersect with religion, which was preoccupied with the Why.

Theologians who endorsed the sharp separation of science and theology in our century could support with conviction the freedom of scientists to investigate nature however they wished. But the new-found mutual respect between scientists and theologians exacted a price from the latter. These theologians found that they were expected to relinquish a role which they had historically exercised. In this new existential theology there was very little concern with nature at all. Nature belonged to the natural scientists, it was now beyond the pale of theology, and theologians found themselves redirecting the goal of their efforts. For example, Karl Barth's understanding of the doctrine of revelation "emphasized an impassable epistemological gulf between revelation and the results of human inquiry of any sort, including science."[3] And in spite of other important differences, what the existentialist theologies of Bultmann and Tillich had in common was, in the words of George Hendry, "that they relate[d] the thought of God not to the world of nature, but to certain aspects of the experience of the self."[4]

But in his Warfield Lectures of 1978 Hendry surely overstated the case when he asserted that nature has virtually been dropped from the agenda of theology for the past two hundred years.[5] Had he suggested that the subject of nature has been fundamentally problematical for theology since the Enlightenment, he would have been far closer to the mark. Once the goal of natural theology, with its dependence on the argument from design, began to lose its persuasive power, the relevance of nature to theology was thrown into fundamental question.

Why did the argument from design eventually lose its appeal to some? With the advantage of hindsight it has become clear that the contribution natural scientists could provide to religion in the program of natural theology lost its effectiveness to the same degree that the mechanical explanation of nature's operation became their dominant concern. The appeal of the argument from design was that it confirmed what the believer already possessed prior to the inquiry into nature's How: that the most fundamental reality of all was describable in terms of person.[6] But the personal dimension had to be added to the blind machine-like operation of nature that was discovered by the scientist. If it was not, a different answer regarding the most fundamental reality emerged. The decisive question was: "Is the Person or is matter in motion the ultimate metaphysical category?"[7] The choice was dependent not on logic but on an act of faith.

This issue lay dormant within natural theology for some time, surfacing only gradually. In England the pursuit of natural theology continued unabated from the seventeenth century until well into the nineteenth. But as natural scientists more and more willingly restricted their task to the discovery of how nature operated, they resorted less and less to nonmechanical categories in their explanations. Even so nonmechanical a notion as Newton's attractive force, which for the great English natural philosopher clearly possessed spiritual significance, was stripped of all personal dimension by Newton's successors. Because the action of force had been subjected by Newton to a regularity describable by mathematical law, attractive force was incorporated into a view of the cosmos that was impersonal and mechanistic.

It was not until the nineteenth century, however, that the fundamental cleavage between the indifferent cosmos of the scientist and the value-laden world of the personal self began painfully and gradually to emerge to clarity. To the extent that the debates between scientists and theologians were understood to be about this issue, the views of theologians were extremely relevant to ordinary members of society. Some refused to concede that there was an unbridgeable chasm between the realities with which the scientist and the theologian grappled. But others boldly declared the irrelevance of nature for questions of human value and attempted to forge on without what John Dillenberger has called the age-old "vertical and depth dimension between man and the cosmos."[8]

Nor was theology the only intellectual context in which the crisis came to the fore. Some philosophers expressed discontent with the manner in which traditional philosophy assessed the dilemma facing Europeans. But the proto-existentialist laments of Friedrich Nietzsche and Søren Kierkegaard were largely incomprehensible to the great majority of their contemporaries, including the so-called educated and reading public. Better grasped was the discontent of the neo-Kantian philosophers, whose attempt to provide a critique of scientific materialism after mid-century was at least appreciated by many even if it remained largely ineffective.

When the question was the meaning of nature for theology, however, the issue could be stated in a manner that confronted everyone. One may have no opinion about the place of nature in the grand philosophical systems of philosophers like Kant, Fichte, or Hegel, to say nothing of neo-Kantian philosophers such as Otto Liebmann or Friedrich Lange. And one may not realize that Nietzsche and Kierkegaard implied much of anything about nature at all. But with theologians it was different. Admittedly some of their efforts were not much easier to comprehend than those of philosophers. But three of the figures highlighted in this volume intended their works for the literate public and therefore stood in close proximity to it. Invariably theologians did get around to addressing concrete issues that gave sharp focus to their contentions. Whereas philosophers did not have to state their position on traditional doctrinal matters that involved nature, theologians did. What they wrote could directly affect what was said from the pulpit in Sunday morning sermons. That defined at least one place where their views could be called to account by virtually anyone.

Although the historical relationship between science and religion in the nineteenth century has been investigated many times, the story frequently focuses on scientists. In this book attention is drawn specifically to theologians and the distinct segments of society to whom and for whom they spoke. If little attention has been paid by historians of science to theologians as representatives of social groups, even less has been afforded the German theologians of the nineteenth century. Past studies of science and religion in the nineteenth century invariably have dealt with the English or American scene. But in the course of carrying out my investigation I have identified four fundamentally different positions to which German theologians came, each of which resonated with different groups of people in the society at large. Each tradition developed its own separate answer to the question of the significance of natural science for theology, and each claimed the allegiance of many Germans.

It has generally been assumed that German theologians were largely uninterested in natural science in the nineteenth century. And yet in Germany more than elsewhere the foundations of the theological research on which the twentieth century would rely were established in virtually all theological disciplines except apologetics. We should not be surprised, then, to find that in fact questions about the role of nature and the knowledge of nature form important points of entry into German theology, and that an acquaintance with this development provides a novel and much needed insight into the changes in German culture between 1800 and 1900.

German Protestant theology in the nineteenth century began its development against the background of profound change. As the century drew to a close German social, political, and economic life was still decidedly feudal. During the tumultuous years of the French Revolution and the ensuing Napoleonic period there was a great deal to which Germans had to become accustomed. In the midst of the political upheaval of the time Napoleon redrew the map of Europe and humiliated Germans by exposing the political impotency of the loosely knit amalgam of many small states that defined Germany. All around there were calls for reform. The abolition of serfdom in Prussia in 1807 was just one of a series of initiatives intended to raise the German-speaking states to a position of greater power and influence in a Europe experiencing rapid social and economic change.

In the intellectual realm there was also profound change; indeed, it had been underway well before Napoleon came upon the scene. By the dawn of the nineteenth century the rise to prominence of an ideology of *Wissenschaft*,[9] while still underway, was nearly complete. If anything, Germany was producing the most innovative thinkers in all of Western civilization. The challenge to Germans in the nineteenth century was to relate their accomplishments in the intellectual understanding of the world to the rapid changes that continued to confront them.

This challenge was particularly acute in theology, where an

older feudal pattern had been under attack for some time. In feudal theology nature was assumed to be totally contingent on God's will. Arthur Lovejoy long ago pointed out that nature for the feudal mind had an instrumental but not an autonomous value; that is, nature could not take on the attributes of necessity and immutability without challenging divine prerogatives.[10] But developments in the natural sciences, especially during the Scientific Revolution of the seventeenth century, brought with them a willingness to view nature as possessing ever more autonomy. This trend continued unabated throughout the Enlightenment, and the methods employed in the investigation of nature spread in Germany beyond the boundaries of the natural sciences into the other *Wissenschaften*.[11]

In the next chapter I will describe four distinct responses to these changes within the German theological community. In three of the reactions theologians attempted to meet the challenge before them through a creative application of the basic assumptions of normal theological science as it had been practiced in the nineteenth century. At least one response, however, contained assumptions contrary to those of traditional theology. These two approaches to the theological task in the rapidly modernizing world of late nineteenth century Germany are dealt with below in Parts II and III respectively.

For theologians who did not or could not question the traditional goal of theological explanation, the way to deal with the increasing body of scientific truth in the nineteenth century remained what it had been before: to demonstrate how a more complete knowledge of nature was compatible with proper theological understanding. There were, to be sure, major disagreements about what constituted proper theological understanding. Because most studies about natural science and religion in the nineteenth century have focused on these disagreements, one gets the impression that these debates exhausted the spectrum of theological response to developments in natural science.

The debates among these theologians are of course important; indeed, a discussion of how they played out in Germany forms a major portion of this book. But if the value of this part of my study is to bring to the attention of English-speaking readers the largely unknown responses to the growing influence of natural science among German theologians who were comfortable in the world of nineteenth century assumptions, a different service is rendered where the far less familiar fourth theological school is concerned. Here it will be necessary not only to explain what was concluded about natural science but also to indicate what it was about this position that marks it as extraordinary as opposed to normal theological science.

For theologians in the fourth group nothing less than a transformation of the theologian's craft was capable of handling the apparent implications of modern science. With the benefit of hindsight we can recognize in this transformation a change in perspective similar to that engendered elsewhere in European society as it moved from the nineteenth to the twentieth century. Other expressions of "the shaking of the foundations"[12] may of course be found in the politics, literature, philosophy, and art of the period. An additional manifestation of the profound challenge to the older, more comfortable perspective of earlier times occurred around the turn of the century in the revolutionary developments within the physical and biological sciences themselves.

The usual techniques employed to meet the challenges encountered in natural science and in theology had evidently failed in the minds of some as the century drew to a close. In theology the transformation entailed, as it did in natural science during the same period, the surrender of a confident nineteenth century worldview in which it had not been possible to doubt the ultimate triumph of human reason in the on-going search for truth. More specifically, the transformation involved the removal of nature from the purview of theology. Theologians willingly acknowledged that nature's truth and God's truth were not identical.

Because this fourth theological option represents *terra incognita* for most readers in the late twentieth century, an extensive introduction to it is called for. Marxist thought has given substantial attention to this transformation within theology in the late nineteenth century, attempting to locate what transpired there within the larger historical epoch of the modern era. This approach typically involves a one-dimensional analysis which is driven by a positivistic understanding of natural science. Yet it is almost alone in its recognition of the existence and potential significance of the rise of de-natured theology.

Ueli Hasler, who has recently addressed the subject, begins by registering his dissatisfaction with several contemporary discussions of the relationship between natural science and religion. He finds little to assist him in Hans Blumenberg's well-known treatment of secularization, in Karl Heim's application of twentieth century physics to theology, in the speculative musings of Teilhard de Chardin, and, finally, in Wolfart Pannenberg's theology of nature.[13] He also consults but abandons attempts to define and explain developments such as the materialization of consciousness, the alienation of nature, and the emergence of an ecological crisis, all of which appear at first glance to hold promise for understanding the abandonment of nature by theology. But the Frankfurt school's treatment of the materialization of social and cultural life based on the notion of "instrumental reason," Christian Link's radical break with the leading premises of modern epistemology as a response to questions raised by "ecological theology," and even Georg Lukács's derivation of the structure of materialized consciousness from the reduction of natural objects and of labor to their exchange value are all seen by Hasler to rely on an objectification of idealistic philosophy and not on the social processes responsible for consciousness formation.[14]

What Lukács does make clear, according to Hasler, is that the relationship between religion and the ecological crisis is not to be understood as Lynn White presents it, namely, as a result of early Christian attitudes about the meaning and place of nature. Hasler sets out to demonstrate that, due to the middle class conception of nature in the nineteenth century, theology, by which he means only what I have identified as the fourth theological tradition of the century, retreated to "an ultimately de-natured and nature-annihilating concept of reality." As a result theology since the end of the century has restricted its interpretation of reality to the self-purposeful subject, hence it has been unable to register an explicit theological protest against the materialization and exploitation of nature.[15]

Hasler wishes, then, to claim that the elimination of the concept of nature from Christian theology is to be understood "as a reaction to the development of the middle class conception of nature between the French Revolution and World War I."[16] In this class-based analysis Hasler leans heavily on Lukács's *Destruction of Reason*, in which Lukács investigated how and to what extent bourgeois philosophy and sociology was embedded in the struggle of the middle class for economic, social, political, and cultural domination over reactionaries on the one side and the workers' movement on the other. He also draws on Dieter Schellong's class study of nineteenth century theology, which, while it does not include nature, focuses on bourgeois understandings of self and the world.[17]

Inherent in the development of capitalism, Hasler argues, lay a challenge to the role of nature in theology that previously had been absent. When John Locke viewed society as something necessary, lawful, and unchangeable, he saw nature as that which was itself determined and which in turn determined social activity. Nature no longer constituted a theater of caprice within which one must powerlessly surrender. That was the feudal concept of nature. Rather, nature had become an arena of determinism on whose regularity one could rely. Locke consequently portrayed feudalism and absolutism as unnatural states.[18] The relation between what Franz Unger calls "normative nature" and "real nature" was inverted from what it had been in earlier times. Whereas in feudalism what Hasler calls "the real in nature" was subject to what was normative from theology, now the normative came from the scientific investigation of real nature.[19]

Whether or not one wishes to explain the impact of natural science on the new theology in such terms, after the seventeenth century the growing importance of empirical investigation in determining the constitution of the cosmos was undeniable. The implications of this change of perspective for theology took, as already noted, quite some time to emerge.

Hasler argues that the autonomy of nature that was produced by developing capitalism proved to be a stumbling block to Christian theology, and that the inevitable response of theologians was to remove nature from theology altogether. The retreat to a denatured theology was therefore the predictable outcome of the shift from feudal to capitalistic society. Hasler asserts that the attack on the religiously legitimized feudal worldview could be successful only if nature was seen to be a world that was lawful to itself, a world that existed by itself, a world that was bereft of God *(entgöttert)*, in sum, a world of things and events that were empirically verifiable.[20] As such, nature stood at humankind's disposal as a medium in which bourgeois humanity became conscious of itself not only as an autonomous subject in opposition to the transcendent authority inherent in the feudal religious worldview, but also as a medium subject to human domination through reason. The bourgeoisie used nature to legitimize its usurpation of religious authority. This Hasler sees to be "the neuralgic point in the relationship between the Christian and middle class view of the world."[21]

What are we to make of all of this? One has the impression here, as one has frequently with Marxist treatments, that too much has been explained. Hasler himself grants that the notion of a middle class conception of nature is fraught with difficulties. And when he argues, not unreasonably, that the apologetic intent lying behind the middle class appeal to nature at the time of the French Revolution could not be expected to be the same as it would be in 1848 or in the Age of Imperialism at century's end,[22] we realize that a good deal of room for maneuvering has been prepared. We should not be surprised, then, when the most diverse expressions of the middle class conception of nature emerge.

Not only does the empiricism on which Locke's endorsement of nature's autonomy rests exhibit the new middle class conception, so too does Fichte's idealism. According to Hasler, Fichte destroyed feudal nature by erecting the mind as the new authority. Hasler cites Peter Ruben to say that "Fichte's hatred of nature is the hatred of the serf against feudally dominated nature."[23] Fichte destroyed feudal nature by interpreting all nature as something that did not exist by itself but was the product of the spontaneously active *Ich*. Fichte's interpretation removed nature from dependence on God and shifted the Deity's traditional role to human beings, thereby further emancipating humanity from transcendent religious authority.

Although Hasler's selection of Fichte in this context is misleading,[24] his examination of nineteenth century theology in light of the middle class conception of nature uncovers much that is useful. But he also unnaturally joins thinkers whose treatment of nature is virtually opposite in emphasis. Schleiermacher, whose theology, Hasler shows, incorporated nature into its very core, allegedly reflects a middle class conception no less than does Wilhelm Herrmann, who, according to Hasler, "annihilates" nature in his later theological work. To see both Schleiermacher's theology of nature and Herrmann's theology against nature as representative of the same middle class conception contributes to the impression that Hasler's position is indeed nonfalsifiable.

Hasler's rejoinder to such criticism lies in his focus on the perceived need after 1848 to hold together both a growing industry and a reactionary politics. The old Lockean bourgeois conception of nature, which sought to legitimize the middle class through a nature that determined social activity, may have worked well during and immediately after the French Revolution. But, he argues, at midcentury this mutual reinforcement between nature and society allegedly broke down. The "natural" had to be overcome, exploited, and made to serve. Neo-Kantian thought, including neo-Kantian theology, emerged as a justification of the autonomy of the social-political (moral) realm over against nature.[25] So according to Hasler, Schleiermacher's theology at the beginning of the century, and Ritschl's and Herrmann's at the end, both represented what he has called in the subtitle of his book "the accommodation of theology to the bourgeois conception of nature."

As interesting as this analysis may be, it is not without difficulties. One should expect that de-natured theology would emerge first and most clearly not in Germany but in England, where increased industrial production developed earliest. But the English theological tradition reveals the opposite tendency. Nowhere is nature more important to theologians than among the English. From Boyle forward the tradition of natural theology and the argument from design were refined among the English more than in any other land,[26] and this tradition continued unabated throughout the political and economic changes that accompanied the English industrial revolution. Nor can one dismiss the absence of a de-natured theology because of the relative lack of reactionary politics. Where is one to find the French de-natured theological school to accompany the ever increasing industrialization of the Second Empire? Apparently no easy association between the middle class conception of nature and the emergence of de-natured theology is possible.

The historian can avoid the difficulties encountered in Hasler's class analysis. If one's account of what happened to German theology in the nineteenth century through its interaction with natural science makes no claim to be a causal explanation, which, of course, the Marxist approach does, then one is free to organize the story in any number of ways. Hasler explicitly denies that the meaning of theology is to be explained "either from the development of internal theological debates or by reference to general historical intellectual tendencies."[27] Although I concur that such treatments—if carried out in isolation from the cultural context in which the intellectual developments occurred—remain empty and unconvincing, I vehemently deny that an intellectual history of theology cannot contribute to our understanding of German culture in the nineteenth century.

What kind of intellectual history is pursued in this book? Above all it is a history based on the assumption that in the society of knowledge,[28] which is made up of competing intellectual descriptions of and responses to the forces shaping human life, no one articulation enjoys privileged status as the proper reflection of human concerns. In this study any coherent intellectual outlook that expresses the convictions and serves the needs of a significant segment of society becomes a component of the mosaic of views that constitute the intellectual culture of an age.

In the case before us, which has as its focus the theological response to natural science in nineteenth century Germany, we must be willing to explore all theological systems that commanded the allegiance of various social groups. We cannot, for example, exclude what might appear to some to be theological responses on the fringe if these positions proved meaningful to substantial numbers of people. My disagreement with Hasler's exclusive focus on the development of de-natured theology is therefore fundamental.

A willingness to accept different theological perspectives as important sources of insight into German society in the nineteenth century does not, however, mean that one presumes to have provided an objective depiction of past intellectual culture. Any schema which attempts to make sense of the confusion surrounding issues of natural science and religion in the second half of the nineteenth century will necessarily be subjective, often reflecting the race and sex of the author, and invariably disclosing elements of his or her own cultural stance. Another scholar's categorization of nineteenth century German theological opinion about nature might well differ from that presented here. My goal is simply to depict the theological response to natural science in a manner that coheres with and informs the understanding of nineteenth century German culture we have from other sources.

How, then, might one approach the complex spectrum of opinion about religion and science that greets the historian of nineteenth century thought? My divisions of the German theological community will be made clear in the next chapter. Here I should like to introduce a preliminary consideration that has informed my study. I have chosen to be guided by a concern which in my view was commonly shared by those living in the nineteenth century. It has to do with a conception traditionally dear to the hearts of both natural scientists and theologians, a notion seen to lie at the heart of both science and theology—the conception of truth. For most people in the nineteenth century the question of truth was the key issue in the relationship between science and religion. It has been useful, therefore, to inquire what the understanding of truth for nineteenth century theologians was, how it changed over the course of the century, and what appeal the various responses to the question of truth enjoyed among diverse societies of knowledge.

It should be stated at the outset that in appealing to different conceptions of truth as a means of shedding light on the developing relationship between science and religion in the nineteenth century I am fully aware that the question of truth has become uninteresting to some contemporary historians and philosophers of science. Among philosophers Karl Popper decides it is better to stop referring to truth; Nelson Goodman holds that it is largely misleading to the task of what he calls worldmaking; Imre Lakatos, if Ian Hacking is right, tried to find a substitute for it; and Hacking himself doubts that the question is important.[29] As for historians, the strong interest in social history among many historians of science has moved the focus of the historical investigation of science away from the content of scientific theory. The question whether a particular scientific theory was regarded as true or false by some past scientific community holds little appeal. The science of the past is of interest to these historians *only* as a reflection of the culture. Many now prefer to inquire about what the language and categories employed by scientists of former times betray about the relations of power and domination among members of the culture.

My interest in the concept of truth is motivated by the conviction that it is instructive to frame a historical investigation in terms important to the subjects of the study. Those in the nineteenth century who were concerned about the relationship between religion and natural science assumed that the question of truth was at stake. Whether or not historians and philosophers in our day find the question intriguing, it *was* part of the mental universe of nineteenth century Germans; indeed, for most of the century there was remarkable agreement among diverse segments of German society about how one should think about the nature of truth. As the century drew to a close, however, this consensus could no longer sustain itself, and a competing understanding of the nature of truth made itself heard.

However Germans a century ago assessed the relationship of natural science and religion, references to truth and to the search for truth abounded. To ignore this dimension of the issue because talk of truth has become problematical or uninteresting among scholars today might cause us to overlook a useful approach to a complicated historical problem. Becoming familiar with alternative conceptions of the nature of truth has been extremely helpful to me as I have attempted to sort out the theological responses to natural science in the late nineteenth century. My hope, of course, is that by bringing the question of truth to bear on theologians in the nineteenth century I might better understand how and why the different positions were meaningful to diverse individuals.

Although the interests of the historian are different from those of the philosopher, I have benefited enormously from the treatment of the question of truth that philosophers have provided.[30] My investigation has convinced me more than ever that historians and philosophers of science have much to offer each other.

The oldest understanding of the nature of truth is that given in the so-called correspondence theory of truth.[31] Here, as the name suggests, truth consists of some form of correspondence between the worlds of thought and things, or, as Bertrand Russell put it, between "belief and fact."[32] In this theory nature is conceived to be completely independent of the mind that knows it. To establish the truth of a claim about nature, one must exhibit the match between the rational categories in terms of which the assertion is made and the apprehension of nature given in experience. In the classical form of the correspondence theory, which perhaps owes its most common understanding to Aristotle,[33] the world is presumed to be rational, although the demonstration of the rationality may be extremely difficult to provide. The truth of nature, however, is presupposed to be accessible to reason, at least in principle. As a student at Oxford the young Isaiah Berlin was initially captivated by this perspective. Later he described it as a Platonic ideal which took its cue from science and which was founded on a correct understanding of the rules that governed the universe. This ideal implied

> that, as in the sciences, all genuine questions must have one true answer and one only, all the rest being necessarily errors; in the second place, that there must be a dependable path towards the discovery of these truths; in the third place, that the true answers, when found, must necessarily be compatible with one another and form a single whole, for one truth cannot be incompatible with another.[34]

When nature's truth is conceived in this manner, one understands that it is independent of the wishes or intentions of the mind that knows nature. Nature itself supplies a constraining influence over human imagination. "The natural scientist does not give in to the belief that he has created the law," wrote the Dutch physiologist Jacob Moleschott in 1867. "He feels in his innermost being that the facts impose it on him."[35] These sentiments could equally well have been voiced by any number of nineteenth century German theologians, who were convinced beyond doubt that they had not created the truth which they revered, but were constrained to believe in it bècause it imposed itself on them.

The correspondence theory of truth, as it was commonly exhibited in the nineteenth century, rested on the requirement that what is properly conceived by the mind is ontologically real. As both Popper and Hilary Putnam have said, it is a theory for the realist, since it allows one to speak of a reality different from the theory.[36] It involved, in other words, a metaphysical assertion. Truth is determined by what really exists. Our knowledge of nature, when it is correct, is the same as the truth of nature.

As already indicated, this perception of the nature of truth was certainly not new with the nineteenth century. It had been present in Western thought in one form or another since at least the time of the Greeks. Given the emphasis on the role of empirical verification that appeared in the Scientific Revolution of the seventeenth century, it expressed well what most scientists and theologians since that time understood the scientist's task to be. The stress was on the discovery of the unknown, conformity to nature's authority, the determining role of fact, and reason's incredible power to copy nature.[37]

For the German theologians the most important challenge to the relation of correspondence in the determination of truth emerged in conjunction with the thought of Immanuel Kant.[38] Precisely because metaphysical claims about nature's truth lay beyond the realm of knowledge, a Kantian could not measure truth by professing to establish a correspondence between thought and things in the manner described. The laws governing all experience in space and time, what Kant calls laws "involved in a nature in general," did not impose themselves onto us; rather, we are the "lawgivers of reason."[39] One year after Kant's death the neo-Kantian physicist-philosopher Jakob Fries made explicit what he felt the master's system meant regarding truth:

> We cannot, as is usually done, speak of truth as opposed to error by saying that truth is the correspondence of a representation with its object. We can only say that the truth of a judgment is its correspondence with the immediate cognition of reason in which it is grounded . . . I call this truth the inner or empirical truth of cognition, because we concern ourselves in this idea only with the unification of all our cognitions into a system of reason.[40]

Drawing on Fries and Fries's student Ernst Apelt, Hermann Lotze later in the century reinforced Fries's contention when he spoke of the strict distinction "between truths which are valid and things which exist."[41]

This conception is consistent with what is known as the coherence theory of truth. There are several varieties of the coherence theory, all of which share the assertion that the truth of a proposition consists not in the correspondence between the proposition and a reality independent of anything that may be believed about it, but in the proposition's coherence with a system of beliefs. "The coherence, and nothing else, is what its truth consists in."[42]

In later chapters I will follow up on the implications for theology of a Kantian understanding of truth, particularly among the neo-Kantian theologians at the end of the nineteenth century. According to Ralph Walker, Kant held to an impure coherence theory of truth because he resorted to coherence only when ascertaining truth on the level of phenomena. This level, of course, is where the investigations of natural science take place. The beliefs he held about things-in-themselves, which were to Kant a necessary part of his system, did not cohere with those he held regarding the phenomenal world.[43]

As I will discuss in Chapter 7, the theologian Wilhelm Herrmann understood all this to mean that the truth arrived at in natural science must be seen as provisional, since it is influenced by the presuppositions which one brings to the attempts to know nature.[44] The truth of a claim or a statement about nature is determined by its consistency with the system of beliefs about nature being utilized. The entire rational system is of course applied to nature, but when this is done it must be openly admitted that we do not know nature in itself to be as coherent as the system is—that is the metaphysical claim of the realist. For the anti-realist Herrmann,[45] the application of our system of beliefs about nature, including our belief in its rationality, is done as a hypothesis. The truth of a proposition about nature is not measured against external nature, as in the correspondence theory, but internally with respect to its coherence within the system of beliefs about nature. When truth is thought of in this fashion invention is emphasized over discovery, the role of presuppositions becomes more consequential than nature's authority, values assume at least equal weight to facts, and the limits of reason are deemed more significant than a naive trust in its power to copy nature.

Because claims regarding truth have invariably been involved in attempts to evaluate issues in science and religion, it is helpful to bring the distinction between the correspondence and coherence theories to bear on the development of German theology in its relationship to natural science in the nineteenth century.[46] What becomes evident is that most nineteenth century German theologians understood the question of truth in terms of correspondence and elaborated their differences within a commonly shared set of assumptions. For some, however, truth as correspondence forced a decision between science and religion that represented an unacceptable set of alternatives. To avoid them these theologians abandoned truth as correspondence in favor of an understanding of scientific and theological truth which, in making no metaphysical claims, was consistent with the so-called impure coherence theory of truth of the Kantian tradition.

The loss and gain for theology was clear. Theology lost its capacity to speak of nature, but at the same time it also denied natural science the right to usurp the old theological prerogative of making assertions about the ontological status of the cosmos. To the extent that natural scientists, especially those caught up with the new Darwinian vision of nature, did not heed this message, which was so carefully expounded by the theologian Wilhelm Herrmann, scientists were portrayed to be as dogmatic as the very priests they delighted to detest.

Through the appeal to the distinction between the correspondence and coherence theories of truth it becomes possible to avoid having to decide whether or not the German debate over science and religion fits or does not fit the infamous warfare motif. Indeed, a quite different constellation of adversaries emerges in the approach taken here. In this study the creationist and the scientific naturalist are not pitted against each other in the manner they have been in our traditional understanding of the nineteenth century debate. Rather, where the significance of nature for theology was concerned, such opposite thinkers as the conservative creationist Otto Zöckler and the pantheistic naturalist David Friedrich Strauss shared something that marked them as advocates of a nineteenth century worldview in contrast to the emerging mentality of the Ritschl school which neither of them understood.

The major subjects of this book lived and wrote for the most part during the latter half of the century when the achievements of natural science made it impossible for theology to ignore the impact science was having on German society. The next chapter is an attempt to orient the reader to the main strains in German theology from earlier in the century that gave rise to the four representatives

I have chosen. Because of its richness and diversity, any attempt to summarize and classify German theology in this period is bound to have omitted figures whom another author deems indispensable to a proper grasp of the century. My selection has been determined primarily by the particular focus of this inquiry into the changing relationship between natural science and religion.

2

The Shape of German Protestant Theology in the Nineteenth Century

HOWEVER VALUABLE theology may be as a vehicle for understanding the past, we must face the fact that nineteenth century German theology is clearly unknown territory for those living at the end of the twentieth century. Some overview of the shape and development of German theology is therefore both reasonable and necessary.

THEOLOGY IN THE GERMAN ENLIGHTENMENT

It is less meaningful to divide the Protestant community in the eighteenth century into conservative and liberal camps than it is to distinguish pietistic from rationalistic styles within the theological debates. After 1800, when the meaning of theological rationalism changed into something quite different from what earlier theologians had understood it to be, the old opposition between rationalism and pietism was displaced.

In the eighteenth century, however, the question ultimately came down to the trustworthiness of reason in theological matters. Orthodox rationalists, as Karl Barth calls them,[1] were convinced that much could be known through reason, and that even those revealed truths that were beyond reason were not therefore necessarily contrary to reason. Pietists, by contrast, avoided reliance on a reasoned systematic theology, trusting their personal experience of "rebirth" and their retreat into religious communities separate from the world to provide a foundation for their faith.

Although it is perhaps tempting to associate the labels conservative and liberal with pietism and orthodox rationalism, respectively, the two sets of terms belong to different times. Rationalistic theology in the early eighteenth century, for example, did not question the fundamental tenets of Christianity simply because it was rationalistic. For most of the century the terms "pietistic theology" and "rationalistic theology" serve as differing theological approaches more than as indicators of loyalty to the basic articles of the Christian religion.

Of course eighteenth century *philosophers* did challenge theology by questioning rather than merely analyzing such fundamentals as miracle and God's superintendence of nature and history. Indeed, rationalistic philosophy directly criticized the validity of the traditional arguments for God's existence that eighteenth century rationalistic theology had endorsed, convincing pietistic theologians more than ever that reason alone was no sure ground on which to construct a sound theology.

Critical rationalism even invaded theology toward the end of the eighteenth century in the thought of the theologians known as "neologists." Contrary to orthodox rationalists, these men were, in the words of Karl Aner, "not content with the logically mathematical possibility of conceiving the content of revelation, but pushed this content to one side as contrary to reason."[2] With the neologists the content of the Christian religion lost its capacity for a distinctly *Protestant* outlook. John Dillenberger and Claude Welch note that for neologists whatever truth remained "was not really different from that which thoughtful men were saying outside the church . . . In these men Protestantism had become a victim of the Enlightenment."[3] There was, then, heresy aplenty prior to the nineteenth century, but it originated as much from secular thought as from theology.

For the older rationalists some truths of religion could not be uncovered by reason, nor could they be understood by reason. Truths such as the mystery of the Trinity, the human and divine nature of Christ, and the unfathomable grace of God were above reason, being supplied by revelation and accepted by faith. Still these truths were not contrary to reason since they were recognizable as essential to Christianity through the use of reason.

A key question of the continuing theological debate in the eighteenth century concerned the possibility of miracle. Here, where it became a question of nature's truth, the orthodox rationalists qualified but did not contradict the correspondence theory of truth. For orthodox rationalists miracle was an undeniable prerogative of the Creator, who, when necessary, could work together with the natural order to accomplish divine purposes. Hence, rational categories simply did not apply completely to nature.

Soon, however, there was discontentment with the prospect that reason's power could not be presumed to extend completely to the natural world. Those whom Albert Schweitzer has referred to as "half-and-half rationalists," and who have been labeled "transitional" theologians by others,[4] downplayed the role of miracles, preferring not to rely on them for evidence of the truth of Christianity. These theologians allowed the supernatural to intrude in the person of Jesus, whose life and redemptive work occupied center stage in the original creation of nature and in the establishing of an order of grace. The transitional rationalists were content with natural religion as long as it was seen as fulfilled in Christ through revelation.

With the neologists, who wrote in the latter part of the century, the move to thoroughgoing rationalism neared completion. For these theologians the correspondence between the rational mind and reality was complete. Even the content of revelation had to be accessible to reason. It was not that revelation did not exist; revelation was still, as it had been earlier, the divine disclosure of truth. But now the content of the revelation could not be mysterious in the sense that it defied reason or remained aloof to reason. Furthermore, reason itself was understood in the narrow sense of logic and common sense, and the truths of reason were seen as practical conclusions whose obviousness arose from inner certainty.[5] Reason came more and more to be associated with an empirical attitude, in which the individual appealed not only to the observation of nature but also to one's own "inner experience." Unbiased observation was assumed to be possible, reliable, and the source of the truths of material and spiritual reality. Through the use of their unique rational gifts, human beings had access to eternal truth.

When members of this neological tradition applied the commonsense understanding of religion to the interpretation of the Bible, what resulted was an inevitable turn away from biblical accounts as literal history. J. F. W. Jerusalem's treatment of Genesis revealed that wherever the biblical record defied common sense, its literal meaning was to be abandoned. When Genesis implied that light was created before the sun, Jerusalem attributed the contradiction to a secular cultural source. Revelation corroborated natural knowledge, it did not stand entirely apart from it.[6]

Hermann Reimarus went even further in his *Apology or Defense for Reasonable Worshippers of God,* which was published by Lessing between 1774 and 1778 in a series known as the Wolfenbüttel Fragments.[7] Reimarus, a professor of oriental languages at Hamburg, brought out into the open what was only hinted at among the neologists: he asserted that revelation was inconsistent with a thoroughly reasonable religion. In Reimarus the primacy of natural religion was established without doubt, and total rationalism had been achieved.

With the appearance of the critical philosophy of Immanuel Kant, however, the ground was laid for interpreting rational religion in a new way. In Kant's careful analysis of the nature and limits of reason the major concern was the vindication of the necessity and certainty of the knowledge of reason, both of which had suffered under the critiques of David Hume. But in order to establish the certainty and necessity of the knowledge of reason Kant found that he had to restrict its scope to objects of spatial and temporal experience. This meant that traditional doctrines about God, free will, and immortality simply could not be objects of the knowledge of reason, since they could not be confined to the realm where reason applied. In *Religion within the Limits of Reason Alone,* written in 1793, Kant examined religion in light of his earlier conclusions about the unquestioned authority of reason in the realm of sense experience and in light of his claim that it is impossible to make God, freedom, and immortality objects of knowledge.

What emerged was a religion which found its origin not in reason but in the human capacity for moral sense. This conclusion was not totally new to the eighteenth century rationalist theological tradition. The neologist J. S. Semler had also located the center of religion in morality, with Jesus and the Bible assuming significance as vehicles through which the moral ideals necessary for the human race were communicated. Kant, however, worked out the implications of this view of religion in conjunction with his philosophical analysis of reason. He concluded that the undeniable moral capacity human beings exhibit when they act receives no assistance from pure reason but nevertheless could only be possible on the assumption of the existence of God, freedom, and immortality. By the century's end, however, few seemed to grasp the significance of Kant's radical separation of religion from the world treated by natural science, and among those who did, few were content simply to accept Kant's defense of religion based on practical reason as the final word.

Theological Trends of the German Romantic Era

Theological Rationalism in the Nineteenth Century

The theological rationalism of the nineteenth century was to be sure born from Enlightenment thought. But in the nineteenth century the new rationalism developed beyond the older more orthodox varieties of the eighteenth century, and it proceeded without heeding the restrictions on reason that Kant had erected. Where in the Enlightenment the assumption of orthodox rationalists was that creeds could be and ought to be articulated and defended by reason, by the end of the century the role of reason had so changed that systems such as that of Christian Wolff and others seemed woefully naive. Armed with a growing confidence in the power and sufficiency of reason, theological rationalism proceeded into the nineteenth century with a boldness that indicated it had not heard, or did not accept, Kant's qualification of the classical correspondence theory of truth in his denial that the conclusions of pure reason were relevant to religion.

Coupled with these traits was the attitude, characteristic of much in the earlier Age of Reason, that one must have the courage to face truth wherever it may lead, even when unpleasant or threatening. And because the implications of Kant's philosophy regarding the accessible and complete correspondence between reason and nature's truth had not yet been made explicit, the nineteenth century rationalists exuded a confidence that came across as cockiness. Rationalism's message became crystal clear. All three of the "burning questions of rational religion" identified by Welch, miracles, original sin, and the special status of the biblical narrative,[8] demanded great courage to address, for in the eyes of the nineteenth century rationalists all three required that central tenets of traditional Christianity be abandoned, and all were identified as inconsistent with rational religion. Rationalist theologians in the early nineteenth century made it their special mission to take the offensive against traditional supernaturalism, and their challenging message was widely heard and broadly influential.

Perhaps the most famous of the nineteenth century rationalists was Julius A. L. Wegscheider, a professor in Halle from 1810 to 1849. Wegscheider's *Institutes* of 1816 represented as much as any single work the rationalist spirit that captivated many theologians and even, according to Barth, "reached the masses of citizens and peasants through the pulpits of countless villages and small towns."[9]

Wegscheider's thought reveals that theological rationalists of the nineteenth century clearly relied on a correspondence conception of truth, for he promoted the notion that the modern age was characterized by a recognition of the correspondence between the laws of human thought and the laws governing the natural realm. This correspondence could not be ignored or superseded in religion. Theology must investigate the origin of dogmas and criticize them in light of modern reason.[10] For Wegscheider humankind's accessibility to truth was therefore obvious and not at all that problematical. His representation of theological rationalism contributed to the boldness, the confidence, and the elitist air by which it was known.

What Wegscheider and the rationalists did not abandon was the teleological function that religion had always performed. History was not for Wegscheider random development, nor was the human race without guidance. Divine guidance in history was evident in the very existence of the Bible and in the person of Jesus, both of which imparted to human beings the message that everyone could reach eternal life through divine knowledge.[11] What was natural could also be goal directed.

Besides Halle the other center of theological rationalism was Heidelberg, whose eight hundred theological students formed the largest contingent anywhere in Germany in the early nineteenth century.[12] In addition to Johann Heinrich Voss, arch enemy of romantic mystics, pietists, Catholics, and, incidentally, aristocrats, Heidelberg held after 1810 the patriarch of theological rationalism, H. E. G. Paulus. Paulus had been at Jena during its rise and fall as the center of German romanticism, but his reputation in theology was made in Heidelberg.

In spite of an attempt to deny that his principal goal was a naturalistic analysis of miracles, Paulus's treatment of the New Testament accounts in his *Life of Jesus* of 1828 was impressive precisely because of the manner in which he explained away miraculous events. Paulus provided believable explanations of what really happened at the feeding of the five thousand, the transfiguration, the resurrection, and other central miracles in Jesus' life. He excused the New Testament writers by arguing that since they did not know the secondary causes of the allegedly wondrous events, they succumbed to the Jewish love of miracle, causing them to attribute events directly to God's action.[13] When Paulus's message was combined with that of the rationalist Old Testament scholar Wilhelm Gesenius in Halle, supernaturalistic interpretations of Scripture were up against formidable opposition.

Paulus died in 1851 at the age of ninety, Wegscheider in 1849 at seventy-eight. But even before mid-century the era of theological rationalism had been displaced. By the 1840s critics of rationalism were everywhere. One was used to the denunciations of conservatives like Ernst Hengstenberg and the neo-pietists. Joining these theologians were critics among the liberal followers of Schleiermacher, whose very conception of religion had arisen in part because of a dissatisfaction with the rationalist tradition. In addition, the rationalists were attacked by the Young Hegelian theologians, who, although openly hostile to theological rationalism, in some ways represented its continuation and fulfillment. Even the arch-rationalist Paulus recognized that the most responsible stance he could assume was to support the same Young Hegelians who so severely criticized him. Schweitzer writes that it was as though Paulus acknowledged that the end of rationalism had come, "but that, in the person of the enemy who had defeated it, the pure love of truth, which was the only thing that really mattered, would triumph over all forces of reaction."[14]

Conservative Theology and Its Context

The heritage of the older orthodox rationalism of the eighteenth century lived on in the nineteenth in the conservative theology of Germany's two major Protestant traditions, the Lutheran and Reformed denominations. While conservative thought emphasized what it always had—an acceptance of the inerrancy of the Bible and an acknowledgment of its authority where basic doctrines of the faith were concerned—conservative German thinkers became occupied as much with internal ecclesiastical politics in the nineteenth century as they did with the defense of faith against its detractors.

As opposed to the theological rationalists these conservatives were known as supernaturalists, since their trust in human reason went only so far. Although they certainly did believe that their faith was rationally defensible, they did not hesitate to draw a line beyond which they would not allow reason to carry them. Sacrosanct from their point of view were fundamental beliefs in the human need for redemption, salvation through the self sacrifice of a fully divine Christ, and, where nature was concerned, God's direct creation and continuous maintenance of the world.

Naturally they constantly opposed rationalists on matters like miracles. The distance between the two groups concerning such things was so unmistakably clear that it was hard for either side to engage the other. The disagreement was not over *whether* the laws of reason matched or corresponded to reality. Where reason could be applied it gave reliable results. The disagreement was, as it had been for the orthodox rationalists from whom the conservatives derived, about the *extent* to which human reason applied to the natural world.

If the lack of common ground made the disagreement between conservatives and theological rationalists as uninteresting as it was predictable, the same cannot be said for the internal debates in German conservative theology. Because of their political significance, these debates assume more importance than otherwise might be ascribed to them.

Among the political reforms instigated by Karl Frieherr vom und zum Stein in Prussia in the first decade of the century was the creation of a state bureaucratic authority for church affairs, even though there was as yet no official Prussian state church. The intent here was to redress the relative autonomy of the local congregation by creating a body to whom the congregations must answer. For want of such many clergy had over the years become apathetic and overly accommodating to the tastes of the day as they encountered them in the local situation.[15] In the wake of the national effort required to meet Napoleon's challenge and eventually to defeat him, the public mood was sensitive to patriotic and religious fervor. The Prussian King Frederick William III may have felt threatened by the romantic political yearnings of some of his subjects for a united German fatherland, but he exhibited no hesitation where Prussia's religious heritage was concerned. On the contrary, the king's support in 1817 of a proposal for the union of the Lutheran and Reformed denominations into a new *Evangelische Kirche* provided him with an occasion to exert his power through what Schnabel has referred to as "absolutist methods."[16] What appeared at first glance to be a concession to the passion for unification being demanded by liberals in pursuit of a more modern Germany turned out on closer examination to have been entirely consistent with the perspective that characterized much of Restoration politics.

In a time when emotions were easily stirred, church union fooled many as long as it involved only the issue of common communion. It did not, however, sit well with conservative German theologians. When in 1817 the king threw his weight behind the union of Prussia's two leading Protestant traditions, he defined a challenge that would occupy conservative Protestantism in Germany for much of the rest of the century.

Church union quickly produced a host of detractors, especially from conservative quarters. Most well known was the Lutheran pastor Claus Harms of Kiel, who, in his youth, had become a powerful enemy of rationalism. Harms set out to provide in his popular writings what he felt was lacking in others.[17] When in 1817 he printed his new ninety-five theses on the occasion of the three-hundredth anniversary of Luther's original declaration, his targets were rationalistic theology and church union. Thesis 75 employed a sexual metaphor to denounce the proposed merger: "Like a poor maid they want to make the Lutheran Church rich through copulation. Do not consummate the act over Luther's bones, for he will come alive, and then—watch out!"[18] Harms opposed the secular government's insertion of itself as the highest arbiter of religious concerns, "a mistake that has to be made right in orderly fashion" (thesis 90), and he defended the rights of congregations to select their pastors (thesis 91). The issues were clearly drawn, producing close to two hundred brochures pro and con.[19]

The substantial opposition of orthodox theologians to church union is an early indication of the identity and strength conservative Protestant theology would exhibit throughout the remainder of the century. The issue became the place of the confession, an important foundation for both Lutheran and Reformed doctrine and church discipline. Merger implied that the historic confessions of each denomination, which represented the rational articulation of each denomination's theological position, would have to be compromised when they differed. That prospect produced a revival of the German confessionalist movement.

While there were both Reformed and Lutheran confessionalist movements in nineteenth century Germany, Lutheran confessionalism was by far the stronger and more visible. Initially union required only a common communion and an open system of church appointments, but in 1822 the king, exercising his prerogative as supreme bishop, decreed a common liturgy. When protests against union increased, drawing liberals into the fray against the autocratic measures of the crown, Frederick William III retreated to a request for voluntary compliance with the new liturgy. Still opposition continued. In the late 1820s the conservative Lutheran journal *Evangelische Kirchenzeitung* found that its failure to denounce union was increasingly causing many Lutheran theologians to distance themselves from contributing to the journal.[20]

Following the three-hundredth anniversary of the Augsburg Confession in 1830 the king became more assertive. When he ordered adherence to the agenda of the new Evangelical church, resistance among many Lutherans stiffened. The king viewed the mounting resistance as political opposition, especially after several congregations formed a "pure" Lutheran church opposed to union. In 1834 Frederick William III acted, forcing some congregations to accept new Evangelical pastors. Such action predictably led to charges of religious persecution, to secret meetings for worship, and to the gathering of die-hard Lutheran subjects in particular regions. In Silesia the "Old Lutherans," as they have been called, came together in considerable numbers.[21] Their strict confessionalist position and their unyielding opposition to church union was defended in the *Zeitschrift für die gesammte lutherische Theologie und Kirche,* which ran from 1840 to 1878.[22]

The confessionalist controversy in German conservative theology illustrates the depth of the conservative belief that the truths of religion had been articulated as correctly and completely as possible in the rational statements of the original confessions of the church. When in the latter half of the century the treatment of nature and the claims of natural science became more explicitly included in the arguments about the confessions, the conservatives resorted to the traditional apologetical approach of employing reason to defend the credibility of the fundamental tenets of the faith, including God's creation and governance of the world, but denying to reason a capacity to apply so completely to the world that it could exhaust an explanation of Christianity's mystery. What this amounted to was a rational defense of the historic confessions of the church.

Friedrich Schleiermacher and the Birth of a Liberal Tradition

After the turn of the century, largely as a result of Friedrich Schleiermacher's response to the intellectual challenges of the day, Protestant theology developed a liberal tradition; that is, Schleiermacher articulated from within theological ranks a constructive reinterpretation of the very meaning of religion. Schleiermacher's positive reevaluation of the essentials of Christianity contrasted sharply both with the traditional theological approaches of the eighteenth century and with the destructive rationalistic tendencies of those Schleiermacher called religion's "cultured despisers."[23] As we shall see, however, in the end Schleiermacher remained committed to harmonizing the truth of religion with that of nature.

Schleiermacher's thought was a defense of the Christian faith against philosophers and even against the neologists, whose modernistic approach to Christianity became enormously popular in the nineteenth century. But if Schleiermacher's work represented a defense of the Christian faith, it was vastly different from the traditional apologetics of the eighteenth century. Schleiermacher realized that the debates between the older orthodox rationalists and pietists were, in the face of the new rationalism, largely irrelevant. No longer could the theological debate focus, as it had in the past, on how the fundamentals were to be apprehended and articulated. Now in need of reinterpretation—indeed a constructive, living, and vibrant reinterpretation—was the very foundation and nature of religion itself. Schleiermacher rejected the notion, dear to rationalists old and new, that religion consisted of metaphysical analyses that produced a set of beliefs.

But Schleiermacher also denied Kant's conclusion that because religion could not have its foundation in knowledge, it had to be identified with the human capacity for moral sense. According to Schleiermacher religion drew its life primarily from that specific component of human consciousness that was neither cognitive nor volitional. Religion was born of feeling *(Gefühl)*, which was prior both to cognition and volition. Cognition and volition both have reference to a mediated external object, but in religion, said Schleiermacher, we encounter a primary human experience.

Here was a view of religion that changed the rules governing the theological debates. For example, according to Schleiermacher it was no longer necessary to forsake Christianity simply because one did not believe in miracles; on the contrary, Christianity truly understood did not rely on a traditional grasp of miracle at all. To Schleiermacher God's presence was evident in nature, but it was evident far more in everyday natural events than it could ever be in miracles.[24] God, frequently synonymous with the Infinite in Schleiermacher's epoch-making *Reden* of 1799, was apprehended primarily in human feeling about the universe, not in knowledge of it.

In his system, however, nature played an important part. Schleiermacher's best-known biographer, Wilhelm Dilthey, writes that "it is impossible to represent Schleiermacher's system without starting with the philosophy of nature and its historical value, for Schleiermacher's system is founded on the truth of this philosophy of nature."[25] Dilthey's claim is supported by the testimony of theologians from Schleiermacher's own day. In 1826 F. C. Baur was under consideration for a vacancy on the Tübingen theology faculty. In a confidential report two Tübingen theologians opposed Baur's appointment because of his endorsement of a "theological view which in consequence of Schelling's nature-philosophy is at the present time propagated especially by the writings of the Berlin theologian Schleiermacher."[26] Furthermore, Hasler has determined that the concept of nature appears more than 820 times in Schleiermacher's two-volume *Christian Faith* alone.[27]

Schleiermacher vehemently rejected what Hasler has called "Kantian dualism" and "Fichtean solipsism" because in both the objective conditions affecting the actions that could change society were blocked out.[28] What he missed in these philosophers he found in the *Naturphilosophie* of the young Friedrich Schelling. Here was a philosophy of the real as well as the ideal, here was support for Schleiermacher's own conviction that the realms of both the physical and the ethical shared an ontological foundation. Like Heinrich Steffens, who also built a bridge from Schelling's *Naturphilosophie* to ethics in the first decade of the century,[29] Schleiermacher responded eagerly to Schelling's representation of the physical realm as an evolving process within human history whose structure was affected by the accompanying self-realization of humanity. Because of his firm belief in a correspondence theory of truth, Schleiermacher felt that there had to be a mutual participation of the ideal in the real and real in the ideal *(das Ineinander von Realem und Idealem)*. This, he was convinced, provided a solid justification for viewing moral action as a continuous dialectical shaping of nature through reason.[30] Natural law and moral law agreed in their productive and formative functions; both captured reality's creative and evolutionary dimensions in such a way that they expressed what was normative for a given stage of development. As in Schelling, the prototype of natural law for Schleiermacher was not mechanism but organism.[31] Contrary to Fichte and Kant, then, Schleiermacher refused to tolerate nature's being cut off from his theological concern for ethics. Schleiermacher resisted the outright destruction of feudal nature by finding a new role for it to play in theology. In Marxist terms, the tones of a society that was no longer feudal but not yet capitalistic found in Schleiermacher's theology vibrations that produced a resonant response.[32]

One might suspect from the thrust of Schleiermacher's *Speeches on Religion* that the central role of feeling for his theology would displace any concern for a cognitive component of religious experience. But Schleiermacher's theology was a far cry from that of the neo-Kantian Jakob Fries, who also focused on the dominant role of feeling in religion and whose thought represents our final theological trend of the Romantic Period. Schleiermacher's commitment to a perspective in which the ideal corresponded with the real did not at all match Fries's emphasis on truth as coherence. We should not be surprised, then, that Schleiermacher defended the possibility of religious knowledge where Fries argued that it was impossible.

It is true that Hegel criticized Schleiermacher because he assumed that the Berlin theologian's reliance on feeling left no grounds on which to base a claim that knowledge was possible in religion. Hegel castigated Schleiermacher's religion based on feeling. "If religion in man is based solely on feeling," Hegel wrote in the Preface to H. F. W. Hinrichs's *Religion in Its Internal Relation to Science,* then "it is correct that this has no further determination than to be a feeling of dependence, and so a dog would be the best Christian, because it has this feeling most intensely."[33]

But it has been shown that Hegel's view of Schleiermacher's system was distorted, and that Schleiermacher accepted the possibility and necessity of knowledge in religion.[34] And those who followed in his footsteps, accepting as he did that there was no lack of correspondence between thought and things, between the ideal and the real, also agreed that one of theology's tasks was to articulate and to make clear the truth of the knowledge it was possible to attain in religion.

To many theologians and to numerous pastors Schleiermacher's theology appeared to be an unnecessary concession to the pantheistic trend of German idealist philosophy. They recognized Schleiermacher's enthusiasm for Schelling, and beyond that they understood that Schleiermacher's participation in liberal politics was not unrelated to his theological outlook. Schleiermacher's vision looked boldly ahead; theirs was directed longingly backward. But in Schleiermacher these theologians had an opponent whose redefinition of religious questions was broadly appealing to the times. Henceforth it *would* be meaningful to distinguish their views as conservative in contradistinction to the innovative views of those caught up in theological liberalism.

If the old sharp antagonisms between pietists and rationalists had already begun to fade before Schleiermacher's *Reden* of 1799,[35] after 1800 there seemed to be no danger that these old antagonisms could rival the new theological divisions that lay ahead. The common enemy for conservatives and liberals was rationalism, and the question facing these theologians in the early decades of the nineteenth century was whether they would fight the enemy with old fashioned means or with new weapons now at hand. Conservatives felt that the new weaponry might well blow up in their faces and inflict wounds so mortal that they would be handing the victory to the enemy. Liberals felt that they had no other choice than to modernize their arsenals if Christianity was to have any chance at all in a world becoming increasingly secular. And on the left stood the enemy, the rationalists. Their primary commitment was to reason, which they understood in such a manner that the object of their research, religion, frequently appeared to be more an incidental than an essential component of their concern.

In 1824 the new rationalist journal, the *Neue kritische Journal der theologischen Literatur,* characterized the two ends of the theological spectrum of the day when the editors declared that they would "ensure a free platform to every scholar who contends with scientific earnestness and propriety, be he a supernaturalist or rationalist."[36] Between these extremes stood the nascent liberal school, which took its cue from the broad new categories of romantic thought. Because of the rise to prominence in philosophy of Georg Friedrich Hegel in the 1820s, Schleiermacher's influence on liberal theology was supplemented in the 1830s with that of the new speculative thinker from Berlin.

The Neo-Kantian Theology of Jakob Friedrich Fries

Prior to the 1820s Hegel's name was far from the most recognizable in German academe. It was not until 1816 when Hegel finally landed an academic post in Heidelberg, having served as the rector of the Nuremberg gymnasium up until then. Far more well known in the first two decades of the nineteenth century were Schleiermacher in theology and Friedrich Schelling in philosophy. While still a *Privatdozent* Hegel had lost out in 1805 to another young philosophy student at Jena in their common bid for the Heidelberg position, the neo-Kantian Jakob Fries. When Fries received a call back to Jena in 1816, Hegel took his place in Heidelberg.

At Heidelberg Fries taught philosophy, physics, and mathematics. But Fries was not interested only in philosophy and natural science, he was also vitally concerned with religion. As a loyal Kantian, he wished to delineate the limitations of reason in order to clarify the place of the nonrational in human experience. Fries stood outside the main currents of thought in the Romantic Period because of his persistent retention of the critical principles of Kant. The prevailing tendencies were either to be unaware of Kant's radical separation of religion from the realm of nature or to call for an overthrow of the Kantian rift between phenomena and noumena.

Fries was nevertheless a child of his time in his unique understanding of the role of feeling in religious experience. Friesian Kantianism is without question romantic Kantianism. This is nowhere more clearly evident than in Fries's book of 1805, entitled *Knowledge, Belief, and Aesthetic Sense,*[37] in which he attempts to set forth clearly the relationship between natural science and religion.

As one might surmise from the title, the goal of Fries's effort was to distinguish three separate modes of conviction, which he was willing to call modes of cognition *(Erkenntnis)*.[38] The first, knowing *(Wissen)*, involved both the immediate knowledge of intuition[39] and immediate knowledge of the laws governing consciousness which we become aware of when we utilize the mind in making judgments.[40] *Wissen* also involved mediate knowledge. When we use judgments and inferences in combination with immediate knowledge to gain new knowledge, the knowledge gained, said Fries, was mediate, not immediate knowledge. It must draw on some external immediate knowledge in order to come into existence, and its certainty arose not from itself, but from the immediate knowledge on which it relied.

This mediated knowledge was scientific knowledge, for natural science, through the use of judgments and inferences, drew on the immediate intuitive knowledge of the senses, mathematics, and the categories of the understanding to acquire new knowledge. Should anyone inquire about the certainty of a scientific assertion, any justification would, according to Fries, ultimately have to hark back to some form of immediate knowledge.

The common understanding of proof was substantially damaged in the Friesian analysis. Proof was only a mediate process that could not of itself guarantee the truth of its conclusion, at least not in the old sense of correspondence with external reality. The immediate knowledge on which a proof depended resulted from either the purely subjective structure of our mental capacities or the non-necessary nature of empirical information. It is perhaps now more clear why, in the section of his 1805 work entitled "Truth," Fries abandoned the classical understanding of truth as correspondence in favor of a conception based on coherence. Like Kant, Fries wanted "to win a place for belief," and he wanted to do it by humbling *Wissen*.[41] "We succeed in defending the rights of belief [*Glaube*] mainly by showing how knowing arises only subjectively in reason."[42]

But having become conscious of the limitations that reason is subject to in knowing, reason could, through the logical process of negation, at least think away the limits. If reason showed itself to be tied to phenomena in knowing, then it at least could formulate a category that was the negation of phenomena. This was the origin of belief. Fries was careful to point out that such a process did not produce any positive notion of what lay beyond the limitations of knowing. The origin of believing in contrast to knowing "must remain thoroughly negative, really having no positive content other than that of the negation of all negations, or the denial of limitations."[43]

But how did believing express itself? If the world of knowing was confined to space and time, the way to negate this restriction was to think space and time away. The negation of these limitations could be captured by the idea of an eternal realm outside space and time. A further characteristic of the phenomenal world, which was of course tied up with its being restricted in time and space, was its rigid necessity. We can think away necessity by imagining a realm in which freedom, not necessity, reigned, and by realizing that necessity was absent in the presence of purpose. Both of these experiences were in fact encountered in our consciousness—we have a consciousness of our freedom, and we know ourselves to be purposeful beings. But while our experience testifies to the presence of a world that stands in marked contrast to the finite world of knowing, Fries was careful to point out that it was not within our power to prove the reality of this world or to say anything positive about it.

Fries was not content, however, to make religion into a merely moral matter. For Fries there was one more mode of cognition, and it was fundamentally different from both knowing and believing. *Ahnung* referred to a feeling—it had no relation to processes born of reason like knowing or believing. Since for Fries feeling was the ultimate source of conviction, it provided the ultimate assent. *Ahnung,* then, was not just a third category tacked on to Fries's analysis. *Ahnung* was a focused feeling, the feeling that the eternal was reflected, albeit in an imperfect and restricted manner, in the finite.

Since we could not, according to Fries, have any concept at all of the realm of the eternal from within the world of knowing, and since we could not obtain any positive notion of the eternal from the world of belief, it being merely the logical negation of knowing's limitations, how could we ever possibly recognize the eternal in the finite? Whatever the answer, we certainly could not expect that the means by which we will recognize the eternal in the finite would remotely resemble those of knowing and believing. We would not meet any concepts or ideas. *Ahnung,* after all, was feeling; it did not belong to reason, but stood behind it. But what feelings?

Drawing on Kant, Fries pressed on to argue that in feelings of the beauty and sublimity of nature we encountered the eternal in the finite. Why should we respond emotionally as we look out from the tops of Swiss mountains? Why was it that flowers were beautiful? Fries would reply that it was only possible because there was purpose, specifically, purpose in itself. Beauty and sublimity were ends in themselves. There must be a realm of purpose for beauty and sublimity to be possible. But the eternal was exactly that—a realm of purpose, the world of believing that was born of reason itself. We are one source of purpose, but in *Ahnung* we encounter it and recognize it in nature. In our encounter with a purpose whose source is outside us, we give life to our belief. Herein lay real religion, and not in the imposed purpose of our misplaced conceptual knowing of organism. "A positive notion of the eternal is possible for us through its relation to the finite," wrote Fries, "but we can lay hold of it only in feeling through the beauty and sublimity of nature."[44] In the end Fries was true to his Romantic Age, for the only expression we could hope to give to *Ahnung* was through poetry and the arts.

Fries's immediate influence in theology was felt in the Berlin theology faculty in the person of W. L. M. DeWette, who saw himself to be Fries's disciple. In spite of the efforts of DeWette on Fries's behalf, Hegel received the call to come to Berlin in 1818.[45] By the end of the next decade the competition for the devotion of theology students was between Schleiermacher and Hegel. Although DeWette continued to reflect Fries's influence, it was not in the theology of the first half of the century that the themes Fries had addressed were picked up. Not until after the revival of interest in Kant at mid-century would theologians once again address the issue of the knowledge of nature in terms similar to those Fries had employed.

Four Theological Tendencies after Mid-Century

Hegelian Speculative Theology

"Hegel and Schleiermacher were the great figures under whose shadow the German theology of the whole nineteenth century developed."[46] With this judgment Claude Welch does not mean to suggest that these two German thinkers influenced their age in the same way, for Schleiermacher was primarily a theologian who was drawn inevitably to philosophy, while Hegel was first a philosopher who found it impossible to avoid religion.[47] If Schleiermacher supplied the impetus for the renewal and regeneration of theology in a modern age, it was Hegel who provided the incentive for the historical study of the origin and development of Christianity.

Hegel's star rose after his appointment to a position in Berlin in 1818. What attracted budding philosophers and theologians to his system was his conviction that reason need not be abandoned in order to apprehend all of the diversity and complexity of human experience; whatever was real was also rational. This was Hegel's way of seeking and achieving the unity all the thinkers of the Romantic Era longed for.

Like Schleiermacher, Hegel owed a debt to Schelling, for it was Schelling who, with the possible exception of Goethe, expressed more effectively than other writers of the time a dissatisfaction with the meaning of rationality as it had been understood from Plato to Kant. In rejecting a static conception of reason, whose categories remained immutable and outside time, Schelling, in Lovejoy's words, "temporalized God himself."[48] God in Schelling was understood as a Being in the process of development, not so much a Being who *is* as a Being who *becomes*. For Schelling God was to be understood through the language and categories of life, not through the attributes of being. "Has creation a final goal? And if so, why was it not reached at once? Why was the consummation not realized from the beginning? To these questions there is but one answer: Because God is *Life,* and not merely being. All life has a *fate,* and is subject to suffering and to becoming. To this, then, God has of his own free will subjected himself."[49]

Like Schelling Hegel too rejected the classical understanding of reason and its implications. In its place he substituted what Barth calls Hegel's "boldest and most weighty innovation," the "heartbeat of the Hegelian system": the assumption "that the concept does not so much exclude the concept that contradicts it, as the fundamental axiom of western logic had previously held, but includes it."[50] Armed with this dialectical understanding of reason, knowledge (including knowledge of God) ceased to be a simple result but became instead a process, a movement fueled by negation toward the identity of the concept and object.[51] And knowledge of God was knowledge of everything, for it was absolute knowledge, knowledge that by its nature could not brook any limitation.

Hegel's dissatisfaction with romantic theology stemmed from his conviction that dialectical reason need not concede to feeling, or to some other nonrational component of the human intellect, the task of completing what reason could not accomplish for itself. Hegel felt contempt for all religion of feeling and for all religion that did not insist that the content of religion be susceptible to analysis and clarification. If one remained content with the mere immediacy of religious experience, then one never reflected on its content, one never had a genuine rational cognition of God. Hegel's thought attracted all those who understood romantic theology to be relinquishing prematurely the ability of humans to arrive at religious *knowledge.*

If the employment of Hegelian reason involved a movement

toward the identity of the concept with its object, the same process was reflected in the development of religion in history. It is from the idea of development inspired by Hegel that his influence on Protestant theology was felt in the nineteenth century debate over the relation between religion and natural science. Historians from both the middle of the nineteenth century and from our own time have underscored the link between Hegel's philosophy of evolution and the emergence of historical criticism in the period immediately following Hegel's death.[52]

Among the theological problems Hegel raised for his age, that taken up by left-wing Hegelian theologians—the significance of Jesus in the development of the Spirit to self-consciousness—focused attention directly on the meaning of historical facts. As in natural science, where belief in the central role of scientific facts was the cornerstone on which most based an unquestioned confidence, so in theology the facile assumption was that historical religion rested on objective facts. While it was true that the commonly received understanding of the historical record had been undergoing critical review for a considerable time, no one before Hegel had implied that the Christian faith could be indifferent to historical facts because the essence of Christianity was to be found in great ideas that were immune to historical disproof.

Historical investigations of the biblical record in the years following Hegel's death moved from "lower criticism" of the text, in which the primary goal had been to determine the most accurate reading, to "higher criticism," in which the Bible was subjected to the same analysis any secular literary work received. In this task no holds were barred, nothing was so sacred that its historicity could not be questioned.

The most notorious application of Hegel's thought to the historical question was David Friedrich Strauss's two-volume *Life of Jesus* (1835, 1836). Strauss appreciated Hegel's teaching regarding the reconciliation of the finite and the infinite in the unity of God and humanity, and he agreed with Hegel that the reconciliation could only take place in history. Strauss preferred to raise the process of reconciliation above a single individual and even above the singularity of historical fact. His study of the life of Jesus depended upon the exposition of what he called myth, the historically conditioned expression of the Hegelian Idea. If Christian reconciliation was not to be made dependent on historical fact, then myths, which were always born of a particular time and place, must not be mistakenly viewed as the core of Christianity.

The *Life of Jesus* of 1835 set out to expose the mythical elements contained in Christian teaching about Jesus. To do so Strauss appealed to the well-established critique of supernatural intervention already carried out by rationalist theologians. Beyond this Strauss argued that accounts in the Gospels should be rejected whenever they were internally inconsistent, whenever they violated commonsense principles of human behavior, and whenever they betrayed the use of literary devices clearly intended to promote the biased intent of their authors.[53] By the time he was through Strauss had rid the New Testament portrayal of Jesus of so much that what remained could hardly serve as the inspiration of faith and devotion.

Yet Strauss wished to assert that the essence of Christianity, when properly interpreted and understood, *could* inspire devotion and commitment. Strauss criticized right-wing Hegelians who, having sought ideal Christian truth beyond the brute facts of history, retreated back to the brute facts as if they were also necessary. In Strauss's Hegelian perspective one remained on the level of the ideal. Strauss tried to persuade his readers that the Christ of history was not a person but an idea.[54]

There were other Hegelian theologians after Strauss who, though they appreciated the claim that the Christian principle could not be simply identified with the historical individual named Jesus, struggled to rescue a sense in which the historical Jesus was more than of merely accidental interest. Ferdinand Christian Baur, founder of the famous Tübingen school of higher criticism, labored to establish a positive historical basis for faith that was not mythical.[55] The most prominent Hegelian theologian at mid-century, Alois Biedermann, agreed with Baur about the necessity of Strauss's negative critique and about the need to go beyond that critique in a way different from Strauss himself. Biedermann was convinced that Strauss had forsaken his Hegelian goal at the point where the challenge was greatest: the establishment of a purely conceptual positive statement of Christianity's truth.

Among the Young Hegelian theologians, however, the one

whose influence was most directly felt in the scientific community was Ludwig Feuerbach. Feuerbach's critique of Hegel in 1839 marked a turnaround from the strongly idealistic position he had held throughout the 1830s. Here and in his famous work of 1841, *The Essence of Christianity*, Feuerbach demonstrated his dissatisfaction with his former approach. In Feuerbach's conversion to what he called "an empirical or historical-philosophical analysis,"[56] speculative theology gave birth to its opposite.

Based on his approach of "unveiling existence" instead of "inventing" it,[57] Feuerbach examined the major doctrines of Christianity in order to demonstrate their tie to real and immediate human needs. In the process he exposed religion as one grand anthropomorphism, a projection of human needs and wishes into the heavens. Feuerbach's intent was, like Strauss's, not so much to destroy Christianity as to preserve it by identifying its essential nature. For Feuerbach religion was, in Karl Löwith's phrase, the "detour taken by man on the way to finding himself. Man first transfers his own essence to a point external to himself before he finds it within himself."[58]

Feuerbach's critique of Hegelian thought and his grand expose of religion not only profoundly influenced the development of dialectical materialism in Marx and Engels, it also served as the conceptual point of departure for what has become known as the German scientific materialism of the 1850s. In the writings of Karl Vogt, Jakob Moleschott, and Ludwig Büchner, Feuerbach's message served as the inspiration for an aggressive campaign to replace unwarranted submission to religion with a commitment to the overriding value of natural science: free and uninhibited inquiry.[59] Thus the scientific materialists ironically obtained their intellectual preparation for Darwin from a one-time theology student who, having turned from theology to philosophy under Hegel, abandoned his master to become, as he described himself, "a natural scientist of the mind."[60]

Not for several years after the *Origin of Species* in 1859 did anyone from the Hegelian school take up the question of the meaning of natural science for theology. Feuerbach wrote nothing that captured attention after his *Theogonie* of 1857, and even that did not create much interest. For his part Biedermann offered only the most cursory observations on the subject. When the Hegelians did get around to concerning itself with Darwin, the response came from the historical-critical school Hegel had spawned. The first representative was none other than Strauss himself in a book called *The Old Faith and the New*. As the title indicated, Strauss was still attempting to characterize the essence of religion, especially as his understanding contrasted with that of more traditional minds.

Orthodox Theology

With the ascent of the romantic visionary Frederick William IV to the throne in 1840, compromise in the confessionalist controversy that had raged for over two decades ensued. The insistence on literalism where the confessions were concerned exerted so powerful an influence that the king decided not to force the issue. Lutherans who demanded that they remain separate from Evangelicals were formally recognized in 1845, although they were not permitted to use the word "church" to identify their Lutheran organization.[61]

It was not long before there appeared theologians who, like the Old Lutherans, wished to retain their adherence to the historical confessions, but who did not wish to claim separate status from territorial churches. These "New Lutherans" were centered in Erlangen University. They were led in the 1840s by Adolf Harless, later by Johann Christian Konrad von Hofmann and Gottfried Thomasius. All three of these men served as editors of the theological organ of the Erlangen program, the *Zeitschrift für Protestantismus und Kirche,* which was founded in 1838 and continued until 1876. The program opposed what it called "destructive rationalism" and "pietistic separatism," but it also asserted that "the church cannot allow freedom under the name of research and science when the scientific investigator . . . separates himself from the higher duty of conscience that imposes a commonality on all individuals who belong to the church."[62] Members of the Erlangen school saw the primary task of theology to be the recovery and articulation of the living truth at the heart of the church confession.[63]

Confessional theology was at its height in the 1850s and 1860s, those difficult years for Germany preceding unification. Paralleling hopes for political unification raised during the Revolution of 1848 was the movement for ecclesiastical unification. In spite of instituting an all-German annual meeting *(Kirchentag)*, the church turned back to a more conservative course following the failure of the revolution.[64] When Martin Kähler looked back to the 1850s from the end of the century, he remembered the dominant presence of biblicists and confessionalists.[65] From the continuing campaign of Ernst Hengstenberg for the authority and inerrancy of the Bible, carried out after 1827 in the *Evangelische Kirchenzeitung* under his forty-two-year editorship,[66] to the attempts of German Lutherans to save the classical doctrines of the person and nature of Christ from destruction by the historical criticism of the New Testament, the need to restore a solid objective foundation to Christianity was everywhere evident among conservative German theologians. Germany participated in what Welch calls "the mood of restoration and recovery" that was strong in mid-nineteenth century Christianity.[67]

With victory over France in 1870 Germany emerged as a unified Reich under Prussian administration, raising the question of the status of the confessional churches in the annexed states. To allow a separate and formal Lutheran church, which Bismarck proposed to do, ran counter to the Kaiser's view, to say nothing of its inconsistency with a spirit of national unity. Bismarck as usual engineered a compromise under which the territorial churches were neither forced into the Prussian union nor exempted from state regulation. In this way he was able to avoid the separation of large confessional groups from the state church.[68]

For the remainder of the century orthodox Protestant theology, up to this point heavily involved in the endeavor to preserve and clarify traditional doctrines of the Trinity, Christ's nature, free will, election, and the like, began more and more to respond aggressively to the challenges facing religion that were so visible in the new age of realism and materialism. As doctrinal debates receded, confessionalist theologians turned to biblically based preaching and apologetics in defense of what they saw as the basic truths of the gospel and its overwhelming power to meet the deepest needs of the human heart. The *Neue Kirchliche Zeitschrift,* which eventually filled the lacuna caused by the cessation of the confessional journals of the 1840s, continued to oppose union but also saw the need to fight the positivism it saw enhanced by the exact sciences.[69] To be sure, the extensive development of liberal theology in Germany during the nineteenth century meant that the situation for German conservatives was very different from that of their counterparts in other countries. From the vantage point of German delegates to the international meeting of the Evangelical Alliance in New York in 1874 the supreme confidence of English and American theologians was sorely misguided.[70] German conservatives knew full well the extent of the challenge facing apologists of traditional Christianity in a modern world with its "modernist" theology.

Two of the major centers of the new biblical theology were Tübingen and Greifswald.[71] Earlier in the century Tübingen had been famous as the location of F. C. Baur's Tübingen school, which in the 1840s supplanted the rational supernaturalistic view of the old Tübingen school with a purely historical, scientific investigation of the New Testament that rejected all supernatural explanation. The new Tübingen school began its decline in the late 1840s, finally dying with Baur himself in 1860. Within three years Karl Köstlin, a former member of Baur's school, could write to another of Baur's former students, Adolf Hilgenfeld, about the new atmosphere in Tübingen: "With regard to your question as to how things stand here, there is little to tell you that is good; the Old Tübingen School is now here again, just as it was before Baur."[72]

But it was Greifswald which housed the conservative theologian who would treat the relationship between natural science and religion most systematically. While J. W. Hanne considered the "ape origin of humans" in a series of articles for the *Zeitschrift für wissenschaftliche Theologie* in 1868, his colleague Otto Zöckler labored from the late 1850s until 1906 to defend biblical Christianity as he understood it against the challenges brought by modern natural science. Zöckler was not alone in his call for German theologians to give more attention to apologetics, but he clearly took the lead among orthodox theologians addressing the issue of the relation between science and religion. As a result his thought provides the best and by far the most complete example of the conservative German theological community's stance on the question.

Mediation Theology

As its name connotes, the essential characteristic of *Vermittlungstheologie* was its refusal to identify with either extreme on the continuum of theological opinion. For almost any set of opposite positions that can be named, the mediating theologians wished to draw from both sides. They appreciated the methods of the rationalists, yet they never intended to abandon the explicit theism of the supernaturalists. They affirmed the role of historical confessions, yet they acknowledged that there could be no absolute formulation valid and obligatory for all times. They defended as necessary a complementary relation between theological scholarship and practical church affairs, but they did not permit one to be excluded or overemphasized. And finally, they respected the conclusions of Hegelian theologians, but they did not revere the speculative human intellect as the final authority to which lesser human capacities, such as feeling or even empirical perception, must necessarily submit.

All of this was characteristic of the life and work of Schleiermacher. Although some have been willing to declare that the mediating theologians formed Schleiermacher's school,[73] others have resisted associating the two. Barth grudgingly agrees with those who say that Schleiermacher established no school, while Welch clearly wishes to view the mediators as theologians who attempted to go beyond Schleiermacher.[74] Even the mediators themselves resisted the notion that they constituted a school. The editors of the most famous organ of mediating theology, the *Theologische Studien und Kritiken,* which ran for over a century beginning in 1828, announced in the opening number that in avoiding slavery to the letter on one side and the lawlessness of fanaticism on the other, the journal belonged to no school and did not wish to establish a new one. In particular, wrote the editors, they sought contributors who in no way saw themselves as members of "mediation theology as a school."[75] What resulted was a loosely knit group of theologians who strove to satisfy simultaneously the claims of *Wissenschaft* and the church.

All these disclaimers to the contrary, there is an important sense in which Schleiermacher's approach dominated the program of mediation theology even after his death in 1834, and it is in this same sense that Schleiermacher's thought represents mediation theology in its relation to natural science. In typical fashion, Schleiermacher neither embraced wholeheartedly what Hasler has called the destruction of feudal nature that was threatened in idealistic philosophy, nor made nature simply the instrumental object of divine caprice in his theological system. Nevertheless, if in Kant's theology religious cognition was an impossible goal, Schleiermacher and those who came after him wished respectfully to disagree.

It is on this point that some members of the mediating school of nineteenth century German theology thought they could distinguish themselves from Schleiermacher himself. Because they did not appreciate the central place nature occupied in Schleiermacher's thought, they credited Schleiermacher only with the groundbreaking insights that it was impossible to reason from knowledge to faith, and that doctrine was to be understood as an articulation of humankind's primary religious consciousness. These mediators saw Schleiermacher as one who remained content with a mere description of that consciousness rather than one who attempted to move from the realm of feeling and faith to knowledge. In other words, some mediators understood Schleiermacher to be conceding that knowledge was impossible in religion, while they were not at all willing to give in to such a claim. They felt it was therefore necessary to go beyond Schleiermacher, not to the extent that they wished to grant religious knowledge the upper hand over feeling, as did Hegelian speculative theologians, but at least to the extent that theology retained an actual cognition that complemented our nonrational experience of religious consciousness. As August Twesten, Schleiermacher's successor in Berlin, put it: Schleiermacher assumed a standpoint "above" the Christian consciousness, but he, Twesten, meant to stand "in" it.[76]

In fairness it should be pointed out that most of the mediating theologians did not pursue their program of reconciling historical Christianity and the modern scientific consciousness by turning their attention to the specific question of the relationship between theological and natural science. Not until the scientific materialism of the 1850s and Darwin's *Origin of Species* of 1859 did the question force itself onto the German theological scene.[77] Even then German theologians, like theologians elsewhere in Europe and America, were not quick to conclude that natural science had become antagonistic to Christianity.[78] Up to the 1870s the mediating theologians were little concerned with apologetics. Their task was not a defensive endeavor; it was a positive and constructive attempt to provide a genuinely critical renewal of theology that would reconcile Christianity and modern culture.

In the early days of *Vermittlungstheologie* German theologians applied their efforts to the major branches of the theological enterprise. Twesten took up the restoration of doctrinal theology, a task he had expected Schleiermacher to address more than he did. Karl Immanuel Nitzsch in Bonn, later in Berlin, used systematic theology to link dogmatics and ethics in a system structured around biblical concepts, an approach Heinrich Hermelink has called "the purest form of mediating theology in the sense of a biblical reinterpretation of confessional writings."[79] The historian of the group was Karl Ullmann, who had studied with Schleiermacher in Berlin and also with the rationalist Paulus and the Hegelian Karl Daub in Heidelberg. Ullmann's work on the sinlessness of Jesus showed the influence of both of his Heidelberg teachers through its combining of the moral Christ with the historical Jesus.[80]

If in the first generation of mediating theologians "a strong biblical orientation was fused with revival impulses and with the influence of Schleiermacher,"[81] in the second generation the central concern was specifically the doctrine of the person of Christ. Earlier mediators, and indeed Schleiermacher himself, had called attention to the centrality of Christology for all theology, but it was Isaak Dorner, the most well known of all the mediators, who made this doctrine the basis of a systematic theology that purported to move beyond Schleiermacher, from faith to objective religious knowledge.

In order to justify his claim that God as an object of religious consciousness could also be an object of knowledge, Dorner argued, as Brandt maintains Schleiermacher argued, that religious faith presupposed a universally valid rational knowledge. In other words, for the God of faith to be possible, knowledge of God must also be possible. This unpacking of knowledge was for Dorner the task of systematic theology working together with philosophy.[82]

Dorner's thinking resulted in a reformulation of the doctrine of the person of Christ in the mid-1850s. In the process of working out the implications of God as an object of knowledge, Dorner was led to question the assumption that God was immutable in every respect. What resulted was a doctrine of a progressive incarnation in which "God as Logos constantly grasps and appropriates each of the new facets that are formed out of the true human unfolding."[83]

Of those theologians whose views are better classified in the mediating school than elsewhere, only a few made reference to natural science. To find an appropriate representative of the spirit of the mediating view who did concern himself with science it is necessary to move from the ranks of the best-known theologians and move among less well known figures. At first glance Alexander Schweizer of Zurich, a loyal follower of Schleiermacher who rejected extremes both to the right and the left by continuing his mentor's commitment to practical theology, appears to be a good choice. But Schweizer came to natural science indirectly by critiquing the enthusiastic reception Darwin received in the popular writing of David Friedrich Strauss. A better choice is Rudolph Schmid, president of the Theological Seminary in Schönthal, who wrote two extensive studies on Darwin in which the general relationship between science and religion was addressed. Although Schmid was not particularly well known, he wrote from a stance that placed him among the mediators, and he intended his work for lay people as well as other professional theologians.

For the majority of the mediators, as for the most important members of the Hegelian speculative tradition in theology, the central issues in the closing decades of the nineteenth century did not involve natural science. As late as 1876, when the Göttingen theologian and acquaintance of Dorner, Friedrich Ehrenfeuchter, addressed the religious crisis of his day in a book entitled *Christianity and the Modern Worldview,* he explained the rise of the anti-religious spirit as a result of the thinking of philosophers and theologians from the eighteenth century to his time. There was only an envious passing reference to "the exact investigation of nature" as a discipline that had preserved its ability to furnish some kind of certainty in a world where uncertainty reigned."[84] For theologians who were used to asking the question, What is it that is true in and by itself?

it was up to the German spirit, as Dorner put it, not to succumb to English empiricism.[85]

Ritschlian Neo-Kantian Theology

If German theologians of the schools identified above had finally begun to address the relationship of natural science to theology after 1870, there was another school just then coming into its own that would also have something to say on the subject. John Groh writes that the rise of the Ritschl school in Göttingen around 1870 signaled that mediating theology was dead.[86] Dorner confirms this modern assessment from his nineteenth century vantage point in a letter to H. L. Martensen, written after the publication of his *System of Christian Doctrine* in 1879. Ritschl, wrote Dorner, was the representative man of the new theological mood that wanted to loose itself from metaphysics and turn to the realm of the practical as the new foundation on which to base theology.[87]

The mediators had for some time enjoyed the favor of state authorities because of their moderate position between extremes, but the new German empire demanded a theology of action to match the spirit of the times. The radical message of this new school about the relation between natural science and religion would prove to be wholly unsatisfactory to all of the existing schools, be they conservative or liberal. The Ritschlians rejected the positions of conservative, mediating, and speculative schools alike, and in the position they assumed on the question of science and religion, they threw down the gauntlet not only to their contemporaries but to future theologians in the twentieth century as well.

The early career of Albrecht Ritschl contained little that would suggest he would one day be the founder of a school. Born in 1822 to an old Prussian family, Ritschl reflected the academic religious interest of both his father and grandfather when he took up university studies in Bonn. Ritschl's father was a bishop in the Prussian Union church in Pomerania, a stronghold of confessional Lutherans whose growing strength and opposition eventually motivated him to retire.[88] Ritschl grew up, then, in what might safely be called an establishment family of the Prussian elite, whose loyalty to the throne allowed for flexibility where ecclesiastical politics were concerned, but whose acceptance of a traditional supernatural Christianity was never seriously questioned.[89]

But Albrecht went off to Bonn in 1839, just as the Hegelian theological tradition, which was suspect enough from his father's point of view, was giving rise to Young Hegelian critics on the left and a supernaturalistic Hegelianism on the right. Bonn was dominated by Karl Immanuel Nitzsch of the mediating school, who, along with other Bonn theological professors, tried to balance theology between the demands of *Wissenschaft* and the more practical needs of the heart. Ritschl read Strauss's *Life of Jesus* around 1840, and, although he wrote to his father that he did not want his study of Hegel to land him in Strauss's camp, he was impressed enough with the need to be *wissenschaftlich* that he complained to his father about the rigid position of the orthodox school. Writing about a follower of Hengstenberg who had charged that his teacher Nitzsch was an unbeliever, Ritschl observed: "If only these people did not want to confuse the area of science [*Wissenschaft*] with that of belief, they might keep their unscientific approach to themselves."[90]

Ritschl did not escape the influence of either Schleiermacher or Hegel as a student or for some time thereafter. Especially in Halle, where he completed studies for a doctorate at the end of May 1843, Ritschl became enamored of the Hegelian approach to theology of Ferdinand Christian Baur. After receiving the doctorate Ritschl spent the remainder of the summer semester in Halle engaged in private studies of Schleiermacher and of Kantian philosophy.[91] But it was Baur's historical investigation of Christianity that engaged him. At Tübingen under Baur's influence he completed in 1846 a critical investigation entitled *The Gospel of Mark and the Canonical Gospel of Luke,* in which he argued for Luke's dependence on Mark as the source for the original gospel narrative.

Ritschl's first academic position was as lecturer back in Bonn at his old university. From 1846 to 1853, when he was promoted to extraordinary professor, Ritschl taught primarily New Testament and historical theology. In the winter semester of 1853–54 he offered, for the first time, a course in dogmatics. This course, plus courses in New Testament and another on theological ethics, Ritschl would offer regularly for the rest of his career in Bonn and throughout his later years in Göttingen.[92] It was in 1864 that Ritschl left Bonn for Göttingen to take up the position in dogmatics previously held by the Hegelian Isaak Dorner.

Although Ritschl's teaching focused on dogmatics and ethics after 1853, a cursory analysis of his publications reveals that, despite continuing to write in these areas, he became ever more heavily engaged in the history of theology. In 1853 he published an article on the Elkesaites, an ancient sect in which magic and astrology played a conspicuous role, and another on fourteenth century mysticism. In 1857 his eventual joining of dogmatics and Reformation history was signaled by an essay on Andreas Osiander's doctrine of justification. In the 1860s Ritschl turned his attention to the study and promotion of Luther, although it was not until after he went to Göttingen that his historical study of the doctrine of justification and reconciliation and his massive history of pietism made their appearance.[93]

The world of secular politics commanded Ritschl's attention briefly while he was still young. In 1848 he endorsed constitutional democracy for Germany and was, as a new theology professor in Bonn, an observer at the Frankfurt Parliament. With the failure of the Revolution of 1848 Ritschl confined his overt political activities thereafter to ecclesiastical matters concerning church union, although he was known to harbor conservative pro-Bismarck sentiments in the 1860s and after the German states unified into a new German empire.

Within clerical ranks he wrote no less than seven articles in favor of church union or against the New Lutheran separatists. Some wish to see this activity as an expression of Ritschl's early held belief that the task of modern theology was to find the true religion that lay beyond the difficulties in the confession,[94] but others point out the definite link between Ritschl's polemical writings against confessionalists and the confessionalists' treatment of his father, to say nothing of their opposition to Ritschl's own understanding of Lutheran Protestantism. "Ritschl's reformation studies must be placed in the context of his continuing animosities toward the confessionalists and the pietists," writes Philip Hefner, "because these studies were urged as weapons against these parties."[95] Hefner notes that Ritschl's pro-union sentiments won him no friends in Göttingen, which was in the thoroughly Lutheran province of

Hannover, and that Ritschl felt severely ostracized by his university colleagues when Hannover was annexed by Prussia after the victory over Austria in 1866.[96] Ritschl, who more than once displayed an insensitivity to the feelings of others, did not help himself in October of 1869 when he identified orthodox Lutheran confessionalism as political anti-Prussianism.[97]

The description of Ritschl's career so far provides virtually no grounds for understanding why a Ritschl school emerged in the 1870s. Clearly what made this possible was the publication of his magnum opus, *The Christian Doctrine of Justification and Reconciliation,* beginning in 1870. For many in the German scholarly community Ritschl's erudite exposition, with its clear focus on two fundamental doctrines of the Christian religion, struck a welcome chord. Ritschl's return to basics spoke to the sense of loss and disorientation that accompanied the "future shock" of late nineteenth century Germany.

It was an age of profound and rapid change. The political unification achieved in Bismarck's new empire suddenly confused age-old identities and loyalties of Germans used to seeing themselves first as Hessians, Bavarians, Badeners, or citizens of some other German state. The swiftly expanding German industrial sector brought with it a new world of machines which, especially in transportation and communication, meant that Germany had begun visibly to share in the "life at high pressure" described by the Englishman W. R. Greg.[98] New social realities spawned by the swelling middle class and a growing workers' movement altered traditional roles in Germany's tripartite class structure. Finally, advances in scientific and technological knowledge had produced an accompanying worldview which did not respect the honored truths of historical religion. Older Germans who could recall life from early in the century could well have cited the English poet John Donne's succinct characterization of life in the wake of Copernicus's disorienting achievement: "Tis all in pieces, all coherence gone."[99]

Amid the insecurity of an iconoclastic modern age Ritschl assumed, without conscious intent, a prophetic role. Admittedly, this image of Ritschl must be justified, since his acquiescence to the new Germany of his fellow Pomeranian, Bismarck, hardly qualifies him as the radical prophet who boldly calls the king to account.

Ritschlism, in fact, has been called a theology for the rising Protestant bourgeoisie of the new empire.[100] To those of a Marxist bent Ritschl's theology represents the shift in the bourgeois conception of nature that accompanied the post-1848 era.[101]

And yet the impact of Ritschl's work was as a call to return to the essential message of Protestant Christianity in spite of all the challenges it had endured. The focus was Luther and the Reformation, and the theme was justification by faith. Contrary to the emphasis on God's intolerance of sin and the human anxiety it provoked, which scholars of the day customarily found in Luther's thought, Ritschl's interpretation of Luther's contribution argued that justification was solely an act of God that in no way depended on humans. God, who was not estranged and who did not require to be reconciled, aimed the act of justification at humans, who did stand in need of reconciliation.[102] Drawing particularly on Kant and Schleiermacher, Ritschl made the central feature of his dogmatics the notion of the kingdom of God, which represented, in the words of Claude Welch, "the purpose which the whole universe is designed to serve." As such it involved both God's highest will for the world and for human beings and a call to ethical responsibility and action.[103]

But it was not solely Ritschl's positive focus on justification that resonated so strongly with some segments of German society. There was a negative dimension to Ritschl's message that was equally if not more important. Ritschl argued that the proclamation of a justification that was divinely initiated and freely bestowed without dependence on the action of humans had been delivered by the young Luther with a particular intent that was later abandoned. What the young Luther had intended, but the older Luther and others since had not been able to sustain, was a break with the epistemological assumptions of scholastic thought. Writing later in his *Theology and Metaphysics,* Ritschl argued that the young Luther had wished not to appeal to the Platonic theory of knowledge, which had held sway in theology up to then and "in which everything is deduced from above by means of general concepts."[104] The young Luther, in other words, had wished to proclaim the truth of the gospel message without requiring that it be made part of a worldly reality whose truth was allegedly accessible to human reason. Ritschl's young Luther had intended to remove metaphysics from theology. In terms of the analytical framework of this study, Ritschl was arguing that the young Luther had intended to reject the correspondence theory of truth.

But neither Luther, nor Melanchthon, nor theologians since had been able to live up to this worthy goal. Even Schleiermacher, Ritschl noted, shared the fundamental error of portraying God through natural theology.[105] Ritschl's *Theology and Metaphysics* was a reply to those who had denounced his attempt, once it was forced into public attention by the emergence of a Ritschl school, to call his age to account on the basis of his view of the young Luther's intent. Because the Platonic epistemology reigned in theology, Ritschl noted, it was held up as the condition on which the integrity and correctness of all theological presentations of Christianity and Ritschl's personal character depended. His opponents accused him of nothing less than devaluing and mutilating Christianity by denying that theology should make use of universal concepts that would square the truth of Christianity with human knowledge in general.[106]

Ritschl's polemical stance against speculation and metaphysics as constitutive for theology must be viewed against the background of the increasingly aggressive natural scientific worldview of the post-Darwinian years. It is most likely the case that Ritschl had not designed his work on justification to be understood as a rejection of modernity or even as a fundamental challenge to it. He spoke of his own theology as having no apologetic interests.[107]

But it is equally clear that Ritschl rejected any worldview in which human beings were helplessly caught in a cosmic web of mindless cause and effect. In the first volume of his *Christian Doctrine of Justification and Reconciliation* Ritschl, writing about Kant's contribution to our understanding of reconciliation, asserted that "the will of every rational being is moral only as it is a general law-giving will, that is, in its autonomy."[108] But it was not until the ninth chapter of the third volume that Ritschl took up what possessing moral autonomy meant where our understanding of and our relation to the natural world were concerned.

Ritschl conceded that because believers were endowed with a sensual nature they were part of the world and were dependent on it. "No one can alter the mechanical conditions of sensory existence . . . Each one is subject for his self-preservation to the laws of mechanism and organism."[109] Ritschl noted further that humans and other living things have managed to control nature to some extent, each in their own way, so as to produce a multifaceted and mutually supported interaction with nature. But this limited mastery of nature only served to expose the many forces of nature that humans could not bind, and therein lay a lesson:

> [Regardless] how many parts of the world one masters through labor, no one dares to believe himself capable of mastering the whole [of nature] in this way, even if he identifies himself in a moment of exalted disposition with the progressive power of human cultural development. But humans [dare to] compare themselves even to the entire nexus of nature when they comprehend themselves through their spiritual feeling of self as a dominion standing near to the other-worldly God, claiming to live in spite of the experience of death.[110]

Ritschl identified the religious worldview as the certainty that one was an object of the care and leading of God in spite of life in a world of suffering, evil, and death.

> As long as one holds to the view that certain restrictions of our freedom [that is, the cause-effect world of nature] are unconditionally evil, one is recognizing [one's] dependence on natural and particular causes, dependence on the world. In overturning the value of evil one attains freedom not only from the individual things from which evil emerges, but freedom over the world in general. For individual evils do not represent merely the relations in which the world limits our freedom; rather, the opposite thought, that one is an object of divine care, means that as a spiritual whole one has in God's eyes a higher value than the entire world.[111]

Ritschl did not claim that this insight was unique to Christianity, nor even first enunciated there. It broke forth in nearly every religion. What Christianity contributed was a worldview within which the question of eternal life was answered. Mastery over the world in Christianity had to be understood as the task of the kingdom of God.[112]

Was Ritschl recommending, then, that one turn one's back to the world? Ritschl rejected the idea that his understanding called for ascetic retreat from the world since all aspects of the mastery of the world, including control of nature, were included "in the task of the kingdom of God." He clearly did not believe that culture, "the intellectual and technical manner of mastering the world," should stand in contradiction to religion. But while culture in this sense should not contradict religion, it was at best an adventitious factor in the religious worldview.

> The domination over the world to which Christianity guides people is not meant in an empirical sense . . . All possibility and probability which is readily credited to natural science is completely neutral in our view of our practical posture in the world of nature, in part because we cannot consolidate that possibility to any degree of real knowledge [*wirkliche Erkenntnis*], and in part because our judgment of ourselves as spiritual persons is never influenced by our knowledge of natural laws. [113]

What Ritschl in effect declared to his age was that the conclusions of natural science were not "real knowledge," that the truth natural scientists claimed to possess was not of a kind that could be said to correspond to reality. As one who openly admitted that he was relying on Kant and Lotze, Ritschl rejected the metaphysical realism he saw being smuggled into the debate over evolution.

But Ritschl did not wish to dwell on this issue. After noting that "collisions between religion and science, particularly natural science, only originate when laws that apply to the narrower area of nature are raised to world laws and used as keys to an overall view,"[114] Ritschl deferred discussion of the matter to another time. But before dismissing the topic he made one final observation about the charge that the teleological and miraculous aspects of the Christian worldview were intolerable for those who limited themselves to a mechanical explanation of the world.

> If tied to this [approach] is the claim that the scientific view of the world gets by without ideas of purpose or the assumption of miracles, that is a self-deception. Miracles in the sense of effects not mediated through laws are assumed in every philosophical or natural scientific theory of the universe . . . Without the concept of

purpose the explanation of organic beings and that of the whole of nature cannot be undertaken at all . . . If one grants no confidence to this concept in the explanation of nature because it designates a condition of spiritual life, specifically the conscious will, which must not be applied to nature, so likewise is a perspective employing an acting cause merely abstracted from our experience of our own will, and must not be related to changes of natural phenomena if the concept of purpose has to be held apart from its explanation . . . The deception that one invites to oneself in religion is also committed in every investigation of nature undertaken by means of a law of an active cause.[115]

Ritschl's reliance on Kant in all this was no less disguised than was his opposition to the worldview of Strauss or Ernst Haeckel. Although Philip Hefner has correctly observed that "it is quite possible to interpret Ritschl's entire theology as an attempt to deal with the challenge of the scientific worldview,"[116] treatments of Ritschl or the Ritschl school to date have not found their way into the work of historians of science or, even more surprisingly, into accounts of the rise of neo-Kantianism after mid-century.[117] Ritschl, of course, preferred not to dwell on the negative aspect of his understanding of religion; he emphasized the detailed theological exposition of the positive features of justification and reconciliation. Furthermore, Ritschl was not a philosopher. He clearly was rejecting classical epistemology, declaring that Kant's contribution had supplied him with the theory of knowledge on which was based his rejection of natural science as "real knowledge." It was left to a confident young theologian named Wilhelm Herrmann, whose *Metaphysics in Theology* of 1876 signaled the creation of a Ritschl "school," to make clear exactly what the implications of Ritschl's theology for natural science were.

The rise of the Ritschl school in the late 1870s finally brought neo-Kantian theology to the attention of the educated public in a way that could not be ignored. Earlier in the century the neo-Kantian Fries may have won a theological disciple in W. M. L. DeWette in Berlin, and his thought may even have given rise to a Friesian school of philosophy in Jena under the leadership of Ernst Apelt and Matthias Schleiden.[118] But nowhere did a Friesian school of theology emerge to challenge that which Hegel produced.

Up until 1870, then, the recognized schools remained the same three which the young David Friedrich Strauss identified in 1837: the speculative school, the believers *(die Gläubigen)*, and the mediators.[119] As mentioned above, the concerns that occupied the attention of these three schools had been biblical criticism, the confessional controversy, and Christology, respectively. More than other factors, the appearance of scientific materialism in the 1850s followed quickly by the publication of Darwin's *Origin of Species* forced at least some theologians to respond to developments in natural science.

Because Darwin's theory was seen to challenge the existence of purpose in the universe, it also cast aspersions on the Christian doctrine of redemption. None of the representatives of the theological schools from the second half of the century wished to abandon either redemption or purpose. Consequently the German theologians frequently, though not always, examined the relationship between natural science and theology in the context of the Darwinian debates.

PART II

Nature Retained

I shall not like to have to concede that where worldview is concerned, an objective solution is impossible.
David Friedrich Strauss

Belief and science are nothing but two ways to one objective truth that differ with regard to direction, but not with regard to goal.
Otto Zöckler

What is correctly thought subjectively is also objectively true.
Rudolf Schmid

3

The New Hegelian Faith of David Friedrich Strauss

IN MOST WAYS David Friedrich Strauss belongs more to Darwin's own time than he does to the era of the Darwinian debates; indeed, there are notable similarities in the biographies of the two men. Born just one year apart, both attended university in the late 1820s and, although both assumed their futures would somehow be involved with religion, neither finished university studies with a clear idea of what he would do.

Here the patterns diverge, for in the very years Darwin was traveling around the world on a journey whose impact would not be felt until some twenty-five years later with the publication of his notorious magnum opus, Strauss was writing and seeing through the press the book that made him the immediate *bête noir* of Europe. Each, however, was guilty of the same offense, for each spoke the unspeakable so persuasively that the only reasonable response was complete and total outrage. Even in the publications from their old age, Darwin's *Descent of Man* and Strauss's *The Old Faith and the New*, both continued to harass people of deep religious convictions.

"To Strauss," wrote Adolph Kohut in 1908, "goes the renown of being the first one of the leading Protestant theologians from the first half of the nineteenth century who boldly and openly, even enthusiastically made Darwin's theory his own."[1] And yet, one might not expect that a theologian who was profoundly influenced by Hegel would end up embracing any variety of Darwinian materialism. It has already been noted that Hegelian theologians at mid-century appeared little interested in the fortunes of natural science. How is it that Strauss, one of the founders of the Young Hegelians of the 1830s, became the first Protestant theologian to adopt a thoroughly Darwinian worldview in the decade after the *Origin of Species?* If Strauss is to serve as a representative of the response to Darwin from the Hegelian speculative school of theology, it is necessary to clarify the precise manner in which Strauss spoke for this group.

To understand why a member of the Hegelian theological tradition did become captivated by issues and controversies in natural science at mid-century, and to determine if Strauss can be used as a representative of this tradition in any sense, we must retrace the path that brought Strauss to the new confession of faith he set down in his last major work.

The Formative Years

The philosopher Eduard Zeller, friend and biographer of David Strauss, tells us that Strauss was born into an upper bourgeois Württemberg family, "such as frequently existed at that period in a small south German town."[2] The son of a retail merchant whose preferences for reading the classics contributed to his lack of success in managing his business, Strauss was the older of the two children who survived from a total of five born to his parents. Weak and delicate as a child, David turned his energies to his studies at school, with the result that he rose to the top of his class in his native town of Ludwigsburg.[3]

Because of his mastery of Latin and Greek, Strauss successfully passed the series of examinations required for entry into one of Württemberg's lower seminaries. Gifted adolescents who otherwise might not be able to join the sons of wealthier parents at a university could, on completing the seminary's program, be admitted to a university for further study of theology. In 1821 Strauss, a tender youth of thirteen, left Ludwigsburg for the seminary in Blaubeuren.

As in the curriculum of his primary school, language and literary studies occupied a major component of the secondary regimen. In addition to Greek and Latin poetry and prose, Strauss also took Hebrew and French. Although his final grade report listed his ability in mathematics to be merely average, his achievements in the rest of his courses, which included logic, psychology, history, and religion, were sufficiently good to place him among the best six students in the class. To this point in his education there was no indication whatever of an interest in natural science or even in the radical philosophy that would soon mark his future. He was following a course that would bring him straight into the ministry.

One might expect that it was Ferdinand Christian Baur, one of his teachers at Blaubeuren, through whom Strauss's defection came about. After all, Baur would soon become the pivotal figure of the Tübingen school of higher criticism of the Bible. There is, however, no evidence that this was so. For one thing, Baur's lectures at Blaubeuren were confined to ancient literature and history. It is true that Baur was enamored of Schleiermacher's theology when Strauss first encountered him in the early 1820s, but there is nothing to suggest that Baur communicated anything of Schleiermacher to his seminary students. It is also true that Baur was eventually drawn to Hegel and philosophy, but this occurred later, and when it did, it came through the stimulation of his former student Strauss.[4] Strauss's loss of orthodox faith appears to have been as much the result of his own study as of the influence of the teachers he had at Blaubeuren or even those he met in the Tübinger Stift, the famous institution of theological training, which he entered on leaving the seminary.

Along with some of his seminary friends Strauss enrolled in Tübingen University in the fall of 1825 to complete the final phase of his theological training. The first two years of the five-year program called for the study of ancient languages and philosophy. Because Strauss and his friends found Tübingen's philosophers to be woefully inadequate and unexciting, they banded together to read philosophy on their own. Here, for the first time, Strauss encountered a formal consideration of the place of nature in the works of philosophers.

Had the group wanted to turn to classical philosophers its members might have begun their study with such great thinkers as Leibniz or Descartes. But because these young and restless minds were free to choose whomever they wished, they preferred to plunge into the works of the more recent German writers who had been redefining the task of philosophers.

First and foremost there was Kant. Ever since the Königsberg philosopher's critiques had burst onto the scene philosophy had not been the same. Now everyone who wished to claim familiarity with philosophical trends had to be prepared to say whether or not they agreed with Kant's rejection of rationalistic and dogmatic metaphysics in favor of what had become known as critical philosophy. Strauss and his friends dove into Kant and found themselves repelled. As he put it later, he had just emerged from adolescence to a more self-assured sense of self-consciousness. It was not the time for him to hear that in spite of his new found feelings of enthusiasm and self-confidence, much of what humans in the past had aspired to know simply was beyond the grasp of mortals. He believed that it was through direct feeling that one possessed truth. His own words summarize it best. "I could not perceive the reason for all the fuss and mistrustful precautions with which Kant approached the knowledge of things. Far from understanding the questions and problems with which this thinker was concerned, I certainly did not understand why these questions had even been formulated."[5]

If the German romantic philosophers Fichte, Schelling, and Hegel were discontent with Kant's surrender of the knowledge of things-in-themselves, Strauss was downright impatient with it. Fichte's *Theory of Science,* with its abstract rejoinder to Kant, made no impression on the budding theologian, and Hegel's newly found fame and influence in Berlin had not yet affected Tübingen. That left Schelling. Among natural scientists Schelling's influence had been waning since at least 1815. Nevertheless, it was Schelling's philosophical system, especially the *Naturphilosophie,* which provided for Strauss, as it had for many before him, a bold and satisfying reply to Kant's qualification of the human intellectual capacity to know the world as it really was.[6]

If Kant's critical philosophy had forced a rethinking of religion within the limits of reason alone, Schelling's transcending of Kant continued to reshape the categories of theology. Schelling's organic universe, in which God evolved along with the world, produced a comprehensive pantheism that was as attractive to the young Strauss as it was radical. As he struggled to escape the insecurity of adolescence and to establish his independence, his penchant for embracing radical ideas began to show itself. Gone for good was the simple faith of his childhood. What would replace it, however, was far from clear.

Schelling himself had been much influenced by the German mystic Jacob Böhme. Strauss soon found that Böhme's "direct gaze into the depths of the divine and natural worlds"[7] answered his romantic yearnings for philosophical certainty more than did Schelling's focus on the intellect. From German mysticism of the past it was but a short step to Justus Kerner and the romantic spiritualism of the present. Kerner, though Strauss's elder by twenty-two years, came from the same Württemberg town in which Strauss was born and lived within visiting distance of Tübingen. For over two years Strauss reveled in romantic mysticism. During this time he became highly intolerant of any rationalism that purported to make clear in concepts what he was convinced could only be grasped directly through mystical union.

At first glance it is curious that Strauss somehow overlooked Schleiermacher during the first years at Tübingen. The great Berlin theologian was at the peak of his fame in the mid-1820s and, as already mentioned, he was an important influence on Strauss's Blaubeuren teacher Ferdinand Christian Baur. Schleiermacher's rejection of rationalism, his emphasis on the centrality of feeling, and his appreciation of Schelling all suggest that Strauss would have found in him the very combination of openness and vitality he sought. Yet Schleiermacher was not a factor in Strauss's development during these early years.

The answer seems to be that in turning to contemporary philosophy Strauss and his friends did not consider the theologians of the day. Not until the third year of their training did the three-year study of theology commence. When Baur, who was now teaching in Tübingen, directed their attention to Schleiermacher, the group turned not only to Schleiermacher's *Speeches on Religion* but also to *The Christian Faith*.

Schleiermacher's impact on Strauss was undeniable, but it was not sufficient to make him into a disciple. Now twenty, Strauss was rapidly developing a polemical personality. No halfway theology would be good enough for him, even if it did convince him that his preoccupation with the occult was irresponsible and not a worthy foundation for the theology of the future. Schleiermacher's mediating position between philosophy and theology satisfied none of the iconoclastic impulses he had inherited from the romantic authors he had been reading. Schleiermacher had wakened him to the need to take the claims of philosophy seriously, but, in accord with his radical personality, he was not content to take them only as seriously as did Schleiermacher himself. To the group of young students Schleiermacher represented "a precarious armistice, and we found ourselves advised to look ahead to the war."[8]

During his last two years in Tübingen Strauss and his friends discovered Hegel and studied the *Phenomenology* carefully. It was tough going, but Strauss eventually grasped Hegel's central point concerning the Christian religion and philosophy. Religion and philosophy might have the same content, said Hegel, but the former needed to express that content in symbolic form while the latter did not.[9] Both pointed to the same truth—the participation of humanity in divinity. But where philosophy spoke of the evolution of mind to self-consciousness, Christianity spoke of the union of the divine and human in Jesus Christ. Of course few understood the profundity of religious symbols, and because of the temptation to remain content with the symbols themselves, fewer still grasped the philosophical truths toward which the symbols pointed.

Hegel's ultimate focus on the nonsymbolic truth of philosophy meant that he never developed an interest in historical criticism, nor did he value the work of those engaged in the quest for the historical Jesus. Whoever Jesus was, the union of God with humans occurred in his self-consciousness, and the gospel's representation of Jesus' person and his deeds gave symbolic expression of the humanity of divinity and the divinity of humanity. To Hegel there was no need to claim or substantiate a literal interpretation of the miraculous aspects of the New Testament account.

As a youth Strauss had never warmed to the alternative explanations of supernatural events provided by representatives of the nineteenth century rationalistic tradition in German theology. Now he found that the more he pursued theological study, the more problems like miracles and a literal reading of Scripture simply faded into the background. In Schleiermacher's work the essential core of Christianity had little or nothing to do with miracles,[10] and his teacher Baur exposed Strauss and the others to the limited conclusions his developing skills in New Testament criticism were bringing him. By the time he came to Hegel, Strauss had no problem either with the deliberate neglect of miracles or the nonliteral interpretation of the Bible. Indeed, Hegel taught him to look beyond the gospel stories to the deeper philosophical meaning being symbolized.

Once Strauss understood what Hegel was saying about the identity of philosophy and Christianity, Schleiermacher's mediating position no longer seemed necessary. But there were several implications of Hegel's system that Strauss and his friends had to address as they completed their work in Tübingen and prepared to assume duties as clerics. For one thing, what were they to preach to their congregations? While Strauss decided that he could use traditional language in his attempts to communicate the message Christianity contained, his friend Christian Märklin disagreed. Both had accepted responsibilities that involved preaching. Strauss spent nine successful months in a small and rather liberal congregation near Ludwigsburg, but Märklin experienced great personal frustration when he chose to express the gospel to his parishioners in the language of Hegelian philosophy.[11]

Strauss spent the following year, 1831, as tutor in the seminary at Maulbron. During the same year he submitted an essay to the Tübingen faculty in order to earn a doctoral degree. His desire was to go to Berlin where he could study under Hegel. Arriving in Berlin in November of 1831, he met Hegel and looked forward with enthusiasm to the study of philosophy under the master. But within a few weeks Hegel succumbed to cholera, forcing Strauss to be satisfied with attending the lectures of Schleiermacher in theology and the Hegelian Carl Michelet in philosophy.

THE LIFE OF JESUS

It was from Schleiermacher's treatment of the life of Jesus that Strauss first got the idea for what would become his own most celebrated work. From Schleiermacher's lectures he learned a new respect for the Berlin theologian, yet the treatment of the life of Jesus, which Strauss had from notes taken by students during an earlier semester, still suffered from the same deficiencies he had originally attributed to Schleiermacher's whole theology. Schleiermacher had "raised himself above Christ the individual, but [he] has not yet attained to the spirit."[12]

In the summer of 1832 Strauss was recalled to Tübingen as tutor at the seminary. In addition to fulfilling his tutorial duties, he worked on his own account of the life of Jesus for the next three years. Immediately he became known in Tübingen for his defense of the new Hegelian philosophy, a position that soon evoked suspicion among some of the philosophy professors. Nevertheless, it was most likely because of Strauss's return that Baur seriously took up his study of Hegel, so that by the beginning of 1835 Baur too counted himself a partisan.[13]

Once the first volume of Strauss's *Life of Jesus* appeared on June 1, 1835, it was not long before the magnitude of the sensation it would cause became clear. Within eleven days questions about his future at the seminary were raised, and by the fall he was gone. And yet, in light of what had transpired in German theology in the preceding years, what Strauss had written in his book had apparently all been said before. As Baur observed to a friend: "In a certain sense one can rightly say that the work actually contains nothing new, it simply pursues a path long ago struck out and followed to its natural end."[14]

Strauss's book, like Darwin's a quarter of a century later, focused public attention in a way that other works devoted to the same subject had not. Both books acted as a crystal introduced into a supersaturated solution, disturbing the tenuous equilibrium of pretended scholarly normality and, with horrifying suddenness, making unmistakably clear the heretical precipitate society had so naively been dissolving. Some initially viewed each work as just more of the same old radical research that European society had been increasingly absorbing since the eighteenth century. It was not long, however, before everyone realized that these books had gone beyond the limits of what could be tolerated.

That *The Life of Jesus* exploded like a bombshell on Germany and, in translation, on European society is indisputable. *Why* Strauss's book created the sensation it did is more difficult to explain. Some factors are obvious. For one thing, the book clearly was written by an author who displayed extensive knowledge of his subject and who possessed credentials to back up his claims. As with any new contribution, theologians could and did disagree with individual points in Strauss's scholarship. But no one could simply dismiss the work as the pretentious ramblings of a theological hack. In addition, the book was a well-crafted literary production, prompting Schweizer later to call it "one of the most perfect things in the whole range of learned literature."[15]

Finally, the political and intellectual context in which Strauss announced his results was ripe for new controversy. Not only was *The Life of Jesus* the summation and culmination of a long tradition of radical theological thought in Germany, but, because of the attempts to reform Germany's congregations, it also came at a time when religious people from all social levels were testing their rights and their powers to resist external directives concerning their theological preferences. To the throne the refusal of Lutheran and Reformed confessionalists to accept the union of 1817 might well have seemed similar to the stubbornness of political constitutionalists who also wanted to restrict royal prerogatives. Strauss's radical message added fuel to the fire, complicating and stirring up even more the debate over freedom of conscience in the struggle between throne and altar.

But none of these factors is sufficient to explain the incredible impact of Strauss's seminal work. For a complete explanation one must turn to the content of the book itself. When that is done one sees that Strauss did indeed go beyond the theologians who had preceded him, at least those who were well known. Strauss was not simply denying the miracles of the New Testament as others before him had done. In the penultimate section of *The Life of Jesus* Strauss confessed that he was guilty of what must appear to be a "monstrous act of desecration":

> Through the results of our investigation up to this point, it would appear that the largest and most important part of what the Christian believes about his Jesus has been annihilated; all encouragement which he draws from this faith has been taken from him, all consolations robbed. The limitless treasure of truth and life, on which humanity has been nurtured for eighteen centuries, seems to have been laid waste, what [was] most sublime plunged into the dust, God deprived of his grace, humans of their dignity, the bond between heaven and earth torn apart.[16]

What had Strauss said that prompted the young theologian to assume such bold self-importance? To begin with, Strauss took sides with those who denied all historicity to the supernatural events of Jesus' life as described in the Gospels. Not that Jesus himself had not lived and died; Strauss conceded that Jesus was a real person. But not only had supernatural events not occurred, there was also no point in seeking a specific natural or historical event of any kind behind New Testament events. They were, in a real sense, fictional elements.

It is interesting to note that Strauss's conversion to nonmiraculous explanations for all natural events did not come about as a result of his contact with natural science. On the contrary, when Strauss turned to natural science he did so in connection with Schelling's *Naturphilosophie*. If anything, Strauss's involvement with natural science supported the penchant for the occult that he developed in 1827. Most likely it was Schleiermacher's theology, with its denial of the resurrection and other miracles, that more than anything else weaned Strauss away from his facile acceptance of supernaturalism as a youth. Strauss himself locates the moment of conversion in 1828, when he composed an essay on the resurrection of the dead. "I proved the resurrection of the dead with full conviction, both exegetically and also from a natural-philosophical point of view, and as I made the last full stop, it was clear to me that there was nothing in the whole idea of the resurrection."[17]

Giving up supernatural explanations was one thing, but what was to take their place? Why and how did such accounts originate? Strauss's book had in large measure been devoted to answering this question, and his conclusion was sufficiently reasonable to make the work notorious. The gospel stories were myths, fashioned in categories drawn from the Old Testament in such a manner that Jesus could be seen as the Messiah.

Here in the first *Life of Jesus* (Strauss would write a second one in 1864) he was careful to state that the mythical origin of the gospel stories was not the result of deliberate fraud. It was not that individual authors had consciously created first-century variations of Elisha's feeding of one hundred men, his curing of leprosy, or the many other parallels between Jesus's life and that of the prophets. Rather, the hopes and fears of first-century Jews gradually imposed on Jesus prophetic qualities that caused him to be associated with men of God who were known in tales from the past. As the association grew the mythologizing process crystallized into stories that soon were taken as historical fact. While no conscious deceit had been intended, deceit was nonetheless the end product.

And yet the deceit that originated through myth was far more pardonable than that which had been created by the theological rationalists. Strauss delighted in pointing out how contrived and unnecessary the treatment of Jesus' miracles offered by Paulus and other rationalists really was. By trying to come up with naturalistic explanations to replace the supernatural accounts of the Gospels, the rationalists had strained credulity just as much as had the supernaturalists, though in the opposite direction. Although Strauss agreed with the rationalists that no miracles had actually taken place, he felt that it was obviously better to adopt a single principle to show how all the miraculous narratives, including those which described who Jesus was as well as those testifying to his power over nature, had come about than it was to have to cast about for a different natural basis for each new miracle.

Hidden in Strauss's attack on the supernaturalists and rationalists was a radicalism that neither opponent could accept. The rationalists may have nodded assent when Strauss faulted supernaturalists for presupposing that God intervened in the lives of individuals through miracles. They may even have endorsed Strauss's goal of a critical research which had no presuppositions, equating as Strauss did the denial of the miraculous with an absence of presuppositions.[18]

But Strauss was doing something different from the rationalists and supernaturalists alike in his exposure of the mythological basis of New Testament miracles. He was, in effect, attempting to reject the straightforward classical correspondence notion of truth in favor of the pure coherence theory of truth put forth by Hegel, his mentor. Later in his career Strauss would return to the correspondence understanding of truth that was so appealing to his age. But here in 1835 he latched onto the radical idea of pure coherence. Hegel's understanding of the truth of an individual claim like the existence of a miracle was not to be determined by its correspondence to a real event in the past.[19] Rather, Hegel wished first to ask whether the claim was consistent with the whole of human cognition. Since a literal belief in the existence of a miracle did not fit coherently with the rest of our understanding of how nature worked, it could not be a true belief.

But for Hegel, and for the early Strauss, that was not the most important result. Since religion and philosophy had the same content, miracles must represent the truth of that content symbolically. The task, then, was to interpret miracles in a way that *was* consistent with the whole of cognition. Miracles understood in this manner could be said to be true.

Of course the average reader of Strauss heard only the negative result. For those who understood the truth of a claim to mean its correspondence to real events in nature, Strauss's reliance on Hegelian idealism did not supply a satisfactory correspondence to reality. Strauss's readers could not get past his apparent rejection of the very need for historical fact. Although Strauss himself might deny it, it was as if he were disputing that religion, or for that matter any human intellectual endeavor, required contact with or correspondence to an external source, such as history or nature, which lay beyond its influence. Strauss might protest that he did not intend his readers to draw this simplistic conclusion, but when he promised to "re-establish dogmatically what had been destroyed critically,"[20] his readers found their suspicions confirmed.

Yet this was what provided the real outrage, for his denial that we have any reliable factual knowledge of history meant that he rejected historical facts as a basis for meaning in favor of an interpretation that made them a reality in the present. He seemed to be saying, far more openly than Hegel, that the world ultimately was what we say it is. Those who relied on an old-fashioned correspondence understanding of truth were simply misguided. The rationalist who presumed he could know what natural phenomenon had really occurred in place of the alleged miracle was just as biased as the supernaturalist himself.[21]

Except for this difference one might have expected that to orthodox minds Strauss and the rationalists would have been viewed as one and the same. After all, both denied that the New Testament portrait was accurate. But Strauss's formal rejection of the primary authority of external reality, especially because it was combined with youth and haughtiness, gave the greater offense and contributed much to the resentment Strauss provoked in his audience.

When, for example, he irreverently referred to some of Jesus' miracles as "sea and fish stories," he did so in the mocking and disrespectful tone that later prompted Barth to characterize the tendency of his life's work as a desire to find the easiest opportunity for striking out against the theology of the church.[22]

To the general reader Strauss appeared at best to be extremely presumptuous. To professional theologians he lacked the deference they expected in a beginner. To take on supernaturalist explanations might have been tolerated, but to value them no more than rationalist explanations was as bold as it was novel. Nor were rationalists and supernaturalists the only objects of his attack; even liberals like Schleiermacher were cast aside when Strauss audaciously denied the authenticity of John's gospel, thereby daring to reopen a question that many theologians assumed had been decided.[23]

The response to *The Life of Jesus* confirmed that Strauss had opened up a new theological option to supernaturalism and mediation theology. The great bulk of the reaction was negative and critical of Strauss, but that was to be expected. What was surprising was the lack of comprehension of the mythological principle Strauss had defined. Supernaturalists restated their supernaturalism, rationalists continued to explain away miracles through natural means, mediators continued to borrow a little from each. As Schweizer notes, "few understood what Strauss's real meaning was. The general impression was that he entirely dissolved the life of Jesus into myth."[24]

In time the problems implicit in Strauss's mythological principle eventually received the attention they deserved. If *everything* was not mythical, then how did one decide what was and what was not? If Jesus was not divine, then who was this person who had so profoundly influenced Western history? In the years of the initial reaction, however, these questions had to wait for a simpler, more bombastic counterattack to run its course.

Strauss had forced the issue of the relation between the Christian message, however it was stated, and the historical Jesus. Supernaturalists counted the resurrection of the historical Jesus among those matters of which they were certain. For them the response to Strauss was easier than for others. His results represented the logical conclusions that followed from the false assumptions of speculative German theology. The only more radical step would have been to say that the gospel stories were created as a deliberate hoax by the New Testament authors.

Ernst Hengstenberg called *The Life of Jesus* "one of the gratifying phenomena in the area of recent theological literature."[25] Like Baur, Hengstenberg did not regard the book as something new, but saw it as a culmination of elements that had been circulating for some time. "Strauss has done nothing more than to bring the spirit of the time to consciousness of itself, of the necessary consequences that proceed from its primary essence." What was the primary essence from which Strauss's disastrous consequences issued? It was Hegelian philosophy applied to the Christian religion. Hengstenberg, anticipating the protests of right-wing Hegelians, asked himself whether "Strauss had perhaps erred in his interpretation of the true sense of Hegel's philosophy of religion." His answer was unequivocal: "That we in no way believe."[26]

Hengstenberg credited Strauss with hastening what he called the separation of faith and unbelief into the polar opposites they were. "Unbelief will more and more cast off the elements of faith to which it clings, and faith will cast off its elements of unbelief. That will be an inestimable advantage."[27] Hengstenberg distrusted philosophy in general, and Hegelian speculative philosophy in particular. As a representative of the new Hegelian direction, Strauss brought to light for him the symmetric relation between two systems of theology which departed from contradictory starting points.

Hengstenberg was not alone, of course, in thinking of Strauss's book as Hegelian theology. On its publication *The Life of Jesus* was commonly assumed to be an application of Hegelian thought to the study of the Gospels.[28] Strauss himself judged Hegel's system to be important to theology because of the distinction it allowed between the theoretical content of religion *(Begriff)* and the representation of that content in symbols *(Vorstellung)*. At Tübingen Strauss struck on the idea of writing a dogmatic theology that would examine doctrines in light of their biblical (symbolic) bases but that would also look at them with an eye to reestablishing them in a purified philosophical form.[29] *The Life of Jesus* was a case study of this approach, in which Strauss asked whether the Gospels were part of the theoretical content of religion or merely a medium of representation which was separable from fundamental religious conceptions. In 1835 Strauss did not question Hegel's assumption that, although separable, the philosophical concept and religious representation were two forms of the same reality.

Some have argued that Strauss's Hegelianism was not an intrinsic part of his theological contribution in 1835, and that it relatively quickly evaporated after *The Life of Jesus*. But in the same work which is cited to document Strauss's acknowledgment of the difference between himself and Hegel on the value of historical criticism (the third volume of his *Streitschriften*), Strauss made explicit that he saw his endeavor to be deeply indebted to Hegel: "My critique of the life of Jesus stands in a close, inner relationship in its origins to the Hegelian philosophy. Already in my university years the most important point of this system for theology seemed to my friends and me to be the distinction between representation [*Vorstellung*] and concept [*Begriff*] in religion, which could have the same content in different forms."[30]

As far as Strauss himself was concerned, the mythological interpretation of *The Life of Jesus* was an application of Hegelian philosophy to theology. The myths of the gospels were Hegelian *Vorstellungen*, symbolic expressions that pointed to a deeper truth. Where Hegel had been interested in how and to what degree the deeper truth was expressed in historically conditioned symbolic categories, and therefore was relatively unconcerned with the question of the historicity and the task of historical criticism, Strauss worried only at the end of the *Life of Jesus* about reestablishing dogmatically what he had destroyed critically. His primary mission was to expose the symbol as symbol; that is, he saw his primary contribution to be the demonstration that symbols were not to be taken literally. The best way to accomplish this was to show, using the tools of historical criticism, that a literal interpretation of the Gospels was impossible. In this way Strauss intended to force theological attention where it belonged—on the spiritual truth behind the life of Jesus. In his letter to the officials at Tübingen, in which he responded to the demand that he show how his views were compatible with his position as a teacher of religion, Strauss described how he had intended to blend Hegelian philosophy and criticism.

> Now if these two designated directions in the present-day theology—the philosophical and the critical—worked in this manner, hand in hand with each other, then whoever has become intimate with both, as I have, must find himself challenged to bring these two directions into association, and, supported by the philosophical conviction of the intrinsically true content of the New Testament history, to allow its historical form to be ruthlessly investigated by criticism.[31]

Writers who argue that Hegel was not of central importance to Strauss point to the difference between the two on the need for historical criticism and suggest as well that there is nothing necessarily Hegelian about Strauss's mythical principle. It must be granted that like others who cut their philosophical teeth on Hegel, Strauss too adapted Hegel's philosophical system to fit his own needs and wishes. One indication of this appears in later works where he abandoned the view that the symbolic character of the gospel narratives, indeed of Christian doctrine in general, *had* to point to the deeper truth also available through philosophical reflection and analysis.

But if he abandoned the representation in favor of the concept, he maintained his commitment to the latter for the rest of his life. Where Hegel insisted on the identity of philosophy and religion, Strauss, perhaps spurred on by his official rejection by the theological community, forsook his earlier commitment to the value of religious symbols and with it his willingness to permit their use among those incapable of assimilating the deeper meaning. Unlike Hegel, Strauss came to believe that religious symbols were ultimately arbitrary expressions that did not necessarily communicate truth. But in abandoning them, he nevertheless retained his commitment to philosophical truth. In his book on the Young Hegelians William Brazill notes that Strauss "assumed as his sole responsibility the duty of ushering in the Young Hegelian apocalypse, of smoothing the passage from the age of religion to the age of philosophy. All of his work had this duty at its core."[32] In carrying out this duty, Strauss clearly saw himself as fulfilling what Hegel had intended.

The Transitional Years

In *The Life of Jesus* Strauss's eventual disagreement with Hegel was foreshadowed in the very place some locate Hegel's strongest influence on the book. In his evaluation of the person of Jesus, Hegel, consistent with his insistence that philosophy and religion were identical, held that the union of God and humanity occurred in the self-consciousness of Jesus. The incarnation was the recognition of the immanence of spirit and a denial of God's complete transcendence; it was the external representation of a truth that philosophy knew purely conceptually. Strauss, too, insisted on what he called the reality of the idea of the unity of divine and human nature; that is, he too accepted the incarnation as symbol. But no longer did this union have to express itself in one individual. "Would not the idea of the unity of divine and human nature be much more a real one in an infinitely higher sense if I understood all of humanity as its realization than if I single out one individual person to be such? Is not an incarnation of God from eternity truer than one closed off in a point of time?"[33] As John Toews has pointed out, this denial of the real experience of the universal incarnation in the historical person of Jesus meant that Strauss had also "denied the Hegelian claim that a philosophical comprehension of that experience accomplished the ultimate reconciliation of the individual and the absolute." Strauss had defined the union of the transcendent and the immanent in totally immanent terms.[34]

If Strauss saw his efforts in *The Life of Jesus* to be an attempt to harmonize Hegelian philosophy and criticism, other Hegelians did not. Those whom Strauss located on the right and in the center could not go along with his reinterpretation of the meaning of the incarnation. Both insisted on the union of God and humanity in the person of Christ, though center Hegelians like Karl Rosenkranz and Konrad Marheineke conceded that there were problems with the New Testament account of miracles.[35] Right-wing Hegelians wished to deny Strauss a place within the Hegelian tradition by making explicit the political implications of his treatment of Jesus. By making clear the parallels they saw between Christ's relationship to humanity and that of the monarch to the state, they were able to turn Strauss's attack on the historical truth and philosophical necessity of a single incarnation into a challenge to the necessity of the monarch as a single embodiment of the state.[36]

Of course dissatisfied Hegelians were not the only ones to respond to *The Life of Jesus*. Strauss took the counteroffensive against them in the second of the three *Streitschriften*, dealing mainly in the other two with orthodox theologians who also had denounced him. The years just after *The Life of Jesus* saw three new editions of the work and an English translation.[37] Then, in 1840, Strauss brought out his study of the history of Christian doctrine in which he joined another Young Hegelian theologian and philosopher, Ludwig Feuerbach, in criticizing Hegel.

In 1839 Feuerbach wrote his critique of Hegel's philosophy. He had come to the realization that Hegel had failed to take into account the primacy of the human *experience* of nature and of the sensuous. Hegel had remained on the level of pure thought and had not genuinely resolved the opposition between the idea and nature.[38] When two years later Feuerbach brought out his infamous *Essence of Christianity,* he saw himself to be "nothing but a natural scientist [*Naturforscher*] of the mind" and his book to be an application of his new perspective, what he called "an empirical or historical-philosophical analysis" in which there were "no *a priori,* self-invented [propositions], no products of speculation."[39]

One year before Feuerbach's critique Strauss saw his chance of remaining within the theological community come and go. In spite of his assurances that he was not out to disturb religious faith and practice and that he would respect the fundamental truths of Christianity, Strauss's appointment to a theological chair in Zurich was turned into an early pension because of the public outcry. This experience, plus Feuerbach's critique, no doubt encouraged Strauss to take an even more radical position. In 1839 he wrote to Märklin that he had abandoned his attempts to harmonize the representation and concept. No longer did he require an external religious representation of the philosophical unity achieved in Hegel's system. In the *Doctrine of Faith* of 1840 Strauss even gave up the religious categories he had used in past writings. Now he spoke of God in abstract philosophical terms as universal personality, not as an individual personality.[40]

Other central doctrines of Christianity were either expressed in philosophical terms having to do with the here and now or they were rejected outright. Always the standard by which dogma was measured was modern *Wissenschaft,* especially as it was expressed in Hegel's system.[41] "Let the believer permit the knower [*den Wissenden*] to go his way, just as the latter does the former. We leave them their faith, so let them leave us our philosophy. There have been enough false attempts to mediate; only a separation of the opposites can lead further."[42]

To the community of orthodox theologians Strauss's move to the left was completely understandable. They had watched the growth of what they called "pantheism" for several decades.[43] What Strauss had done was to make crystal clear where such thinking led. Those theologians who revered the philosophies of Schelling and Hegel because they believed them to contain important new insights for religion and theology had been taken in by the use of religious language and religious categories. Strauss showed where such thinking led, and right-thinking theologians could brook no compromises with it.

Strauss's rejection of the Mosaic account of the creation in Genesis in the *Doctrine of Faith* was tied explicitly to advances in astronomy and geology. Once these sciences had developed, it was no longer possible to permit the earth to have been created before the sun or even to see the heavenly bodies as devices by which time could be measured on earth. Strauss further pointed out that geologists now required more time than was captured in twenty-four-hour days. Finally, Old Testament criticism had shown that there were separate creation narratives in Genesis 1 and 2, and that they had to be understood as myths without any historical foundation.[44]

The notion that all of humanity, with its rich racial diversity, had originated from one pair struck Strauss as unscientific. Some natural scientists had resorted to "artificial hypotheses" here rather than abandon the tradition of the church. For himself, Strauss could not accept monogenism, since it was to him obvious that the differences between the Negro and the Caucasian could not be attributed to the effects of climate and lifestyle.[45] Strauss went on to argue that the view according to which the first humans were the result of the immediate activity of God was "the negation of the standpoint of natural science and of science in general."

> Science must not allow divine causality to enter immediately into its ranks at any point. In science God as such has not . . . created humankind . . . According to science all organic beings are produced originally from the inorganic. In particular, there is no doubt that our planet has acquired its present state gradually, that it was uninhabitable for organic beings in primitive times, and that these have originated gradually without having had ancestors, that is, through dissimilar reproduction [*durch ungleichartige Zeugung*].[46]

Citing Karl Burdach and Gustav Carus, Strauss went on to describe an evolutionary schema in which a primal formative power *(Bildungstrieb)* worked its influence on the seeds of lower and higher organisms, including human beings, to bring them eventually to maturity.[47] Strauss's goal was to demonstrate that the formation of the first humans was a natural process, the result of the mixture of matter under certain conditions of temperature, electricity, and other factors, and that therefore the human race had not begun with a single pair.[48]

In spite of these radical sentiments about the origin of humankind tucked away in its pages, the *Doctrine of Faith* was anticlimactic to Strauss's earlier writings. Furthermore, attention was drawn in 1841 away from Strauss and to Feuerbach, who, in *The Essence of Christianity*, exposed Christian doctrine and with it God himself to be the result of the projection of human needs and characteristics. Feuerbach was easier to understand than Strauss, and his message was as clear as it was sacrilegious. Had anyone been unsure of the dangerous implications of left-wing Hegelianism, there could be no more doubt. Feuerbach's complete rejection of the reality of a transcendent realm represented a ruthlessly consistent modern theological science. Strauss, with his Hegelian way of expressing himself, still seemed to stop short of embracing a totally naturalistic view. But there was nothing tentative about *The Essence of Christianity*. "Feuerbach," writes Horton Harris, "had dotted the *i* which [Strauss] as pioneer had failed to do."[49] Now it had all been said.

The work of Feuerbach emboldened Strauss to be even more critical of theology and more content with philosophy. But it was most likely his total rejection by the theological community that embittered him. No one of note stood with him; even Baur had tried to distance himself by emphasizing minor differences with Strauss to the exclusion of the fundamental agreement they shared.

Strauss's dismissal from Tübingen, the collapse of a promised appointment in Zurich, and the silence and even attacks of former friends cut deeply. For almost two decades following *The Doctrine of Faith* Strauss turned away from the science of theology that had made his name so recognizable. His life between 1841 and 1860 was further afflicted with two more personal failures, one in his marriage to the opera singer Agnese Schebst, which ended in separation, and the other in his election in 1848 as a representative to the Württemberg provincial Parliament, where his political conservatism surprised and angered those who had backed him.

Strauss eventually did come back to theology. Among the literary researches he engaged in during the 1840s and 1850s were several biographical studies. In 1857 as a result of publishing a two-volume work on Ulrich von Hütten, the sixteenth century defender of Luther, he felt a stirring within himself to do battle for the cause of truth once again. At least one other factor also encouraged him in this direction. Strauss became angry in 1858 over the attempt of some to censure his friend from seminary, the pastor Ernst Rapp, because he openly entertained Strauss in his home. His reentry into the theological scene occurred in 1864 with the publication of a new, completely rewritten book on the life of Jesus. The first study of 1835 had been expressly intended for theologians; this time the title carried the phrase "For the German People."

Strauss seemed to be indifferent to the view theologians might have of his new study. His audience here, as it would be from now on, was the educated and intelligent laity. In 1864 Strauss assumed them to be everywhere, and his characterization of them presumed that they were all really on his side, even if they sometimes did not wholly realize it.

> Is it still, among all people, only those who are to some degree educated and thoughtful who know the long open secret, that no one believes the dogmas of the church anymore? One thinks that one believes—that I concede. But really believes? That I deny. For no one is the Apostles Creed or the Augsburg Confession an appropriate expression of religious consciousness anymore. No one believes any longer in any of the miracles of the New Testament (not to speak of those in the Old), from the supernatural conception to the ascension. Either one explains them naturally or conceives them as legends. And if this is the case with the thinking

laity, the situation, as we have seen, is no better among the clergy. And to what end are these subterfuges? To what end the hypocrisy before others and before oneself? Is it worthy of a man in his relation to religion to resort to it like a cowardly and artful slave with half-truths and empty excuses? Why not come into the open?[50]

It did not take long for the reader to realize that the Straussian leopard had not changed his spots. There were of course some new dimensions to this second *Life of Jesus*. Strauss took into account, for example, works on Jesus published since his own monumental study of 1835,[51] and this time he included an investigation of the historical relationship among the individual gospel narratives.[52] Strauss attempted to reconcile who Jesus was with the enormous influence he had exerted on human history, concluding that he was among the greatest religious personalities who had ever lived. But the rejection of everything supernatural, the denial of the resurrection and other miracles, and the retention of his theory of mythologizing to explain how the supernatural elements had become part of the New Testament accounts, all were stated here as they had been in 1835.[53]

It was naturally impossible for the 1864 *Life of Jesus* to have an impact anything like the one the first edition had. Since 1835 the air had been filled with radical theology. In addition to Feuerbach's call for the Christian world to be replaced with philosophical humanism, there were the bitterly atheistic attacks on religion of Bruno Bauer and Karl Marx, the intellectual anarchism of Max Stirner, and the critical studies of the Bible of F. C. Baur and the Tübingen school. Each of these radical thinkers had been influenced early on by Hegel, and each felt that he had carried Hegel's ideas to a logical conclusion that lay beyond the master's own stopping point. It is in this context that we can begin to understand why Friedrich Nietzsche, in criticizing Strauss's wholesale adoption of a Darwinian perspective in his last major work, sees him still to be Hegelian.

The Old Faith and the New

In *The Old Faith and the New* Strauss completely and thoroughly embraced Darwin. On the face of things it would seem either that Strauss had abandoned his Hegelian idealism for crass materialism, or that he misunderstood Darwin. Such a clear-cut choice may represent the dictates of logic, but it does not take into account the historical development of Hegelian thought after the master's death. In his own way Strauss illustrates the same inability to retain the pure form of the coherence theory of truth Hegel had introduced as others whose early training was indebted to the Berlin philosopher.

Feuerbach's critique of Hegel, on which Marx drew, emphasized Hegel's failure to impart to sensation a primacy equal to that of thought. With no reference to a thing-in-itself at all, Hegel's system seemed to many in the middle decades of the nineteenth century to have lost contact with the real world. Marx, of course, addressed this deficiency through the creation of a dialectical form of materialism, while Feuerbach and Strauss opted for a variety of objective idealism in which the real world conditioned truth as much as did the nature of thought itself. The German scientific materialists of the 1850s were linked to Feuerbach's message in the minds of most,[54] and yet the program of these materialists retained the very idealistic flavor its adherents so frequently tried to denounce.[55] In a similar fashion Strauss's work of 1872, *The Old Faith and the New,* could not shake the Hegelian speculative tradition in which Strauss had been reared.

What *The Old Faith and the New* confirms is that Strauss embraced in his last work the realism that accompanies a correspondence understanding of truth where nature is concerned. He proclaimed that an alleged multitude of people, whom he identified as "we," knew that the universe corresponded to their rational and unbiased understanding of it. Because his philosophical system included a recognition of the primacy of the sensuous aspect of human experience, he was now as bold in his declaration of how the cosmos was to be understood as he once had been regarding the life of Jesus.

In his diatribe against Strauss and his book Nietzsche identified Strauss as Hegelian because of his insistence that one could go beyond the limited domain of sure knowledge which Kant had defined.

> The quite incredible fact that Strauss has no notion how to derive from Kant's critique of reason support for his testament of modern

ideas, and that everywhere he flatters nothing but the crudest mind of realism, is among the most striking characteristics of this new gospel, which presents itself moreover solely as the arduously attained outcome of continuous historical and scientific research and therewith denies any involvement with philosophy at all. For the philistine chieftain and his "we" there is no such thing as the Kantian philosophy. He has no notion of the fundamental antinomies of idealism or of the extreme relativity of all science and reason. Or: it is precisely reason which ought to tell him how little of the in-itself of things can be determined by reason. But it is true that people of a certain age find it impossible to understand Kant, especially if in their youth they have, like Strauss, understood or thought they understood, the "gigantic spirit" of Hegel, and have also had to occupy themselves with Schleiermacher . . . It will sound strange to Strauss when I tell him that even now he is in a state of "absolute dependence" on Hegel and Schleiermacher.[56]

If Hegel ran roughshod over the Kantian caveat en route to absolute knowledge, Strauss had followed him as a youth. What Nietzsche hated in Hegel's system he hated even more in Strauss's announcement of the arrival of the Young Hegelian apocalypse—the presumption that Germany had finally arrived at the end of the search for truth. Strauss represented what Nietzsche dubbed a "cultural philistine," one who, reveling in self-satisfaction over Germany's recent military victories and the establishment of the new empire, delighted to proclaim boldly to the world what German culture had come to understand.

Nietzsche was not deterred in his denunciation of Strauss by the latter's explicit acknowledgment that his views were no more than a confession of a new-found faith. In the fourth edition, which appeared just three months after the first, Strauss inserted an Afterword as a reply to the many voices of criticism his work had called forth. Here he emphasized that his book was a confession, and that he would leave to others the task of showing that he had not misunderstood certain particulars of Darwin's theory. "Incidentally, in the title of my work I deliberately opposed a new faith to the old faith, not a new knowledge. To construct a comprehensive worldview that should replace the equally comprehensive faith of the church, we cannot be satisfied with what can be demonstrated strictly inductively. We must variously add to this whatever presuppositions or consequences result for our thought from this foundation."[57]

Nietzsche treated Strauss as the founder of a new religion. He mocked Strauss for presuming that anyone would be interested in what he happened to believe, knowing full well that Strauss felt justified in his presumption because he was really offering them truth and not mere belief. The whole tone of Strauss's work goaded Nietzsche, for humility before the great questions of life was not part of Strauss's stock in trade. Nietzsche felt he had to expose Strauss's ignorance and presumption by identifying him as a cultural philistine who refused to be told that life's meaning was more complicated than he allowed:

> The person [Strauss] hates most of all, however, is the one who treats him as a philistine and tells him what he is: a hindrance to the strong and creative, a labyrinth for all who doubt and go astray, a swamp to the feet of the weary, a fetter to all who would pursue lofty goals, a poisonous mist to new buds, a parching desert to the German spirit seeking and thirsting for new life. For it *seeks,* this German spirit! And you hate it because it seeks and refuses to believe you when you say you have already found what it is seeking.[58]

Nietzsche's fury was perhaps out of proportion to what one might have expected; certainly Strauss himself was bewildered by what he called "the actual motive for his passionate hatred."[59] He did not see that it was the "gigantic Hegelian spirit" so visible in his book which so offended Nietzsche. As far as Strauss was concerned *The Old Faith and the New* was his personal summing up of his life's work, "a reckoning of [his] household, for [he] would not be a householder much longer."[60]

To prevent the book from becoming dull, which Strauss felt had happened in *The Life of Jesus,* he decided to write freely, without the detailed argument of the scholar. He worried about the project and wanted to flee from it, comparing himself to the prophet Jonah who fled before the Lord. Once underway, he wrote it quickly, knowing full well that his approach would be criticized.[61] It was to be Strauss's considered and final word on four basic religious questions: Are we still Christians? Have we still religion? How are we to understand the universe? and How do we order our life?

If ever Strauss's views on such issues could have expected a fair hearing it was in 1872. Just prior to the appearance of *The Old Faith and the New* his lectures on Voltaire had been warmly received. His open exchange of letters with the French theologian Ernst Renan during the war with France, in which he defended Germany's need to unify and criticized France both for trying to stand in Germany's way and for starting the war, proved to be the most well-received words that ever issued from Strauss's pen. France had altered its constitution three times since the fall of Napoleon I and Germany had never thought of interfering, Strauss wrote. Now France was trying to interfere in Germany's internal affairs.[62] But *The Old Faith and the New* soon quashed Strauss's brief moment of adulation.

Already in the Introduction Strauss found a way to antagonize his readers even before taking up the questions he wished to pose. Those whom he had identified in the second *Life of Jesus* as the thinking people who knew in their hearts the secret of Germany's widespread disbelief became in *The Old Faith and the New* an "innumerable multitude" no longer satisfied with the old faith. They believed change was necessary because the old faith was not consistent with knowledge nor with present-day political reality. Strauss conceded that the majority of this number wished merely to adjust certain tenets of the old faith; for example, some felt they could keep Jesus or the Pope while denying them divinity and infallibility, respectively.[63] Strauss himself identified with the minority who, he said, constituted a "we against the world" who thought logically about religious matters regardless of the hallowed doctrines that might be challenged in the process. The message came through loud and clear: Strauss claimed to speak for those people who were unbiased, honest with themselves, and who valued logical reasoning about their religious position.

As Strauss launched into his treatment of the first query—Are we still Christians?—the clash between the Genesis account of creation and modern science as Strauss understood it became immediately clear. As a vehicle for determining the answer to his question Strauss chose to examine the Apostles Creed, the beginning of which declared God to be the maker of heaven and earth. Strauss challenged his reader either to accept the six-day creation narrative literally, in which case one declared *ipso facto* that one did not care what modern science said, or to abandon it. What he could not see were the attempts to explain away problems such as the creation of the sun on the fourth day after three days and nights had already passed.[64] For his own part, of course, Strauss assumed that the notion of a world-creating deity was a primitive conception common to many religions and worthy of our respect as such. In the case of Genesis, the ancient story had been petrified into a dogma which, because it had been formulated long before Copernicus and modern geology, was therefore bound to conflict with later conceptions if taken literally.[65]

As for the atonement and life everlasting for the redeemed, the two other propositions Strauss identified as part of the Christian creed, the former, according to Strauss, had its origins in Jewish sacrifice, which in turn had evolved from primitive human sacrifice. The exclusivity and inhumanity of the latter, which could condemn innocent unbaptized children to eternal punishment, was no longer tolerable.[66]

Long before Strauss got around to answering whether he and the ones he claimed believed as he did were still Christians, the reader knew what the answer would be. Before pronouncing the inevitable "no," Strauss explained how the old faith had been eroded through the development of historical and scientific research. In the eighteenth century the negative results of rational analysis meant that it was no longer possible to defend supernaturalism. The whole rationalist tradition, which Strauss had always despised and opposed, he here conceded to be an attempt at an honest explanation of what really had happened.[67] Schleiermacher, who had died in 1834, Strauss praised for abandoning miracles but criticized for his stubborn adherence to the divinity of Jesus.[68]

If time had exposed Schleiermacher's "God in Christ" to be a mere phrase, Strauss wanted some of the credit. His own work had been focused directly on Jesus; now in his old age Strauss reported anew that so little positive could be known about Jesus that we are not justified in using him as an exemplar. "The Jesus of history, of science, is strictly a problem; but a problem can neither be an object of faith nor a pattern for life."[69] Without the foundation, Strauss wrote, it was impossible to fix up Christianity so that it was palatable to the modern age. The Christ of faith was a Christ of legend. "We cannot seek support for our action in a faith we no longer have."[70] "In short, if we speak as honest, upright people we must acknowledge we are no longer Christians."[71]

If the answer to this first question was predictable and straightforward, a response to the second—Have we still religion?—was more problematical. Since Strauss was confessing his new faith, one might expect that he would have to rescue religion in some guise or other from the cutting edge of his critical spirit. And yet what could that religion look like? In this second section of the book Strauss continued the prolegomena to the content of his confession, which came only in the final two sections.

Hints of the pivotal role that humanity's relationship to nature would play in Strauss's religion (and therefore the vital function natural science would provide in giving expression to it) were evident from the outset. "One thing, at all events, is certain: that the brute, destitute of what we term reason, is devoid of this capacity [for religion] also."[72] Strauss concurred with those who held that religion was born of the conscious fear of nature. Nature was indifferent to human wishes, an alien power with which nothing could be done. The sole recourse available to a species afflicted with the painful consciousness of its mortality was to invest nature with its own attributes, to recreate human society, with its hierarchy of domination and submission, among the gods.[73]

Having arrived at the question of God, Strauss quickly moved his discussion through humanity's original polytheism to Judaic monotheism and finally to what he called "our modern monotheistic conception of God."[74] Both Judaism and Christianity, Strauss wrote, had emphasized the personal side of the modern conception, while philosophy had concerned itself with the other side, the absolute. History had shown that the two were incompatible.

Strauss defended Copernicus from the charge of having pulled the rug from under the Hebrew-Christian God. Copernicus, he argued, left God's abode untouched. Strauss wrote as if it was the later awareness that space was infinite in extent that raised problems with Western religion, a notion not present in the system of the devoutly Christian Copernicus. He used the occasion to illustrate, without identifying it, the old Hegelian distinction between religion and philosophy. The understanding, he wrote, knew God to be omnipresent, but the imagination could not avoid locating and limiting God in space.[75] Having long since forsaken Hegel's own insistence on the identity of philosophy and religion, Strauss here was setting the stage for his philosophical expression of faith.

Before he allowed himself to spell out the particulars of his position, Strauss felt obligated to address what he called the "somewhat outmoded scientific artillery of the so-called proofs for the existence of God."[76] Contrary to expectation, Strauss did not dismiss them out of court. He smugly pointed out that the cosmological argument, according to which God was the first cause, did not result in a *personal* God. But Strauss did not say that the whole line of reasoning was faulty—only the inference of a personal God was faulty. When, however, he came to the teleological argument for God's existence, which inferred a designer from the design and purpose evident in nature, Strauss, anticipating what he would write about Darwin, blithely rejected its validity. Nature was like instinct, he wrote. It was displayed in apparent obedience to a conscious aim, yet it acted without any such goal. With the same confident air he denied that the moral law in the human conscience had its origin anywhere other than in nature and society.[77]

As for his own view of God, Strauss turned surprisingly to Schleiermacher, who in his posthumous work on dialectics provided "the collective result of all of modern philosophy with regard to the conception of God."[78] Strauss quoted Schleiermacher to say that the ideas of God and the universe, while not identical, required each other to be conceived. On the philosophical level they remained the empty ideas of unity and plurality, respectively, but any endeavor to fill them in, such as conceiving God as a conscious Ego, brought them into the realm of the finite. From this Strauss concluded that "what is left as the genuinely highest idea is that of the universe . . . To get beyond it will never be possible for us; and if we nevertheless try to conceive a Creator of the universe as an absolute personality, we are convinced by the above before we start that we are merely dealing with idle fantasy."[79]

Strauss paused only long enough en route to his explicit answer to the question about religion to explain the evolution of the idea of immortality, and to reject it in favor of the assumption that spiritual existence can only coexist with material. Without a personal God and without personal immortality, Strauss conceded, many would argue he had revoked all claim to religion. But for a second time in the book Strauss reversed his youthful antipathy to Schleiermacher, this time drawing on the famous characterization of religion that had long ago established Schleiermacher as the father of liberal German theology: religion was born of the human consciousness of absolute dependence. Strauss qualified Schleiermacher's conception with Feuerbach's insight that religion also involved the human need to react against the dependence, for absolute dependence would be crushing.[80]

And so Strauss declared that he still did have religion. But as Strauss described the feeling that lay at the foundation of his new faith, one reader at least had to wonder: "How far does the courage bestowed on him by the new faith extend?"[81] Strauss's answer revealed his unquestioned confidence in nature's rationality.

> For us, that on which we feel ourselves entirely dependent is by no means merely a raw power to which we bow in mute resignation. Rather it is simultaneously order and law, reason and goodness, to which we surrender ourselves in loving trust. And more, since we perceive in ourselves the same disposition to the reasonable and the good which we believe we recognize in the world, and since we find ourselves to be the beings by whom it is felt and recognized, in whom it is said to have become personal, we feel ourselves related in our inmost nature to that on which we find ourselves dependent. At the same time we discover ourselves to be free in the dependence; and pride and humility, joy and submission, intermingle in our feeling for the universe.[82]

Hegelian speculative theologians had been critical of Schleiermacher's understanding of religion because, they said, in resorting to feeling he had abandoned all hope of specifying the content of religion in cognitive terms. With his endorsement of religion as the feeling of absolute dependence, Strauss appeared by the end of chapter 2 to have laid himself open to the same charge. But in dealing with the third of his questions—What is our conception of the universe?—Strauss set out to state in cognitive terms as much as he could about the content of his faith.

Strauss began by attempting to explain exactly how one proceeds to formulate a comprehensive conception of the cosmos. The old Hegelian confidence that reason was up to the task did not fail him, for arriving at a conception of the world that confronted us was possible, he assured the reader through "an inferential process."[83]

What now distinguished Strauss from Hegel, more properly from Schelling, was the absence of a dialectical dimension in reason. Strauss no longer conceded, if he ever had, that the knower helped to shape the cosmos that was being known. For the Strauss of *The Old Faith and the New* the truth of the cosmos was not dependent on human wishes or ideas. After declining to elucidate how our different "modalities of receptivity for impressions" could be combined with "the objective causes of these impressions," Strauss merely asserted that the latter "more and more separate themselves into groups . . . till at last this whole richly partitioned and orderly system of our present conception of nature and the world is formed."[84] Strauss did not question his description of the impressions that "separated themselves into groups," because, as he went on to say, these groups "represent mere aggregations of externally coordinated objects, but are intrinsically united by forces and laws."[85] That the laws were knowable Strauss never doubted. That they were not yet known in their entirety he conceded. It was because of the human ignorance of the present that he was required to state his results as a confession of faith. Thanks to the emergence of modern science, however, one had the impression that the days of faith were numbered.

The first particular Strauss felt must characterize his conception of the universe was its infinity in time and space. This attribute, Strauss was careful to point out, applied only to the universe as a whole. Individual parts of the universe, like the earth or the solar system, had been shown to have a beginning and therefore must also have an end. "But if we contemplate the universe as a whole there never has been a time when it did not exist."[86]

Strauss credited Kant with founding the modern scientific method of explaining how portions of the universe originated. In his theory of the heavens Kant resorted to purely mechanical principles, Strauss observed, and deliberately excluded a creator acting with deliberate aims. To his credit or blame, Strauss did not simply refer to the nebular hypothesis of Kant and Pierre-Simon Laplace on the assumption that his reader knew what he was talking about; he went on to provide a summary of its contents and even included recent developments, like spectral analysis, which had a bearing on it.[87] The earth, of course, was one of the planets which was flung from the rotating sphere of nebulous matter and began to cool. Strauss clearly believed no life was possible in these early stages of the earth's history; only younger strata contained remains of organic beings. Strauss raised the obvious question openly: "There was once no organic life on the earth, and later there was; it must, consequently, have had a beginning and the question is, how?"[88]

In his answer to the question Strauss betrayed that he saw the matter not, in fact, as a question of a new faith in opposition to an old one, but as new knowledge conquering old. "Here," he wrote, "faith inserts a miracle."[89] God was said to have originated life through direct action. Even Kant, who required only the preexistence of matter to explain the origin of the solar system, desired that living things could be accounted for through matter alone.

Strauss granted to Kant that a complex form of life like a caterpillar could not somehow emerge directly from inorganic matter. But what of the simplest organism, which, Strauss noted, we now know to be the cell? Darwin had felt obliged to resort to miracle to account for the first cell or even several different cells, but Lamarck felt no such compunction. Darwin's French precursor referred to spontaneous generation to explain the origin of the first life, and Strauss, noting that the conditions on a primitive earth would have been much different from those now extant, concluded that the transition from inorganic to organic was bridged in this way. The elimination of vital force assured that the transition was a natural not a supernatural process, for life was not a new creation. Rather, preexisting matter and force had merely been brought into another kind of combination, and that had been caused by the physical conditions of primeval times.[90]

Strauss cited Emil DuBois-Reymond's opposition to vital force in the text of his book, but because the latter's famous lecture on the limits of natural knowledge appeared almost simultaneously with that of *The Old Faith and the New*,[91] Strauss had to deal with its claims in the "Afterword as Foreword" of the fourth edition. Naturally he was delighted to repeat DuBois-Reymond's explicit rejection of anything supernatural in the appearance of life on earth.

But some of the critics of Strauss's book were quick to emphasize that with regard to the prospect of a scientific explanation of consciousness DuBois-Reymond came across as a humble man of science, in marked contrast to the arrogant proclaimer of Germany's new theological vision. In the second volume of *The Old Faith and the New* Strauss joined materialists like Rudolf Virchow and Karl Vogt in denying the existence of an immaterial soul. It was, he wrote, a hypothesis unsupported by a single fact and explained nothing.[92] After attempting in very confused and unpersuasive fashion to assert that the interaction of mind and matter had to proceed in conjunction with the law of the conservation of energy, he sensed he had gotten in over his head. He tried to get out by taking the offensive. "I will be told that I am here speaking of things I do not understand. Fine, but others will come who will understand them, and who also will have understood me."[93]

Strauss was not used to being cast in the role of a dogmatic theologian, the very people he saw himself opposing. His only recourse was to portray DuBois-Reymond's declaration that humans would never know how to explain consciousness scientifically as at best premature and, by implication, at worst wrongheaded dogma. Strauss confessed he saw no difference between this question and that of the origin of life. He insisted that science had the right to try to replace faith, with its miraculous explanation of the origin of mind, with knowledge that would make divine intervention unnecessary.[94] This way our understanding of the relation between mind and matter would be based on induction and would be as free as possible from the illusions of dogma. It was the prejudiced belief in a nonmaterial conscious soul that prevented Descartes from acknowledging the existence of animal souls, thereby making impossible any recognition of kinship between humans and animals. Without this prejudice, Strauss continued, the theory of descent along with the doctrine of natural selection would force on him the idea that souls came into existence as the gradual result of certain material combinations.[95]

It was Strauss's appeal to Darwinian evolution as the foundation for his answer to the question, How do we conceive the universe? that crystallized opposition against him. "I have become interested in Darwin since his theory has become known," he wrote to Carl Gustav Reuschle, a professor of mathematics in Stuttgart. "I read his major work when it came out and have since seized everything referring to this theme."[96] But even more rewarding than the *Origin* was Ernst Haeckel's *Natural History of Creation*. Of course he did not shrink from accepting the physical evolution of humans, which he called "the *sauve-qui-peut* not only of the orthodox and tender feeling world, but also of some of the otherwise tolerant and open-minded people."[97] By early 1869 he could write to an acquaintance that it was Darwin who first freed humans from the notion of creation. "We philosophers always wanted to get out, but Darwin was the first to show us where the door was."[98] This was the emphasis that greeted the reader of *The Old Faith and the New*.

In *The Old Faith and the New* Darwin was portrayed as the first scientist to replace a miraculous and superintending agency with natural forces where this long-developing series of organic gradations was concerned. Darwin was not the first to recognize that descent had occurred; Goethe and Lamarck both openly had acknowledged descent. Lamarck in particular had anticipated Darwin's theory of transmutation, according to Strauss, as a way of explaining the production of complex life forms from simple ones. But what was missing in Lamarck's version was the How.[99] Darwin's theory, wrote Strauss, was still incomplete and imperfect, but it marked out the direction for the future with certainty. Others had proclaimed miracles dead, but they could never persuasively get rid of the miraculous agency. Darwin's virtue was that his explanation through natural forces "opened the doors through which a happier race to come will throw out miracles once and for all."[100]

Strauss could understand why those he called "the orthodox" would oppose Darwin. What puzzled him was the reaction of the press, which loved to ridicule Darwin's theory by deliberately overlooking the many intermediate gradations between apes and humans. Newspaper writers were hardly solidly on the side of the orthodox, but what they apparently did not realize, Strauss noted, was that there was no real middle ground on the issue. Unless they were willing to assume that nature produced complex life forms directly by spontaneous generation, a prospect Strauss dismissed as neither persuasive nor viable, they had to choose between the two other options available. Strauss was not sure that they had thought the matter through. "Do they know," he queried, "that their only choice is between the miracle—the divine hand of the Creator—and Darwin?"[101]

The exposition of Darwin's theory began with an explanation of its difference from Lamarckian use and disuse. Strauss asserted that the world looked askance at Lamarck's account, and that Darwin's approach was much more down to earth. Strauss explained that Darwin's involvement with the selective breeding of pigeons to procure offspring with desired characteristics was possible because organic types, for all their immutability in the whole, were mutable in their parts in a manner that permitted the deviations to be inherited. By artificial selection human beings had produced varieties "concerning which it is in the last analysis just a semantical dispute if one refuses to recognize them as a new species."[102]

The move from artificial selection to natural selection was explained in social terms. Where did Darwin get his notion of "struggle for existence"?

> It is significant where the Englishman has sought and found it. He did not have to search for it at all, since he had before his eyes the action and the astonishing effects of this principle all around him in his homeland. He needed merely to transfer it from the human world to the household of nature: competition. Darwin's "struggle for existence" is nothing but the broadening into a natural principle of something we have long known as a social and industrial principle.[103]

Although he did not emphasize the dimensions of chance in Darwin's scheme Strauss was clear about its presence. He dismissed Lamarck's explanation of the acquisition of horns in the bull through the love and habit of butting, and he cited Goethe with approval to say that the primary question for the future was how the bull had come by the horns with which he could butt, replacing the assumption that the horns were given the bull so that he could butt.[104] Developing Goethe's example further into a Darwinian account of the acquisition of horns, Strauss explained that if a herd of primeval cattle defended itself by butting its attackers, then any bull who possessed "an incipient horny accretion" would have the best chance of survival. Other bulls without it would not tend to survive to propagate the species. Since unquestionably there would be at least some offspring who inherited the advantageous characteristic expressed in the parent, gradually over time the species would be dominated by those with this enhancement to their survival. Once the characteristic was present in both sexes, a completely horned species would be formed, especially if the female gave preference to the males with horns. Darwin's theory of sexual selection, Strauss added, supplemented his theory of natural selection.[105]

It is not possible to tell exactly how much Strauss required Darwinian chance and how much he allowed acquired characteristics. He spoke of a bull "developing the incipient horny accretion" and of it somehow being "transmitted" to the female, yet clearly only some members of the species were favored with the characteristic while others were not. Although, as is well known, Darwin too allowed for instances of the inheritance of acquired characteristics, Strauss was not at all as careful as Darwin was in his explanation of natural selection.

Like Darwin, however, Strauss did raise problems which he then set aside. Isolation of individuals with a certain trait, he remarked at the end of the first volume, was absolutely necessary for artificial selection. But in the wild such isolation did not seem to occur, and one could expect the descendants of such individuals to revert to the type of the original species, thus impeding descent. Strauss was unaware of the difficulty this problem had caused Darwin.[106] He handled it by simply asserting that the isolation of favored individuals *was* possible in nature, citing examples from the work of the German naturalist Moritz Wagner.[107]

Having finished his exposition of Darwin's theory in general, Strauss broached the subject of humanity's past. As expected, he accepted the bones that had recently been found in European caves as the remains of humans, assuming that they served as evidence of an early brute state.[108] That humans developed from this level of animal existence was not in the least offensive to Strauss. He found it difficult to comprehend that people would willingly accept the idea of the incarnation of God, but that they could not abide the thought of an evolution of animals into humans.[109] Once again Strauss appealed to a social metaphor to drive his point home. Some people, he wrote, preferred a count to a citizen, even if the former had become impoverished through wanton living while the latter had worked his way up through talent and energy. Strauss sided with the citizen.[110]

Before he came to the last of the four questions around which his book was composed, Strauss attempted to summarize the conception of the universe at which he had arrived. Darwin, he repeated, had removed design from our explanations of nature. Quoting the physicist Hermann von Helmholtz to say that Darwin had shown how adaptation could be effected by blind natural law without the interference of an external intelligence, Strauss denounced teleological arguments based on design as against the spirit of modern science.[111] The conception of the universe that was consistent with the spirit of modern science was a cosmos of causal necessity. As long as people allowed the existence of a personal God, Strauss declared, individual purpose and the general purpose of the world would be referred to Him.

Strauss preferred to conceive of the human need for purpose in Feuerbachian terms. To the extent we speak of purpose in the universe, we speak subjectively, at best referring to what we want to recognize as the result of the cooperation of active forces in the world.[112] One had to get away from seeing the cosmos solely from the human perspective—that was a major drawback of the old redemptive scheme in which the universe existed for humans. As there had once been a time when humans did not exist, so that time would come again. Viewed in this way, Strauss urged, the purpose of the universe did not lie in what had been accomplished in the development of life; rather, the purpose of the universe was attained and continued to be attained in every moment of its existence.

> Thus necessarily everything the earth in the course of its development has produced by itself and, as it were, brought before itself—all living and thinking beings with all their efforts and achievements, all the building of states, all the works of art and science, not only those that have disappeared without a trace, but also those that have left behind no memory in anyone's mind—must perish, since

with the earth naturally its history also perishes. Either the earth has failed its purpose here in that nothing has come of its long duration, or the purpose does not lie in something that is supposed to endure, but is achieved in every moment of its developmental history.[113]

Having come to realize that whether or not this was the best of all possible worlds, it was the only possible world, Strauss was impatient with those who tried to discredit his system by calling it materialistic. The contrast between materialism and idealism was merely a quarrel over words, he said. Each perspective complemented the other, each led to the other. Their common enemy was dualism. "We must allow one part of the functions of our being to be ascribed to a physical and the other to a spiritual cause, but all of them to one and the same, which permits itself to be seen either way."[114]

In this vein Strauss reprimanded natural scientists and philosophers for the way they viewed each other. He detested the overbearing tone philosophers took toward the materialism of natural science no less than the abuse natural scientists heaped on speculative philosophers. Natural scientists needed the guiding hand of philosophy to obtain a proper understanding of concepts at the heart of their work, ideas like force, matter, and causality. Likewise, virtually everyone now conceded that philosophers needed to become acquainted with the results of natural science. In the wake of the *Origin of Species* each community was beginning to recognize its reliance on the other, for the Darwinian theory was "the first child of the as yet secret marriage of science and philosophy."[115]

When it came time for Strauss to answer his final question—What is our rule of life?—he proceeded with the same confidence that marked his solutions to the riddle of the universe to this point. Gone was any sense of humility that one might have expected in a mere confession of faith. The reader might have suspected that in a universe devoid of purpose there could be no reason why one course of action should be preferred over another. But the impression now given was that Strauss had a corner on truth and his recommendations, through which he proceeded to justify German middle class values as those that were consistent with a "modern" view of the world, flowed from his enlightened state.

It was true, Strauss conceded, that a divine origin of law imparted to it a sanctity that was hard to match. He identified a foundation for the human "ought" in the characteristic sociability of humanity's brute ancestors. Human beginnings did not originate from the hand of God in the context of an earthly paradise. The real origins of humanity defined a scenario exactly the reverse of the biblical portrait. Humans arose out of a brutish state from which they had continuously ascended. But even in this primitive state human ancestors had begun to counter the animal qualities of competition and egotism with a social impulse that compensated for their relative frailty in the struggle for existence. Laws, customs, and duties emerged with necessity out of the unique human solution to the problem of securing a place for the human species within the animal kingdom.[116]

Strauss did not hesitate to see this development as progressive. Citing Moritz Wagner, Strauss declared that the most important result of natural science in general was "the great laws of progress that ruled in nature."[117] Although he was not willing to conclude that this progress had necessarily reached its zenith with the appearance of humankind, he did recognize, with echoes of his Hegelian past sounding in the background, that in humans "nature had not wanted just to go straight ahead, she wanted to go beyond herself."[118] Humans therefore had an obligation to acknowledge those parts that were animal but at the same time to rule the animal through their higher faculties and to ennoble nature.

As Strauss proceeded it became clear that the ennoblement of nature involved some very specific conclusions. Although his analysis of the (animal) sexual instinct led him to the conclusions that adultery was but one among several justifiable causes for divorce and that with the progress of culture new factors could work together to make marriage impossible, it did not occur to him that one of the rules of life he had come to resulted less from the dictates of his new universe than it had from the torments of his own failed marriage.[119]

Strauss did not hesitate to range widely in his estimation of how those for whom he spoke, the famous "we," should order their lives. In a tone that must have made Nietzsche cringe, Strauss defended the middle class and hereditary property as the indispensable bases of morality and culture.

I am middle class [*Bürgerlicher*] and am proud to be such. The middle class, one may rant at it and mock it from both sides as much as one wants, remains nevertheless the core of the people, the center of its customs . . . Property is an indispensable foundation of morality [*Sittlichkeit*] and of culture. It is the proceeds of work and the incentive to work. But to be this it must be heritable. Without that earning would deteriorate into the raw search for pleasure.[120]

After expressing his admiration for Bismarck and Helmuth von Moltke, whom he ranked in influence with Goethe and Alexander von Humboldt, Strauss went on record against universal suffrage (the poor paid no tax and should not be given the vote), in favor of capital punishment, and in favor of the removal of all connection between ecclesiastical and civil ceremonies.[121] No climax marked Strauss's concluding remarks, only an invitation to read two appendices to the volume in which Strauss set down his thoughts on Germany's great poets and composers.

Final Outrage

Strauss should not have expected anything other than the renewed outrage that greeted *The Old Faith and the New*. In spite of his earlier experiences Strauss had never really gotten used to being an intellectual outcast. He could not understand why he was treated so viciously by the public while others, whose views were not unlike his own, escaped vilification. All he ever wanted to do was to speak the message of his heart, but that was somehow forbidden to him. Earlier, in the aftermath of his *Life of Jesus* and *Doctrine of Faith* during the 1840s, he had written to Baur that he could do nothing but calmly observe what little literary fame he had achieved crumble away.[122] Again in 1868, while completing his work on Voltaire, he complained to his boyhood friend Friedrich Vischer that the great French critic had lived in better times than he, that no one understood him or even wanted to, and that there was nothing left for him to do but "to be quiet and die."[123]

And yet he clearly must have expected something different in response to *The Old Faith and the New,* for once again he was genuinely surprised by the vitriolic outburst his final book called forth. True, his pen had produced another sensation—the book went through six editions in six months. But once again he was denounced by his enemies and, even more painfully than in 1835, once again he was abandoned by those he thought were on his side.

Public discussions of *The Old Faith and the New* almost without exception disapproved of it. While the orthodox enjoyed a certain smug satisfaction that the dreaded theological critic had finally been totally unmasked, liberal thinkers were eager to separate themselves from him. In the "Afterword as Foreword" of the fourth edition, Strauss complained bitterly about the tone of the criticisms. Even though the claims of his *Life of Jesus* had since been confirmed and quietly adopted by educated people, he was being treated as if he were just a beginner. According to the alleged liberality of the day, he continued, it was no longer supposed to matter what a person believed as long as his conduct was exemplary. Strauss pointed to his explicit reinforcement of middle class morals in *The Old Faith and the New* and concluded that liberality was just a pious phrase.[124]

In fact, Strauss had become accustomed to being left alone during the previous two decades of his career; he did not expect to have his work corrected as if he were a boy.[125] In the face of such treatment by his critics, however, Strauss remained defiant. "The time of understanding will come, as it came for *The Life of Jesus*," he wrote. "Only this time I will not live to see it."[126]

What hurt the most, of course, was that even his friends, whom he knew to have convictions similar to his own, refused to defend him in public. Vischer, friend and correspondent for a lifetime; Eduard Zeller, colleague from Strauss's Tübingen years and professor of philosophy in Heidelberg; and Kuno Fischer, well-known neo-Kantian philosopher with whom Strauss had become friends following Fischer's dismissal in 1853 from Heidelberg because of his alleged pantheism—none came forward to support him. Among themselves the friends justified their silence as the best they could, for if they were to make public their hesitations about the excesses of Strauss's polemical work, they would simply be giving more ammunition to his enemies.[127] Vischer complained to the theologian Alois Biedermann that among Strauss's pronouncements on ethics and politics "there were many sentences that I would have attacked vehemently, even mocked, were the book not by a friend."[128]

The differences in the perspectives of Strauss and his philosopher friends were due to Strauss's blurring of the distinction between faith and knowledge. The message of Strauss's book was deliberately and clearly identified by its author as a statement of faith, a tactic by which Strauss hoped he could deflect objections that philosophers were sure to bring. He knew, for example, that they would say that he had not dealt satisfactorily with the thorny and subtle epistemological problems inevitable in the articulation of a worldview. But Strauss could reply: "These sentiments represent merely what I believe. I do not claim that they are either complete or perfect."

Had Strauss been willing to associate his exposition with what Nietzsche called "the extreme relativity of all science and reason,"[129] his confession of faith might not have given so much offense. But, as has been noted above, there were passages in the book in which he had opposed not a new faith, but new knowledge to an old faith. The pure coherence theory of truth which had lingered in the background of his first *Life of Jesus* was gone. Now he believed his description of reality was true because it corresponded to the world as it really was. Strauss imparted the very presumption of certainty that Nietzsche so detested in the book. Like many others of his time, Strauss assumed that if a claim resulted from empirical observation and induction it was unprejudiced as opposed to the biased illusions of the old faith. As he wrote to Fischer, in spite of the contradictions in its subjective dimension, "I should not like to have to concede that, where the worldview is concerned, an objective solution is impossible."[130]

Strauss's friends, while hardly willing to go as far as Nietzsche, saw the matter very differently from the author of *The Old Faith and the New*. A decade before Strauss died Zeller revealed his concern about the uncritical acceptance of empirical results. In a letter to Kuno Fischer in March of 1864, he wrote how delighted he was that Fischer had taken up the challenge to defend philosophy from the presumption of the growing empiricism of natural science. Although Zeller agreed that philosophy had been discredited among natural scientists earlier in the century, he urged that it was all the more necessary for philosophy to oppose the excesses of an empiricism that had become arrogant.[131] Ironically, Strauss himself, in words whose tone was far different from that of *The Old Faith and the New,* had once also defended philosophy against the empiricism of natural science with arguments similar to those of Zeller.[132]

If Strauss had blurred the categories of faith and knowledge, he had also confused his reader about the allegedly spiritual foundation for his new faith. His friends, to say nothing of his enemies, believed that his embrace of materialism had eliminated the possibility of a spiritual dimension in the universe. Strauss's stoic acceptance of a universe indifferent to everything humans required to have meaning brought less satisfaction to others than it apparently had to him. When Strauss asserted that all purpose in the world was merely the result of our subjective perspective, then elsewhere in the work declared that we recognize in the world the same disposition to the reasonable and the good which we find in ourselves, it was too much for Vischer. To Biedermann he wrote: "What is treated with complete lack of precision and with marked contradictions is the concept of inner purposefulness, of development, and the question of materialism and idealism. He does not establish where the trust for the future in the universe is supposed to come from."[133]

Some friends felt personally wounded. Prior to the publication of the book, Strauss had made the acquaintance of the theologian Biedermann and a pastor from Zurich named Heinrich Lang. In spite of Strauss's self-confessed susceptibility to materialism,[134] these two liberally minded men began to see Strauss as almost one of them. When *The Old Faith and the New* appeared, however, they felt betrayed, and in Lang's case angry. Even though he received a warm letter from Strauss in July of 1872, Biedermann soon thereafter wrote to Wilhelm Vatke: "I would have given a finger from my right hand to prevent Strauss from writing that ominous book."[135]

To one of the three men who did defend Strauss in writing, the young Theobald Ziegler, Strauss had made justified claims. He had restored unity to a worldview that had been fractured by the rapid and enlightening growth of natural science. "To free us from this disunity, to dissipate the romantic mist and provide us with a unified and simple worldview that is really accepted as modern—this is, I think, a thoroughly justifiable goal. Strauss took it up, and therein lies the great service of *The Old Faith and the New.*"[136]

Carl Reuschle, a professor of mathematics with whom Strauss

had corresponded about his newly found enthusiasm for natural science, also celebrated Strauss's achievement in a work of 1874 entitled *Philosophy and Natural Science*. Strauss, wrote Reuschle, occupied the first place among philosophers who acknowledged that their discipline could no longer be studied without a preliminary acquaintance with natural science. *The Old Faith and the New* was a prolegomenon to the new philosophy.[137] Strauss's was a "philosophy of the universe" in contrast to Hegel's "philosophy of the idea" and Spinoza's "philosophy of substance."[138] Reuschle could agree with what the conservative E. G. Steude of Greifswald had to say about the path that led from *The Life of Jesus* to *The Origin of Species* to *The Old Faith and the New*: "If David Strauss was once the authority, it became Charles Darwin until *The Old Faith and the New* made it possible to name both in one breath."[139]

In Germany in the 1870s, of course, one would also name the radical evolutionist Ernst Haeckel. Given Haeckel's defense of monism, it is hardly surprising that he had nothing but high praise for Strauss's book. In his famous book *The Riddle of the World* Haeckel called Strauss "the greatest theologian of the nineteenth century," crediting him with courageously proving the irreconcilable opposition between modern science and the antiquated Christian worldview. Haeckel also certified that there was indeed a large group of people who constituted the "we" Strauss had identified. Whether they acknowledged it or not, *The Old Faith and the New* was indeed the honest conviction of educated people of the present who realized that there was an unavoidable conflict between Christianity and natural science.[140]

Ziegler later asserted that Helmholtz, Moritz Wagner, and other German natural scientists had privately confirmed that the natural science in Strauss's *The Old Faith and the New* was correct.[141] That many scientists had not openly shown "the cards of their confession" had been noted by Haeckel, and Strauss concurred in a letter to him that it was not always advisable for them to do so. After all, Strauss, whom many of his day thought of as a modern Judas Iscariot, knew from personal experience as well as any nineteenth century writer what the consequences of speaking the message of one's heart could be.

Strauss died in 1874, two short years after his notorious book had appeared. Biedermann's willingness to have sacrificed a finger from his right hand if it could have prevented the publication of *The Old Faith and the New* is sufficient proof that Strauss did indeed represent the sentiments of a growing number of people. Although there were undeniable personal motives that drove him, Strauss clearly thought that he had honestly dealt with the challenge to theology that natural science had presented. In his conclusions Strauss did not believe that nature had been lost to theology; on the contrary, nature and natural science were for him unavoidably in the middle of any attempt to articulate a new faith. But if that new faith was acceptable to more and more people among the German laity of the new empire, it was neither acceptable to nor influential for the rest of the German theological community.

4

Otto Zöckler, the Orthodox School, and the Problem of Creation

IF GERMANY SUPPLIED the first theological champion of a Darwinian worldview in David Friedrich Strauss, it also produced the earliest substantial critic of Darwin. The Greifswald theologian Otto Zöckler wrote earlier and more about theology and Darwinism than any other theologian in Europe, England, or America.

Unlike Strauss, however, Zöckler's encounter with Darwin came at the beginning, not the end of his theological career. Strauss was a theologian already quite well known in both professional and public circles when his *Old Faith and the New* confirmed impressions long since formed in the minds of Germans everywhere about their radical and irreverent countryman. Zöckler, by contrast, was just beginning his theological work when Darwin's *Origin of Species* provided him with the means he needed to establish a name for himself.

Were it not for his massive *History of the Relations between Theology and Natural Science* (1877–1879), few historians of science in this century would ever have heard of Otto Zöckler. And yet natural science was at the center of Zöckler's life. During his lifetime Zöckler published more than thirty articles and reviews, five major lectures, and four books dealing directly with the relationship between science and religion. His first treatment of Darwin appeared in 1861, making his response to evolution by natural selection uncharacteristically early compared to most members of the professional theological community, who delayed for nearly a decade.[1] It is no wonder that Victor Schultze named Zöckler "the first theological authority in the area of natural scientific knowledge."[2]

The Grooming of a Theologian

Zöckler was born on Pentecost Sunday in 1833 while his father was preaching in the Hessian town of Grünberg, twelve miles southeast of Giessen. There had been Zöcklers in Grünberg since the Thirty Years' War, and their bent for becoming clergymen had long been apparent. By the 1830s Otto's father, Johann Konrad Zöckler, had gravitated slowly but surely away from the highly popular theological rationalism of the teens and twenties to the biblically centered Lutheranism that flourished under Ernst Hengstenberg and others. He was an intensely curious man who loved matters of the mind and who delighted in sharing his knowledge with others.

After several moves the family settled down in nearby Laubach, where Konrad Zöckler established a private elementary school in which Otto received his early training. Although a young Hessian nobleman attended the Zöckler school, Konrad also made it a point to include some of the poor youth of the community in his educational venture.[3] Otto early showed signs of interest in natural science. His room at home became "a veritable natural history cabinet," housing in addition to his butterfly and insect collections a self-built electrophor and a planetarium.[4]

At sixteen Otto went off to the gymnasium in Marburg, home also of the university that rose to fame later in the century as a center of neo-Kantian thought. In the gymnasium Otto encountered August Vilmar, who, along with Hengstenberg, Theodor Kliefoth, and F. A. Philippi, was among the leading spokesmen for strict biblicism and uncompromising Lutheran confessionalism from the 1830s to the 1860s. It was Vilmar who imbued Zöckler with his need for religious certainty, an attribute he carried with him throughout his life.[5] Barth observed that "Vilmar is the type of zealous pastor who has always known what he wanted to say so well that he has never really considered anything else."[6]

Vilmar left his position as director of the Marburg gymnasium in 1851, the same year Zöckler left Marburg to enroll at Giessen University. The tone at Giessen was much different from that set by Vilmar in the Marburg gymnasium. Zöckler found himself drawn not to the liberally oriented Protestant theologians but to the more theologically conservative Catholic reform thinkers Anton Lutterbeck and Leopold Schmid, the first a theologian and the second a philosopher. Both had ended up in the philosophical faculty following the collapse of the Catholic faculty of theology in Giessen. From Schmid in particular Zöckler inherited the strong conviction that secular knowledge did not threaten religion but stood in harmony with it. It was Schmid who set Zöckler on his life's work, the apologetic defense of Christianity against materialism and atheism.[7]

Zöckler remained at Giessen from 1851 to 1854. In March of 1854, having just successfully passed the examinations for candidates aspiring to become gymnasium teachers, he formally requested the faculty of philosophy to confer on him the doctoral degree without the usual examination and disputation. Although he obviously intended at the time to go into secondary teaching, it was not long before things changed.

In all likelihood it was the extended tour of German universities that Zöckler undertook from October of 1855 to March of 1856 that confirmed for him the desire to change his career goal from secondary teaching to becoming a professor of theology. At twenty-two his visits to Erlangen, Berlin, Halle, and Göttingen brought him into contact with theological luminaries such as Hengstenberg, Nitzsch, Franz Delitzsch, and Dorner, awakening in him the dream of following in their footsteps. But he had not completed a dissertation, and his idea of returning to Giessen did not elicit a warm reception in the theological faculty there.[8] Zöckler weighed his prospects and decided to enroll just after Easter in a one-year course at the *Predigerseminar* in Friedberg. The course was required of anyone entering the Lutheran clergy; hence Zöckler would at least have that option as a back-up if his other plans failed. If he was successful in his bid to join the Giessen academic community the practical training he would receive would serve him well as a teacher of future divinity students.[9]

Zöckler thought he would be able to prepare in Friedberg for the examination he had to pass before he could be accepted as a *Privatdozent* in Giessen. Then in May he learned that he would have to complete a dissertation before he would be allowed to take the examination. Perhaps the theology faculty in Giessen was trying to discourage him, for although he met the requirement in the summer with a dissertation on the meaning and usage of the Greek word *elpis* (hope) in the New Testament, he was not invited to take the examination until mid-November. Once past this final hurdle Zöckler wasted no more time, beginning academic activities immediately in the winter semester of 1856–57 in Giessen with lectures on the New Testament, followed by a comprehensive treatment of church history.

Zöckler threw himself into academic work with a vengeance. In addition to his teaching duties as a *Privatdozent*, he attended as many lectures and seminars as possible in natural science. "From [a course on] the letter to the Hebrews, which he taught," writes his son Theodor, "he hurried to botany, then lectured on church history, and immediately hastened to experimental physics and later even to zoology and chemistry."[10] During these years he began to lose his hearing, an affliction that had plagued others in his family. By all accounts Zöckler bore this burden well throughout his life. In keeping with his pious demeanor, he chose to see the delay in the onset of his hearing loss until after his student days were past as a special dispensation from God. Out of gratitude Zöckler immersed himself in the serious academic responsibilities before him, avoiding all temptations of Giessen's worldly life.[11]

Throughout his life Zöckler preferred historical research to all other. Even in his apologetic writings it was in the history of apologetics that his unique talent came through most clearly. His first publications following the dissertation set the tone for much of what was to follow. First came an essay on Luther's (and therefore Lutherans') understanding of nature, followed a year later by an eight-hundred-page study entitled *Theologia Naturalis: A Sketch of a Systematic Theology of Nature from the Standpoint of the Believer* (1860). Both works betrayed Zöckler's interest in nature, and both contained clear articulations of the conservative Lutheran perspective.

Writing against the background of the scientific materialism rampant in his time, Zöckler revealed that he had become aware of the glaring omission in Germany of any established tradition of natural theology. Given his interest in natural science and religion, it is not difficult to see why Zöckler's attention might easily have settled on natural theology. But it was more than that: the eager young Zöckler had caught a vision of how he might make a name for himself in the process. *He* would fill the gap in German theology, *he* would take up the challenge Hengstenberg himself had thrown down in 1856, precisely at the time Zöckler was changing his mind about his future career. Hengstenberg's words, quoted four years later by Zöckler, left no doubt about the need for German theology to develop an interest in natural theology:

> Materialism's appearance has made us aware of the significant deficiencies in the church in the area of scholarship. The modern theology of feeling is totally foreign to apologetic efforts. We are far less equipped than the English with valid proofs of the truth of revelation and its individual teachings. An overreaction to rationalism has made us lukewarm toward natural theology, which in older times was seen as the necessary underpinning of positive theology. These gaps must of necessity be filled.[12]

Lacking the moderation that comes with experience, the enthusiastic *Privatdozent* plunged headlong into his new-found task. Yet even the eager twenty-seven-year-old theologian sensed that he may have asked too much of his audience. Confessing that he was aware his treatment was overly extensive and not easily read, Zöckler defended himself by raising a bold claim for a theology of nature: "A science that . . . lays claims to become associated with Christian doctrine and moral theory as the third equally essential factor of Christian teaching . . . cannot have its main characteristics sketched passively or its major materials represented in but a few fleeting strokes."[13]

Zöckler drew on his dissertation to establish his case for theology of nature (a designation he preferred to "natural theology") as the third major branch of theology. Based on his research on the place of hope in the New Testament, Zöckler argued that just as faith was the foundational principle for doctrine and love was the basis of moral theory, so hope was the principle of theology of nature. He no doubt assumed his reader would make the obvious connection with Saint Paul's reference to faith, hope, and love in I Corinthians 13.

Zöckler in fact referred to this scriptural passage directly, but when he did it was to emphasize that human knowledge was but a dark reflection of real truth.[14] He was conscious that his use of natural knowledge depended on its tentative symbolic nature. In this he was toying with notions well ahead of his time. While Zöckler seemed to be agreeing that the knowledge of natural science could never be said to provide a distortion-free representation of nature, his message focused not on scientific knowledge as deficient in this respect but on the possible role our knowledge of nature could play. Although merely an imperfect symbol of reality, knowledge of nature nevertheless was sufficiently linked to reality to be able to point beyond itself toward larger truth, the clear unreflected vision that Saint Paul promised in I Corinthians 13:12 would one day obtain: "For now we see through a glass, darkly; but then face to face."

An examination of Zöckler's *Theologia Naturalis* may tempt the reader to dismiss his treatment as so much theosophical mysticism. Indeed, Zöckler himself worried that readers in his own day would have just such a reaction. Beyond his examination of the traditional proofs of God's existence, Zöckler discussed ways in which divine attributes were symbolically visible in nature itself. The atmosphere, for example, was described as a demonstration of God's dominion, in which the air represented divine omnipresence while light and color stood for divine omniscience.

But Zöckler was struggling in the work with something more profound than he could really handle. Had he ever been able to come to a place where he explicitly admitted what was clearly implied in his schema—namely, that truth in natural science did not depend on a precise correspondence between reality and our minds, that in fact we deal only with distorted representations of nature—he might have felt confident that he could make his new "theology of nature" persuasive to his age. But neither he nor the great majority of those of his time thought of truth in terms different from correspondence. For them the results of natural science were neither tentative nor inexact.

With his notion of hope-based speculation Zöckler played with the possibility that he was onto something. Hope, wrote Zöckler in the first volume of a projected two-volume work, was the only thing humans brought on their own to establish their link with the divine, since faith and love were gifts.[15] Without hope the examination of nature resulted in the universe of materialism, devoid of all purpose. Hope was the driving principle of Christian natural theologians, what Zöckler called the natural (as opposed to supernatural) foundation for our consciousness of God.[16]

In this first attempt Zöckler gave hints that he understood his results were meaningless to those who did not share his starting point; those who did not, would not, or could not have hope. This emphasis was different from many classical works of natural theology, where the message seemed to be that it was not so much an agreement about starting points as it was the proper rational evaluation of nature that determined the value of natural theology. Zöckler gave evidence that he knew his system would only have value to believers,[17] that it would not be convincing at all to those whose presuppositions were different from his. His conviction was, however, that if he could demonstrate that his reading of nature enabled humans to hope, then his revelational standpoint *(offenbarungsgläubiger Standpunkt)* would perhaps become persuasive.

This position was also defended in Zöckler's article in the *Jahrbücher für deutsche Theologie,* published in the same year as the *Theologia Naturalis.* A review of the most recent natural theology of the English, it was intended to complete the historical survey done in the *Theologia Naturalis* itself.[18] Zöckler argued specifically that the English approach, based as it was on Newton's achievement, related merely to the external side of human experience and therefore was of use in countering only the crudest atheism. Zöckler concluded that while there were still scientists in England who, like Newton himself, combined a deep commitment to science with a strong belief in God and his word, there was evident nevertheless a deistic tendency in England to separate the world from its creator and knowledge from faith.[19] English natural theology, therefore, did not suit the German theologians' need of a foundation for redemptive history that would counter the refined pantheism, materialism, and deism everywhere evident in Germany. German natural theology, Zöckler declared, had to provide for itself illustrations of Christology, eschatology, and the ethical attributes of God.[20]

The *Theologia Naturalis* was largely unwelcome in the German theological community of 1860. Zöckler did receive a favorable public notice in Hengstenberg's *Evangelische Kirchenzeitung,* an organ that Zöckler himself would one day direct, and he was praised by the anatomist Rudolph Wagner and the physiologist Rudolph Leuckardt. But in the main the work was judged to be unscientific, "medieval," and thoughtless.[21] In addition, Zöckler's presentation reflected a definite ecumenical flavor, which he had no doubt learned from his old Roman Catholic teacher Schmid. But when Zöckler wrote against "all false *Unionsmacherei*" in favor of a true union and peaceful settlement among the churches, he was hopelessly trying to resist the disunity being encouraged at the height of the confessionalist controversy.[22]

Writing in 1908 Victor Schultze commented on the program of the *Theologia Naturalis* that "of course it no longer needs to be substantiated that the book was wrong in conception and execution."[23] Schultze had in mind its mystical quality, as seen in what he called the unfortunate influence of Baader, Schubert, and other theosophists. Perhaps the second volume of Zöckler's *Theologia Naturalis* never appeared because many in 1860 had the same objection that Schultze voiced a half century later. But Zöckler's abandonment of his "theology of nature" was also due to a reformulation of his conception of his task. He in no way gave up his interest in the relation between natural science and religion. What did change was his understanding of the use of scientific knowledge in apologetics, from something *symbolically* illustrative of the doctrines of the Christian faith to something that stood *factually* in harmony with the assumptions of religious belief in general. With this shift in emphasis back to an explicit and straightforward correspondence understanding of truth Zöckler became less an illustrator and more a defender of faith.

THE ENCOUNTER WITH EVOLUTION

In Zöckler's 1860 review of natural theology Charles Darwin was mentioned in passing along with Alfred Russell Wallace, Joseph Hooker, and the anonymous author of the *Vestiges of the Natural History of Creation* as a "renewer of the Lamarckian development or transmutation hypothesis."[24] A year later Zöckler devoted a lengthy fifty-five-page essay to the theological significance of the species question, giving special attention to the views of Darwin and Louis Agassiz. As was his wont, he began with a historical review of the debate over the alleged origin of new species, which, he said, rested on the relatively recent claim that there had been a successive development of ever more perfect plants and animals from less perfect forms.

In what must rank as one of the earliest post-*Origin* systematic treatments of the species question by a theologian, Zöckler demonstrated his impressive command of the literature. His historical survey of what he called "the debate over the species question" began with Benoît DeMaillet's *Telliamed* of 1748 and included Georges Louis Leclerc Buffon, Lamarck, Isidore Geoffroy St. Hilaire, Jean Bory de St. Vincent, Lorenz Oken, Georges Cuvier, the anonymous author of the *Vestiges,* and others, ending finally with the real subject of his interest, the contrast between Louis Agassiz and Charles Darwin. These two figures, he said, represented the two most important and most admired voices of the last half century.

Zöckler introduced Agassiz as an opponent of Baden Powell's philosophy of creation in the *Unity of Worlds* of 1855 and of Powell's contribution to the Oxford *Essays and Reviews* of 1860. Powell had argued that species were permanent over long periods, but he seemed to allow that changes in external conditions could be felt within a species. An alteration in conditions might, for example, lead to the extinction of one species while not affecting another species at all, yet a third "might undergo change."[25] Powell rejected the view that the original representative of all species had been created by God in final form. That, he said, was due to the influence of Hebrew cosmogony; indeed, Powell refused the conclusion that the secret of the creative process was susceptible of the treatment of inductive science and that, strictly speaking, a philosophy of creation was possible.[26]

In representing Agassiz's contrary position, Zöckler relied heavily on the first part of the 1857 *Essay on Classification.* Agassiz's basic goal was to establish the idea that the classification of every truly natural and genuinely scientific system of natural history was nothing more than the thoughts of the Creator translated into our human language.

Zöckler wanted his readers to see that Agassiz was not simply prejudging the matter. First he referred to Agassiz's stature as a scientist of the first rank within the scientific community.[27] Next he tried to establish that Agassiz's view was unbiased because it depended "on the objective value of natural scientific concepts of species, genera, families, classes, and kingdoms or groups."[28] In chapter 2 of Agassiz's work, reported Zöckler, he made clear the "objective basis and the real character of the dominant ways the animal kingdom is divided."[29] There Agassiz asserted that Cuvier's four grand divisions of the animal kingdom had acquired "objective validity" among almost all investigators and that they too denoted nothing other than "the same intellectual conception that originally united the animals in the thought of the Creator."[30]

Although Zöckler continued to use similar language to describe Agassiz's understanding of species as "definitely bounded objective units," he indicated his discomfort with Agassiz's inadequate representation of the correspondence between thought and things that existed. For Agassiz species were ideal entities that would exist in the mind of God even if there were no empirical representatives of them. True, Agassiz was careful to list empirical conditions that accompanied the existence of an individual species (geographical distribution, food supply, life span), but these in no way predetermined the nature of species.

Zöckler's discomfort did not necessarily arise because of the idealistic flavor of Agassiz's system; rather, he objected to what Agassiz's system implied for theology. Agassiz did not insist that animal species, including the human species, were each created as an original pair. The sexual pairing of members of an animal species did not occur, according to Agassiz, simply to preserve the species. The preservation of the species was guaranteed by the creative decree; it possessed real existence *before* the first sexually produced representative of the species. Agassiz believed, therefore, that when members of a species made their first appearance as the result of divine creative action they did so in large numbers. Zöckler quoted Agassiz to say that "pines originated in forests, heaths in heathers, grasses in prairies, bees in hives, herrings in schools, buffaloes in herds, humans in nations."[31] It was the last phrase, "humans in nations," that bothered Zöckler. Agassiz's support of polygenism made his otherwise permissible view of species unacceptable. Zöckler saved his critique of Agassiz's view for later articles; here he merely reported it. But the reader could see his raised eyebrows through his insertion into his text of exclamation points and question marks.

When Zöckler turned his attention to Darwin, it was with the spirit of respect that Zöckler always displayed toward those with whom his disagreed. Darwin's *Origin of Species* was judged to be far more circumspect and scientifically significant than the other recent work on transmutation of species in Zöckler's review, Hudson Tuttle's 1859 *Arcana of Nature*. Twenty-five years in the making, Darwin's book, according to Zöckler, "contained comprehensive knowledge of almost all areas of organic nature, eminent sagacity, and a brilliant talent of representation." It had "the character of a scientific achievement that was as attractive as it was epoch-making."[32] It was no wonder that Darwin had already won several converts from the older ranks of the scientific community and promised to acquire many from the younger.

Zöckler proceeded to provide German theologians with a thorough, chapter-by-chapter exposition of Darwin's book, especially as it contrasted with older transmutation theories. As in the discussion of Agassiz, Zöckler's treatment was eminently fair. Although he could not occasionally resist appending a footnote listing contrary evidence to a conclusion Darwin drew, his review of the *Origin* was remarkably thorough and straightforward. He rehearsed Darwin's argument that as domestic variation resulted mainly from selection by humans, so nature "by means of a certain unconsciously selective activity" produced incredible variety in the wild, making it difficult to differentiate species from varieties and enhancing the probability that differing species were originally varieties of one primitive species.[33] He followed Darwin's documentation of "the struggle for one's own existence, from which proceeds the unconscious and naturally necessary [*naturnotwendig*] striving, passed down from parent to offspring, to preserve faithfully the distinguishing characteristics, however small, that have been attained and that assist in adaptation to external nature."[34]

Although Zöckler acknowledged that Darwinian selection was a naturalistic process resting on chance, that by means of it favorable variations were preserved and unfavorable ones lost, he could not resist drawing a comparison between its role in Darwin's system and the place of "contrivance" in William Paley's classical exposition of natural theology. In both cases, Zöckler observed, the system stood or fell with acceptance of the central claim. In fact, Zöckler argued, Darwin's natural selection was really just Paley's contrivance loosed from the presupposition of a personal creative cause and translated into the pantheism of impersonal natural force.[35]

Zöckler was not threatened by Darwin in the way many theologians of the time would later show themselves to be. Like Charles Hodge in America, Zöckler saw the fundamental issue to be one of starting points. His goal was to establish that the decision about the species question was dependent not on the scientific *details* of one view or another but on the *presuppositions* that were brought to the debate. Because of this Zöckler was not intimidated by the impressive detail of Darwin's book. He could report its strengths on the same page he pointed out its deficiencies. He listed faithfully Darwin's answers to anticipated objections in chapter 6 of the *Origin,* he called the chapter in which Darwin incorporated animal behavior into the system "especially rich," and he explained Darwin's rejection of a simplistic acceptance of the sterility of hybrids, all with only an occasional qualification in a footnote.[36]

This same careful and fair exposition of Darwin's position was carried into the chapters in the *Origin* dealing with geological and morphological evidence, the former subject lying close to Zöckler's heart. He noted that Darwin's attitude toward the fossil record assumed, like Charles Lyell's, that it provided an imperfect history of which we have only the last part. He pointed out how geology, morphology, and embryology could be used to support biological transmutation, noting, for example, that "the frequent appearance of purely rudimentary organs and atrophied or freak organisms were favorable to the developmental hypothesis in Darwin's view."[37] Zöckler summarized no less than nine consequences of the revolution Darwin wished to institute that would be constructive results for natural science. Finally, Zöckler portrayed Darwin's vision of the future as one of undisturbed progress in which the mental and corporeal properties of living things were continuously perfected.[38]

Darwin commanded more space than did any other figure because of the immediate theological implications Zöckler saw in the English scientist's system. This is not to suggest that Agassiz's work did not involve theological implications, for in his polygenism and his acceptance of a prolonged age of the earth Agassiz was as far from Zöckler's so-called revelational standpoint and as close to pantheism as Darwin.[39] Zöckler criticized Agassiz for unnecessarily proliferating the number of species, but he argued that the leading idea of his system—that an "objective or natural" system embodied the thoughts of the Creator—was of indispensable value for all theological treatment of nature.

By contrast Darwin's system was suspect because of its highly subjective nature. The understanding of species, classes, and other classificatory categories resulted from subjective judgments arising from the reflective activity of human beings, and it was as incompatible with the biblical account of species as it was with theism in general.[40] Gone was any appreciation of "seeing through a glass darkly." "In this rigorous claim of the objectivity of natural historical systematics and of the fixed character of species lies essentially and especially the genuinely theistic element of Agassiz's worldview."[41]

Zöckler did not, however, simply assume Darwin's results could be ignored. His objection to Darwin, beyond the subjective character of his categories, centered on his method. Zöckler argued that Darwin committed the same fundamental mistake Agassiz did when he assumed that the laws of the past were identical to the laws of the present, a presumption that negated all differences between past and present. Although he admitted that one could never establish such an assumption empirically, he argued that we could infer from the phenomena of growth, be it of plants, animals, states, or nations, that things proceed rapidly in the beginning and soon stabilize with age.[42] Hence Darwin's confidence in Lyell's geology to furnish him with incredible amounts of time was misplaced.

Furthermore, Darwin's notion of "creation by law," in which the mysterious and unimagined element of the creative process was accounted for through exact research, was as impossible to carry out as it was to confirm spontaneous generation empirically. Such claims could not be established scientifically; hence the biblical interpretation had no reason to defer to scientific treatments. Since scientists like Darwin could not compel compliance with their conclusions through the force of scientific reasoning, Zöckler could not imagine why anyone would opt for Darwin's view. There was too much to lose. "Whenever the concept of creation is negated and exchanged for a spontaneous generation from blind natural forces, then the concept of a personal living God different from the world also falls, and as a result the existence of a supreme moral world order to which conscience testifies is explained to be pure illusion with equal necessity."[43]

In his 1861 review of the species question Zöckler set down the position he would defend during the rest of his life. A fundamental assumption guided his treatment of the debate, for, as the title declared, his examination of the question was done "according to its theological significance." His purpose was clear: he would attempt to determine which views it was possible to entertain, even hypothetically, on the assumption one was a theist. All questions could of course be investigated, but not all results were compatible with a belief in a personal God.

The specific issues were for the most part clearly drawn. They included standard questions such as: Were species real (objective) units, or were they arbitrary abstractions of the mind? Did all species originate through divine fiat at creation or had some made their first appearance subsequent to the original creative activity of God? Could the appearance of species ever be accounted for through the action of unconscious, purposeless natural forces? And finally a question on whose answers all the others seemed to wait—how long had species existed?

For all these questions Zöckler assumed that there were unambiguous answers and that the answers were both consistent with and affected by the fact of God's existence. Of course answers merely consistent with a theistic viewpoint *could* nevertheless be wrong. And answers inconsistent with theism could be right if God did not exist. If, however, God did exist, then the results of science, in order to be true, would have to reflect their consistency with God's presence in the universe. For Zöckler that meant natural science, when done *correctly,* provided objective truth.

Zöckler's position in the 1861 essay on the species question represents a transition from his earlier theology of nature, in which the tentative results of the necessarily incomplete knowledge of science could at least provide hope to those whose belief welcomed support, to his later aggressive apologetics, in which Zöckler happily exposed the errors in scientific conclusions that had to be false because they were inconsistent with a theistic perspective. The 1861 essay was for the most part a neutral report of recent views. Only at the end did Zöckler express the emphasis intermediate between his early and mature positions, an emphasis that never completely disappeared from his subsequent apologetical efforts. The biblical view—by which Zöckler meant acceptance of species as real entities, all created by divine agency in a historically recent act—was not only consistent with theism, but, said Zöckler, it also "need not retreat shyly and humbly from the arena of scientific debate and leave the field to one or another of its opponents."[44]

Activities in the 1860s

For the remainder of the 1860s Zöckler directed his apologetical efforts primarily to the crucial issues of the age of the earth and origin of the human race. From his vantage point the debate largely hinged on one's handling of these questions. If the earth had existed for only a relatively short period of time, then all other suggestions about the gradual transmutation of species became academic.

Zöckler was not solely interested in apologetics, however. By his own estimate, apologetics had not been pursued in German theology as it had been elsewhere, and that meant that Zöckler would have a hard time establishing himself as a theologian if he focused his research exclusively in this area. In 1861 German theologians seemed as uninterested in natural science as they had been since early in the century, but they were also remarkably unresponsive to the growing scientific materialism of the 1850s. What was perceived as a crude and sensationalized polemic against religion was left for pastors and preachers to counter. In addition to the general reluctance German academics displayed toward involving themselves in such low-level popular debate,[45] German theologians did not give evidence of any sense of urgency concerning the claims of natural science *vis à vis* those of religion. Their attitude assumed that if one really understood religion, one would realize that there was no problem between science and religion.

As a *Privatdozent,* then, Zöckler was not about to impress the faculty at Giessen by writing long reviews of the works of English and Swiss naturalists. "It was understandable that the faculty became more cautious regarding its *Privatdozent,* even though his academic success was good."[46] A still better confirmation was required to legitimize the young scholar. Fortunately, Zöckler's interest in history included more than the history of natural science. He also was fascinated by the history of the church, a thoroughly traditional and respected field in the theological community. It was his sober and scholarly study of asceticism, published as Zöckler neared age thirty in 1863, that brought him recognition from his peers and a promotion to extraordinary professor of church history.[47]

In his inaugural lecture Zöckler identified his sympathy with the Lutheran confession. "The true purity, fullness, and freshness of practical Christianity," he declared, "is almost exclusively found among those for whom the confession of the church still serves a divinely sanctioned norm for their action in the service of Christ."[48] But while Zöckler clearly embraced the Augsburg Confession, he always did so with a great tolerance toward those who adhered to other articulations of the faith, even toward Roman Catholicism. In his tolerance of diverse formulations of the Christian faith, there is evident an echo of Zöckler's instinctive appreciation of the imprecision of language. Admittedly he felt forced to abandon his earlier attempt to interpret natural science symbolically, demanding instead that natural science be "objective." But he always retained an ecumenical, or, as he himself preferred to call it, an "irenic" spirit where union and confessions were concerned. He once intimated that to defeat the enemy of a Christian worldview a common effort between Protestantism and Catholicism would be required.[49]

In spite of his continuing work in church history Zöckler hardly abandoned his interest in apologetics. In 1865 he founded, with Rudolf Grau, what became the journal of German apologetics, *Beweis des Glaubens*. Originally co-editor with Grau, Zöckler later assumed the editorship by himself, retaining it until his death in 1906. Praised by conservative Protestants and Catholics alike,

Zöckler's journal provided a monthly forum which, as the *Katholische Zeitschrift* approvingly noted in 1907, "concerns itself solely with the defense against attacks on the one hand and the deep grounding of the contents of faith on the other."[50]

By 1866 Zöckler had published over fifteen articles, three lectures on various subjects, and three books, including a 476-page study of the life and work of Hieronymous. In the midst of all this activity the thirty-two-year-old theologian fell in love with Charlotte Geist, daughter of the director of Giessen's gymnasium. The two were engaged in 1865 and married the following year. They were hardly home from the honeymoon when Zöckler received a visit from a representative of the Prussian ministry offering him a position either as an extraordinary professor in Halle or an ordinary professor in Greifswald. Since 1848 Zöckler had viewed Prussia as the power which would unify the German states, but his fellow Hessians harbored ambiguous feelings about Prussia during the 1866 war with Austria. Zöckler tried to get Giessen to raise his salary as a means of avoiding a call to Prussia, but when he was unsuccessful the uneasy relationship that had always existed between Zöckler and the Giessen theology faculty came to an end. Zöckler chose the offer with the higher rank and became a member of Greifswald's theology faculty.

The provincial university in Greifswald was a center of orthodox biblical theology. When he accepted the call, Greifswald had approximately four hundred students, only seventeen of whom were studying theology. The theology faculty, dominated by Old Testament and New Testament scholars, had become stagnant.[51] Zöckler, a conservative young church historian and apologist with a growing reputation as an author interested in natural science, helped to revitalize Greifswald, especially after the end of the war with France in 1870 when he was joined by another young church historian, Hermann Cremer. Through his commentaries on the Old Testament, which began in 1865 as part of J. P. Lange's ongoing Bible handbook, Zöckler became known to Lutheran pastors. When his name also cropped up following 1869 as a co-editor of the *Allgemeine literarische Anzeiger für das evangelische Deutschland,* a venture that Cremer joined in 1871, the impression arose that here was a young theologian with ambition at an institution that was not at all hesitant about its strict biblical stance.

Throughout the remainder of the 1860s Zöckler settled into his role as the leading spokesman on current issues in natural science from what he identified as the "strictly orthodox side."[52] From one vantage point Zöckler's extensive writings on the species debate with its accompanying questions in geology and anthropology command immediate attention because he represents one of the earliest professional theologians to deal systematically with the issues Darwin forced upon them. Carefully reasoned theological analyses of Darwin would not appear in England and America for more than a decade after Zöckler's first statement in 1863.[53] Before Charles Hodge's *What is Darwinism?* of 1874, Zöckler would compose five major studies of various aspects of the debate.

The question arises, if nineteenth century German theology was marked by a lack of interest in natural science, how does it come about that the first serious attempt by a theologian to come to grips with the Darwinian revolution emerged in Germany? The explanation of this curious state of affairs involves, of course, Zöckler's uniqueness as a theologian, particularly the insight he possessed into the severity of the threat that Darwin's theory posed for traditional Christian theology. It stems as well from other more general factors that characterized the reception of foreign science in Germany.

All his life Zöckler was a voracious reader. His own publications are filled with references to works of all kinds, including those whose reputations survived into the twentieth century and a host of those which did not. One of the functions he filled as editor of the *Beweis des Glaubens* was to provide reviews of developments in science and in apologetics at home and abroad. Zöckler's early review of Darwin, then, is not surprising, since he apparently prided himself on being up to date with the latest publications.

But there is more to the explanation than merely the likelihood of Zöckler's reviewing Darwin. Intellectual trends that were well established in Britain did not immediately rise to a comparable level across the channel, nor did the reverse occur. Radical German theology of the 1840s, for example, did not cause a sensation in England until the *Essays and Reviews* of 1860.[54] The delay in transmission meant that a scandal such as that caused in 1844 by the publication of the *Vestiges of the Natural History of Creation* did not occur in Germany as it had in England. There it was greeted as a revival of Lamarck's outlandish theory of development and denounced immediately by scientists and clerics alike as not worth reading. When it arrived in Germany in 1851, however, it was a contribution to the debate over scientific materialism, placing it in the same context in which Darwin's *Origin* was received.[55] Zöckler, then, as a young theologian naturally trying to make a name for himself in apologetics, reviewed Darwin's book as one of the latest challenges of scientific materialism, a movement that had been underway since Karl Vogt's *Physiological Letters* of 1847. His continuing publications on the subject reflected his attempts to establish a German apologetics that would answer the challenge of materialism, as Hengstenberg had urged.[56] While still in Giessen Zöckler assessed the theological options represented in the debate on origins. The title of his 1864 essay, "On the Doctrine of Creation: The Theistic Concept of Creation in Its Battle with the Theories of Materialism, Pantheism, and Deism," made clear what the three rivals to theism were.[57] Of all the doctrines of the church, the explanation of how the world came to be had the advantage of clarifying what might otherwise remain obscure in these alternatives to orthodox theism. Deism and pantheism, for example, both accepted the existence of God. The theist could even accommodate them on such doctrines as God's providence, provided the deist or pantheist acknowledged "at least God's relatively miraculous actions in nature." The theist in return might concede that rule by law in nature was God participating in and with nature, not God arbitrarily working against nature.[58] But similar accommodation was not possible on the doctrine of origins. When pushed regarding creation, the alternative systems quickly revealed thoroughly objectionable implications.

The materialist was the easiest to discredit, for materialists, when consistent, really could not speak about origins at all. Because they denied the existence of immaterial spirit, and because matter for them was eternal, the world had never been created. Zöckler referred to the most consistent of the scientific materialists, Heinrich Czolbe, who, in rejecting all creative process as superstition and suprasensual mysticism, was forced to conclude that life too had always existed on earth and that geological argument to the contrary was fallacious.[59] Most materialists, Zöckler pointed out, were not as consistent as this. Ludwig Büchner, Vogt, Darwin, and others adopted some form of a developmental hypothesis according to which matter possessed a formative power that functioned as the cause of the world's present condition. English materialists like Darwin and the author of the *Vestiges* represented a half-materialistic, half-pantheistic position which permitted mention of a "creator" at the point where pure mechanism ran dry, but the German scientific materialists generally avoided such references.[60] Zöckler did not critique materialism beyond simply exposing what he assumed was its obviously ridiculous foundation.

Pantheism Zöckler introduced as the opposite of materialism in that it emphasized the primacy of force or idea, which formed the basis of matter. In some pantheistic systems, like those of Fichte and Hegel, the idea was totally without substance. In others, such as those of Spinoza and Schelling, there was a primary union and indifference of nature and spirit which separated into independent factors at creation. What was essential to recognize about these pantheistic systems, according to Zöckler, was that in none of them was there an act of creation in which nature, as an entity subordinate to a higher God, appeared because of God's intent. The key was the distinction the pantheists made between creation, which they did not accept, and origination, which they characterized as the self-realization of God. In pantheism transcendence was impossible; in fact, Zöckler commented, Fichte called creation "the basic error of all metaphysics and religious doctrine."[61]

Not surprisingly, most of the pantheists who dealt with the doctrine of recent and nonrecent creation alike, came from Schelling's school.[62] Zöckler reviewed several examples of such systems, showing how in them the creatures and works of creation were produced "not by God, but out of God." In the work of Christian German the human race originated not in an original pair, but naturalistically in a primitive form that was inclined to both good and evil. Humans possessed no individual immortality but strove as a race to develop toward the victory of the good.[63] Carl Gustav Carus depicted humans as the fourth and highest level of life on earth, which originated from a metamorphosis of the substance of the planet itself. Since the process was wholly naturalistic, it claimed to be capable of explaining the divine origin of the human spirit in Zöckler's view.[64]

Finally, Zöckler turned to deism, which, he concluded, was

inconsistent and unclear. In Germany deists were frequently admirers of the philosophy of Johann Friedrich Herbart, who opposed the pantheistic identification of material nature with God. Many natural scientists turned to deism as a way to retain the existence of a personal creator while at the same time insisting on the autonomy of natural law. The challenge lay in the nature of God's creative role. Zöckler's real problem with deism was best expressed later in the decade when he was discussing the creation of living organisms. There he said that God could not have left the task of stamping his image on living things "either to blind chance or to secondary natural causes, but rather he had to perform and carry it out himself."[65]

Zöckler described Herbart's replacement of creation *ex nihilo* with a divine creation of the substratum of empirical reality in time (but not in eternity) as an unfortunate link to the eternal matter of materialism. In such a system God became nothing more than a superficial ordering principle, fundamentally opposed to matter. Popular deism, by contrast, accepted creation from nothing and then moved directly to the rule of law, excluding the possibility of God's active participation in the primitive formation of the world. This left deists with no weapon with which to oppose materialism, even forcing some to toy with pantheistic notions of a world soul.[66] Deism, then, possessed elements of theism, pantheism, and materialism, but in its desire to avoid the godlessness of the last, swayed back and forth between pantheism and theism. To Zöckler pantheism and theism represented the only viable options: either the world was created or it developed by itself.[67] Deism, in trying to have everything, ended up with nothing.

In fact, Zöckler's review of contemporary theistic works on creation showed that even they did not always escape pantheistic overtones. When those with admirable theological and scientific motivations began to explain exactly how the mechanism of law could be harmonized with God's purposeful design, or to demonstrate the significance and proper use of inductive reasoning, they sometimes slipped into pantheistic categories that gave too much autonomy to nature. Zöckler did not hesitate to point out such deficiencies in those who wished to call themselves theists.[68]

The only acceptable position on creation was to see it "as a product of the free triune self-determination of the personal God," effected through God the Son.[69] While Zöckler did endorse the work of numerous theologians of his time, surprisingly including that of Richard Rothe,[70] and while he had demonstrated how easy it was to slip almost unintentionally into pantheism, he in no way wished to excuse those theologians who compromised what he understood to be the biblical doctrine of creation. In 1869, as part of a lecture series that Zöckler, Cremer, and others gave, entitled "Apologetic Lectures on Some Important Questions and Truths of Christianity," Zöckler closed his discussion of the history of creation and natural science with a sound denunciation of "the school of theologians . . . that shares as much as they can the pantheistic-materialistic denial of both the beginning and end of our earthly course." He recognized that the members of this school saw the issue to be relatively unimportant, and that in conceding the field to natural science they hoped to strengthen belief in the central, Christological aspects of revealed truth. But, Zöckler declared, their results gave the lie to their claims, for the manner in which the historical Christ was handled by these theologians denigrated the message of the Scripture, proving that without solid roots and healthy foliage the trunk of a tree could not escape fatal injury. His concluding words put it succinctly: truth must be one, it cannot be portioned out in opposing sections. "The doctrine of creation is of foundational significance for the whole truth of redemption, in particular for belief in the person and work of Christ. One cannot be a Darwinist or materialist in the area of the history of creation and still be a Christian too."[71]

THE AGE OF THE EARTH AND THE ORDER OF CREATION

In other publications of the 1860s Zöckler was more specific about the biblical position on the individual questions of the age of the earth and of the human race. The latter issue was of course tied to the former, for if the creation of the world occurred in the near past, then a distant origin of the human race was logically impossible. In a sense everything hung on one's ability to oppose the claims of the geologists, one group of whom stood out from all the rest. "The apologist for the biblical doctrine of creation really has but one opponent," Zöckler wrote. The foe was "the geological quietism of the Lyellians." "If this one main enemy is defeated, the entire edifice of materialistic hypotheses and systems loses its apparent scientific underpinning only to collapse in a heap like a house of cards."[72]

In contrasting the biblical view to that of Lyell, Zöckler did not deny that in the former the earth was at the center of the universe and immovable. But he did not therefore assume that a biblical view committed him to a defense of the Ptolemaic universe. On the contrary, the achievements of seventeenth century science, because they represented objective truth to Zöckler and therefore stood in harmony with revealed truth, formed a sure foundation for all further investigation of the nature of the universe. That the writers of Scripture shared naive geocentric conceptions was unimportant to Zöckler, since they belonged simply to the genuinely human form into which God clothed his revelation. What value would a more exact description of the heavens have held for Moses, queried Zöckler.[73]

The Kant-Laplace nebular hypothesis was a different matter. As another theory that provided Darwin's transmutation theory with the huge amounts of time it needed, Zöckler was necessarily opposed to it. He listed standard scientific objections such as the "wrong-way" revolution of the moons of some planets, the rings of Saturn being a swarm of meteorites rather than a rotating gas in transition to a final orbit, and others.[74] To these he added the argument that the well-developed environment of the earth could not have come about by chance. The nebular hypothesis was consistent neither with scientific facts nor the scriptural account. It appears to have escaped his notice that Lyell too opposed Laplace's developmental scheme.

Lyell, of course, was interested in the history of the earth, and his conclusions were based on his understanding of the earth's geological past. Unlike Lyell, Zöckler had to interpret geological evidence with reference to biblical events. Zöckler suggested that there were three possible routes to harmonize a six-day creation with geology. As Zöckler described the three options his own position gradually became clear. To defend it Zöckler went to great lengths to take note of the recent developments in geology and paleontology. He was even not above drawing on the uniformitarianism of his arch enemy Lyell to support his case.

The first possibility had a long tradition, going back to such historical luminaries as Tertullian, Gottfried Wilhelm Leibniz, and Thomas Burnet. Even some Catholic and Lutheran theologians of Zöckler's own day still found it satisfactory. Zöckler identified this position as a "hypothesis of the antigeologists" because no geological development was said to have occurred between the creation and the Noahic Deluge. The formation of various rock strata had been occasioned by the Genesis flood itself. Zöckler rejected this view because, except for a few finds from the very youngest rocks, all the plant and animal fossils possessed an undeniable prehuman character. To locate them or even the coal deposits, which formed from the gradual sinking of massive layers of plants in the period after the creation of humankind, Zöckler felt, could be accomplished "only with the most extreme scientific caprice."[75]

Second, one could opt for the so-called restitution theory, in which geological development took place prior to the six days of labor described in Genesis. According to this interpretation, which was made possible among scientists in England by William Buckland, in America by Edward Hitchcock, and in Germany by Andreas Wagner, an immense time passed between God's creative decree in Genesis 1:1 ("In the beginning God created the heavens and the earth") and the first day of God's creative activity in Genesis 1:3 ("And God said, let there be light"). During the interim there occurred the struggle between Satan and his fallen angels with the kingdom of light. The six days of creation represented a restitution of order to the revolutionary chaos the earth had suffered during the long struggle. This explanation not only permitted the earth to undergo extensive geological change but also procured immense periods of time in which it could occur. On the surface it appeared to accomplish for the nineteenth century what Tycho Brahe's explanation of the planets had done for sixteenth century astronomy; namely, through one adjustment in the traditional view it allowed room for both a literal interpretation of the Bible and the results of the latest science.

But Zöckler did not find this compromise theory acceptable. It implied, he said, that there had been more than one destructive revolution, requiring "complete recreations of the earth's surface and its inhabitants."[76] Both the Mosaic account and the most recent natural science argued against such an assumption, however, for both portrayed a basic continuity in the primitive development of our planet that had been interrupted only occasionally by violent change. Moses clearly gave no indication that his account referred to a creation that repeated an earlier action of God. Lyell, too, supported continuity. His achievement was to show that these allegedly frequent catastrophes which destroyed all organic life on earth had no historical basis.[77] But Zöckler was in no way giving in to Lyell's claim of long, drawn-out continuity in geological development. Continuity need not imply immense time; indeed, part of Zöckler's objection to the restitution hypothesis was that it provided enormous amounts of time when none were needed.

To explain his own view, which would not use biblical events either as a *terminus post quem* or a *terminus ante quem,* Zöckler wished to harmonize the creation with the geological epochs of science in what he called the "concordance hypothesis." Later in 1880, when he again sketched his understanding of geological history for a theological journal, he distinguished his concordance theory and the older one of Cuvier. One could not make a direct identification of six rock formations with the six days of creation as Cuvier had tried to do. The number of days was unimportant in itself; what was significant was the correlation between the order of divine creative work and the order of the earth's development. Zöckler called his concordance theory an "ideal" concordance.[78]

What was clear from the outset was that Zöckler did not accept six twenty-four-hour days of creation. As a theologian he criticized those who interpreted the Hebrew word for day, *yom,* as anything other than "a period of years."[79] This opened up the possibility of showing once again that the objective results of geology could be harmonized with the biblical account. He claimed to follow Lyell and others regarding the division of the major geological epochs into five, all of which more or less blended from one into the next. Zöckler preferred to demarcate these epochs through reference to "days" two through six of Genesis 1, the five creative periods of God's activity. Rock formations, fossil remains, and other bases for classification did not provide phases of development that were as cleanly demarcated as the divisions of the Genesis narrative, although Zöckler did attempt to correlate recent geological work with the Old Testament narrative.

On the first day God created light, which did not have reference to the earth's geological condition. The second day, when God separated the water from the land, could be correlated with the earliest geological epoch, one in which the most ancient rocks, granite and gneiss, were formed. In these strata no fossils had been found, in spite of the conjectures of Vogt, Büchner, and others that some life could have perished before the rocks were formed. Furthermore, those fossils attributed to these strata had been shown by Roderick Murchison to belong to the next classification, where the earliest fossils were to be found.

When fossils did make their appearance they did so suddenly and in a multiplicity of species. This, observed Zöckler, was a major problem for Darwin, whose system called for the gradual introduction of species.[80] Genesis recorded that on the third day God created grass, herbs, and trees. With the presence of plant fossils the so-called Paleozoic ("oldest life") period could be distinguished from the Azoic ("no life") era on the one side and the Mesozoic ("medium life") on the other. Zöckler labeled the coal and "lower sedimentary" rocks here transition rocks, noting that the English geologists subdivided them further into the Cambrian, Silurian, and Devonian systems. As one progressed from the Cambrian to the Devonian, the latter of which contained sandstone formations, the number of fossils increased to the point where one could even encounter some fish.[81]

The fourth day introduced strain into Zöckler's harmonization of the days of creation with geological epochs, since it was given to the creation of the sun and the moon. Zöckler suggested that this period could not be correlated with any sharp division in the geological record; rather, it corresponded to a transition from the upper levels of the Paleozoic to the Mesozoic periods, where, in addition to remains of fish, the amphibians could be detected. Because the sun, which Zöckler implied had been present since the creation of light, was removed from the cover of a watery atmosphere, a real diversity of plant life could now flourish. Corresponding to the transition from lower to higher forms was a transition from primarily coal to mid-level sedimentary rocks.[82]

The fifth day included Agassiz's age of reptiles. No mammals appeared as yet; the "beasts of the field" of Genesis came first on the

138 | Nature Retained

Zöckler's understanding of geological history

"Day" of creation	Rock		Period		Fossil	Date
	Classification	Kind	General	Specific		
2	Primitive	Granite Gneiss	Azoic		None	
3	Transition	Lower sedimentary Sandstone	Paleozoic	Cambrian Silurian Devonian	Plants Invertebrates Some fish	

Otto Zöckler | 139

4	Secondary	Middle sedimentary	Mesozoic	Permian Trias	Fish Crustaceans Amphibians Diversified plants
5	Tertiary	Upper sedimentary		Jura Chalk Mammoth Cave bear	Water animals Birds Dinosaurs
6			Cenozoic		Mammals Mastadons Cave animals Humans
				Diluvial Reindeer	4000 B.C. 2700 B.C.

Note: The days of creation do not correspond exactly with one rock classification or kind and may coincide with more than one geological period.

sixth day. Zöckler referred to the geological epoch corresponding to day five as the Tertiary Period. "The biblical declarations about the fifth work of creation agree best . . . with the petrified contents of the upper sedimentary layers or the younger sedimentary formations."[83]

Finally, mammals and then humans were created on the sixth day, which corresponded to the transition from the Mesozoic to the Cenozoic ("recent life") periods. Zöckler emphasized that the Flood of Noah came *after* the appearance of humans, that the Flood was not a new creative epoch but occurred as an "appendix to the last one."[84]

Nevertheless Zöckler did refer frequently to the Diluvial Period, sometimes associating *das Diluvium* with the Ice Age.[85] Just prior to this epoch, in the late Tertiary Period, was the Mammoth Period, or as he also called it, the Cave Bear Period. This was what some referred to as the Stone Age because human cultures had been found from this prediluvial time whose tools were not made of metal. In the late Diluvial Era Zöckler located the Reindeer Period, after which tools of bronze and iron had been found. Zöckler, however, denied that three individual ages of stone, bronze, and iron could be separately identified.[86]

Since the most recent geological period, according to Zöckler, was the Diluvial, humans originated in the Tertiary Era just prior to it. Zöckler's estimate for the age of the human race was approximately six thousand years; that is, he concluded that humans were created somewhere around 4000 B.C.[87] The overwhelming majority of the space he gave to defending his position did not so much contain positive arguments in favor of such a recent origin as put forth a rejection of the older estimates of scientists.

The Origin of the Human Race

Zöckler garnered the specific evidence that had been used by various individuals to extend the Tertiary Period, in which human remains had been found, backward in time. The human jaw found in Florida, which originally had been estimated at 135,000 years of age, had recently been discarded as sound proof of human antiquity. Likewise the Mississippi skull, supposed to reach back some

57,000 years, was far too similar to present-day American Indians to be of interest to any English or German geologists, including Lyell, in their attempts to establish an ancient age for America's original inhabitants.[88]

Zöckler cast doubt on the methods used for geological dating; for example, he cited the various qualifications that prohibited a simple and straightforward correlation between the depth of a find and its age, and listed the many disagreements among geologists regarding the dating of several individual cases.[89] When he came to archaeological discoveries from what he called the Mammoth and Reindeer Periods, which for him occurred just before and after the biblical deluge, he conceded that recent discoveries of the mid-1860s in the caves of Belgium left no doubt about the presence of human remains that indicated a low level of cultural achievement. Because he rejected the attempts of Lyell and others to extend the Tertiary Period back in time millions of years, Zöckler denied that such peoples represented humans at an early stage of evolutionary development.

How then to account for the existence of various barbarian cultures without questioning the accuracy of Old Testament history? Zöckler accomplished this feat with a two-pronged program. On the one hand he called into question the assumptions of natural scientists about the *rate* of geological development during the epochs before humans appeared, and on the other he explained how barbarous levels of culture could have come about subsequent to the original paradisiacal state of humanity.

Throughout his works in the 1860s and after, Zöckler never abandoned his position that the rate of geological development was not constant. It was not that Zöckler believed coal deposits had *not* been formed by the solidification of plant remains through pressure, but that he believed this process took place much more rapidly than generally supposed.

> The presupposition that the earth's formative process has taken place according to the exact same laws of development ... governing the changes of the surface of our planet at the present time *involves an egregious error of a principal kind.* For it makes one and the same the facts of creation and of preservation, whose essential difference is taught not only by theology, but equally well by the

totality of all natural analogics . . . Where is the organism that does not develop itself more quickly and is not far more easily altered in the period of its origin and growth than later during the time of its more mature existence and quieter perseverance?[90]

Nor did Zöckler confine himself to examples such as plant growth or the development of the animal embryo. Inorganic entities, such as metals and crystals, were malleable and easily changed when young but become rigid with age. Even the settling of cement walls during the building of a house exhibited the same fundamental truth. Zöckler cited Bernhard von Cotta, whom he identified as no great friend of biblical geology (though he was a passionate defender of vulcanism in objection to Lyell's actualism), as he attacked the assumption that there had not been geological development. Zöckler believed there had been a time when the earth had exhibited vastly different primitive conditions, and it had undergone development to the state in which it was observed today.[91]

In Zöckler's way of thinking, the creative phase of the earth corresponded analogously to the first four to five years of human life, when the most rapid steps in mental and bodily development occurred. Then followed the years of childhood and adolescence, when, despite backward steps, a gradual course of physical and mental development ensued leading to the transition to mature adulthood.[92] Contemporary experiments in natural science had found that under the incredible temperatures that existed on the primitive earth development would have been fantastically rapid. Zöckler cited the work of German paleontologists who had claimed to produce coal-like substance from vegetable matter under heat within six years, thereby throwing into question the assumption that hundreds of thousands of years were required.[93]

The second defense against a prolonged past for humans involved an explanation of barbarian peoples who, for many paleontologists and anthropologists, represented humans in an early stage of their evolution. Zöckler first noted that the evidence of such cultures indicated that they were very few in number, a fact he thought corresponded better with a recent origin of the human race than would be expected had humans been developing for a long time. Only three skulls from the Mammoth Period and four from the Reindeer had been discovered on which paleontologists could base their conclusions, and, except for the Engis and Neanderthal skulls, most were in poor condition.[94]

Zöckler's main argument, however, centered on the anthropological evidence. The cultural life that these humans experienced, he asserted, could not have been the half-animal, half-human existence that was claimed for them by so many. A host of facts spoke against it. For example, linguists such as Wilhelm von Humboldt, Franz Bopp, and Max Müller held that instead of the small number of simple roots alleged by those who argued that human language evolved from animal sounds, the number was at least one thousand, and the oldest linguistic forms known were not simple and elementary but complex and already developed.[95] Moreover, the earliest historical evidence of Egyptian, Chinese, Babylonian, and Hellenistic culture was already highly developed in 2000 B.C.

> This pre-historical primitive period of our race cannot be thought of as a time of wild bestial brutality and centuries of barbarism, but strictly as a time of childlike innocence and developmental need . . . The forefathers of our race, the patriarchs of the ancient cultures of the Near East and southern Europe, could not have been apes or half apes in the sense of the Darwinian doctrine of creation, but simply early people with the simplicity, naiveté, and narrowness of children . . . From a state of complete barbarism and primitive wildness there could not have come a drive for art like that of the Egyptians in the days when the great pyramids originated or a morally well-ordered family life like that of the Hebrews at the time of Abraham. In short, it is . . . the Golden Age of the human race that we must think of as preceding the earliest historical time.[96]

What then of those archaeological finds which contained undeniable evidence of human cultures that were obviously less developed than those Zöckler described as part of humanity's Golden Age? Zöckler conceded that Eduard Dupont had found evidence which "left no doubt that during the Reindeer Period . . . numerous peoples of rather crude cultural level occupied our western European lands."[97] From especially the paintings of these peoples it could no longer be denied that around the time when the great glaciers had largely covered middle and western Europe there had been exceptions to the progressive development from the original state of humanity Zöckler had so carefully depicted.

But to use these finds as evidence of the evolution of humans was, in his view, an error. Not only did it ignore the numerous reasons why Zöckler claimed the fundamental differences between humans and apes could not be overlooked, but it helped to point out why he devoted extensive attention to an apparently esoteric theological controversy of the mid-nineteenth century. The debate between monogenism and polygenism first broke out in Germany in the 1850s in the encounter between Rudolph Wagner and Karl Vogt.[98] Zöckler's 1861 attack on Louis Agassiz's defense of polygenism placed him squarely on the side of Lutheran orthodoxy.

Polygenists exhibited a variety of motivations for their position. The anti-clerical Karl Vogt was delighted to bring polygenism to the support of his materialistic claims, while Louis Agassiz in no way felt that polygenism undermined the supernatural origin of humans. Still others were simply trying to find a plausible explanation for Genesis 4:14, which indicated that when Cain fled to the land of Nod after murdering his brother, it was already populated. Zöckler did not fail to point out that still another motivation of polygenists might be to find a justification for the existence of a hierarchy among the races. In one version Adam was seen not as the founder of the Jews, but as the father of the Caucasoid race and the spiritually highest representative of humanity. And, according to Zöckler, some North Americans classified the Negro race with apes in order to justify slavery.[99]

Although Zöckler condemned the equation of blacks and apes as unchristian, he himself was hardly above racist judgments. One of his primary arguments for a common origin of all human races was that it was the entrance of sin into human history that made them possible. The different races represented the degeneration and corruption of original humanity as created by God. Once under the stain of sin, not only did the original monotheistic religion degenerate into varieties of all sorts and the original language become confused at the Tower of Babel, but other physiological differences were eventually brought about through cultural practices. Zöckler cited how cultures had over time imposed such physiological characteristics as flattened skulls, noting that much deeper causes of physical misformation were the psychical processes of hardening the heart and degradation of character. An evil moral nature led to ever worsening wildness and an ugliness of external appearance.

Prevailing racist attitudes in nineteenth century Europe provided Zöckler with a means to prove his assertion of racial degeneration, for in citing examples of cultures that had been more and less successful in resisting sin and its effects he claimed to establish that civilized Europeans such as the English and Portuguese, presumably of the highest moral fiber, had been able to resist climatic influences in tropical areas like Africa and the East Indies better than the inhabitants, thereby providing evidence of the causal relationship between the degree of moral culture and racial features.

> That many generations have been able to preserve their original skin color and the quality of their blood forms just as important an indirect proof for the correctness of this assumption as, on the other hand, is furnished directly by the consciousness of many Negro tribes in Africa that they were originally white and must, to a certain degree, carry their dark skin as a result of their inner moral blackness or as a symbol of the dark nature of their hearts.[100]

Clearly, then, what Zöckler applied to races fit primitive cultures as well. To Zöckler barbaric humans of his Reindeer Period as well as the cannibalistic tribes that still existed in Africa both represented the consequences of the fall of humankind. A common origin of the human species was closer to the literal meaning of the Bible's first pair and an indispensable component of Zöckler's foreshortened schema of geological history. If the races had been separately created, Zöckler would no longer be able to explain the origin of early primitive cultures without resorting to some kind of developmental scheme. It was no accident that polygenists like Agassiz invariably held that the human race had been around for a very long time.

Once, in order to defend his biblical literalism against the restitutionists, Zöckler invoked the assistance of Lyell's uniformitarianism. Now he claimed that Darwin's theory also aided him, this time against what he saw to be the dominant polygenist view. "Since the Darwinian hypothesis began to make popular in natural scientific circles the idea of an undivided origin of all or almost all organisms, the number of investigators has become rather significant who not only hold the appearance of all human races from one original pair to be possible, but who assume it to be certain."[101]

As the 1860s came to a close Zöckler added a philosophical

analysis to the arsenal of his arguments for the biblical interpretation of the origin of life. People did not realize, he intoned, that eliminating the concept of creation as a divinely intended act not only committed one to a materialism that undermined morality, but also shrouded in doubt any gain to science through this tactic. Perhaps an appeal to creation was not necessary for the daily needs and interests of society, but any knowledge that claimed to separate itself from belief by basing itself exclusively on what was empirical was not competent to address great questions like those of origin.[102] Zöckler referred to but did not develop the notion that knowledge involved belief. He seemed more comfortable with the position that knowing and believing were two different paths to the same truth which, because they were complementary, must both be taken. His assumption that truth must be understood as correspondence reinforced the conception that it existed apart from us and that our task was to devise means to reach it. "Belief and science [*Wissenschaft*] are nothing but two ways to one objective truth that differ with regard to direction but not with regard to goal. True belief and genuine knowledge can hardly contradict each other; rather, they everywhere supplement and summon each other."[103] And yet there were hints that the two paths, though they must end at the same objective truth, were not equal. Belief was more primary and made knowing possible. "Belief, the immediate apprehension of truth through divinely illuminated reason, gives rise to knowledge, the mediated apprehension *of the same truth* through reason struggling for light. [Belief] is the support, the point of departure, the indispensable basis and presupposition of reason's operations at every level of its struggle forward."[104]

But if belief was a point of departure, did it not disallow the objective result Zöckler demanded? If belief was so fundamental, could it not mean that truth as correspondence must be abandoned in favor of truth determined by consistency with the starting assumptions?

Although Zöckler argued that belief was unavoidable, he did not conclude that it colored the conclusions based upon it. Truth did not depend on human belief; rather, the existence of independent truth required that the belief chosen as the starting point itself be true. Zöckler did not see this requirement as involving circular reasoning. In fact, Zöckler pointed out, it was because belief could not be avoided that two forms of false belief, skepticism and superstition, abounded. Skepticism was the attempt to avoid belief by embracing total unbelief. In refusing to acknowledge that belief could ever be escaped, it represented nothing more than a sickness of the mind that produced error and new forms of superstition. The latter was the misguided disposition to bring what was inaccessible to the sense world into the scope of belief and knowledge.[105]

Although these distinctions were clear to Zöckler, they were not to many of his contemporaries. The difference between superstition and true belief turned out, of course, to differentiate for Zöckler the biblical view from the half-knowledge and illusory knowledge (*Halbwisserei* and *Scheinwissen*) of those who questioned it. But in Zöckler's mind only the proper balance of biblical belief and inquiring reason converged on objective truth. Everything else was a distortion, which, through analysis, could be uncovered. Exposing false science was one of Zöckler's self-appointed missions.

Creation Past and Present

Having made clear what he perceived the fundamental issues to be and where he stood on them, Zöckler endeavored in the 1870s to test his position against that of natural scientists past and present. Whereas after 1881 his role became that of a commentator and reviewer of developments relevant to the growing debates about evolution, the decade of the 1870s marked the era when his attention as a theologian was drawn in the most focused fashion to the relationship between natural science and religion. But even then Zöckler hardly confined his efforts to this apologetical task. His interests in church history and systematic theology, for example, continued unabated throughout his entire career.

Zöckler conducted scientific inquiries into the religious stance of natural scientists with an eye to the specific questions raised by Darwin. His surveys of scientific opinion took several forms. As a historian he exerted the greatest effort, as might be expected, to examine the views of scientists from the past. He did not, however, shy away from an analysis of the variety of opinions in his own day.

Not only did he classify the different positions represented among contemporary natural scientists, he identified what he thought were characteristic tendencies of individual nations as well. Always he wanted to show that there were plenty of scientists of the first rank whose religious stand was not unlike his own. His purpose was apologetic; that is, Zöckler saw himself to be winning a place for his position in the contemporary debate, not to be excluding all views other than his own. His goal was to identify the assumptions behind the alternative views. He knew and made clear to others the assumptions he himself brought to the debate, and he wished to make just as visible the corresponding starting points of those with whom he disagreed.

Of the historical studies the most impressive was Zöckler's massive *History of the Relations between Theology and Natural Science with Particular Reference to the Story of Creation*. Other than a 96-page sketch in a popular German journal, Zöckler claimed to know of no other attempt to trace the treatment of the six days of creation through the history of natural science.[106] The title may not have been quite as revealing as John Draper's *History of the Conflict between Religion and Science* had been a few years earlier, but it was nevertheless clear from its special reference to the creation story and from a glance at the table of contents that Darwinism had played a major motivating role in Zöckler's decision to write the study. His work culminated with a large 248-page section entitled "The Present, or, Relations between Theology and Natural Science in the Age of Darwinism." But from the beginning Zöckler defended writing his history with special reference to Darwin and the creation story.

> In reply to the possible objection that our singling out the doctrine of creation as a special illustrative means to illuminate the subject is arbitrary and in a certain sense insufficient for a complete solution of our task, two considerations are certain. First, creation dogma is of fundamental significance . . . Beyond that the particular richness in the relations that [the history of the doctrine of creation] provides for several modern questions and tendencies of the day, not those of a strictly scientific interest, but those of a general ethical and religious, even political-social significance, anticipates the positions taken on the history of creation. The course of our historical

investigation must of necessity conclude with a critical discussion of the coherence of these questions of the day, namely the Darwinian hypothesis and the controversies with earlier views in the history of creation that come out of it.[107]

Not only did Zöckler overwhelm his nineteenth century readers with more than 1,600 pages of historical information and detail, but he also challenged the popular tendency in his day to describe the relations between science and religion as those of combatants engaged in warfare. In response to "a certain school of English and North American historians" Zöckler made clear at the outset what his attitude in the work would be.

> Under the relations of theology to natural science we do not understand merely the hostile points of contact between the two. It would perhaps be more timely and our work would raise more interest in the eyes of many had we wished to write a history of just the *conflict* between theology and natural science . . . Such an approach would probably be able to win the approval of many in Germany too, especially given the *Culturkampf* that is still going on . . . A conflict-history [*Conflicts-Geschichte*] would be popular, but it would not be true.[108]

Zöckler indicated that he would leave the popular route to A. D. White and others. White's *History of the Warfare of Science with Theology,* which appeared one year before Zöckler's first volume was published, was singled out as a work full of gaps, so arranged to cast shadows on the church and light on natural science.[109] As for himself, Zöckler would write a more objective historical account, one that did not dwell solely on the restrictive influences of religion on science and that included along with these episodes the ways religion had promoted science over the long haul. Zöckler's assumption was that the past contained both good and bad where the relationship between science and religion was concerned, and that any responsible treatment would have to reflect both.

To some extent Zöckler is an exception to the claim that it was the theological far right in combination with a small segment of the scientific community which fostered an image of religion at war with natural science in the aftermath of the *Origin*.[110] Zöckler clearly identified himself with what he called "the strictly orthodox side" of Christianity, yet in his rejection of "conflict-history" he displayed a characteristic more typical of those who were attempting to reconcile Darwinism and religion. Historians of science have largely agreed that for many scientists and theologians in the late nineteenth century, the relationship between science and religion cannot be described accurately in the language of warfare. But for theologians such as Charles Hodge, who, like Zöckler, took their stand against Darwin on the basis of a hard-line biblical position, religion and Darwinism were locked in a life and death struggle.[111] How is it, then, that Zöckler looked past this imagery? The answer lies both in Zöckler's hermeneutical principles and in what Zöckler understood Darwinism to be.

In the last analysis Zöckler and Hodge were of similar minds. Both, for example, believed that there should be harmony between science and religion, for God's truth applied to both. But Zöckler went farther. He rejected outright the position of those who so separated natural science and scriptural accounts of creation that neither could any longer make demands on the other. Unlike Hodge, Zöckler never left the impression that the biblical record was immune to developments in natural science. He explained that if scientific evidence ever cast disfavor on what the Bible taught, "then to be sure it would be serious."[112] Zöckler was obviously convinced, however, that the Genesis account had nothing to fear from natural science. His faith that there was a necessary harmony between the book of nature and the book of revelation was confirmed, he wrote, by his studies of the history of science. When he went on to suggest that "pernicious philosophical doctrines with their harmful moral principles and aspirations" had "crept in under the cover of the pretense of natural science," it prompted A. D. White to pencil in an exclamation point in the margin of his copy of Zöckler's *History*.[113]

Zöckler's position allowed him to attack aggressively the liberal view that the important purpose of the Genesis account was simply to affirm that the world had been created for humans, who represented its reason for being. According to this interpretation the investigations of science were so isolated from religious truth that it was necessary to keep the two realms apart. Zöckler maintained that such separation turned religion into mere poetry. "One values history first according to its truth content and only secondly for its beauty," he wrote in 1880.[114] The evidence that allegedly contradicted the biblical narrative existed, said Zöckler, "only in the preconceived opinions and postulates of the modern chronologists."[115] Some years later Zöckler complained bitterly about the hands-off attitude that more and more theologians were taking about natural science. "Our relations to modern natural science would in fact be severed, every attempt at caring for them further would be useless if the solidarity of natural science with belief in descendance were so all-fired complete and wrapped up. In reality, however, this is not yet the case."[116]

Another reason why Zöckler did not reflect Hodge's bellicose attitude toward Darwinian scientists was that, unlike Hodge, Zöckler did not always identify Darwinism with natural selection.[117] Even when he did focus on natural selection, Zöckler did not interpret Darwinism to be so rigidly anti-teleological that it could, as Hodge argued, be equated with atheism.

In the overwhelming majority of cases Zöckler's references to Darwinism conveyed a concern with the alleged fact of descent. Because of this perspective Zöckler did not credit Darwin with a fundamentally new insight that had not been known prior to 1859. In his historical studies, for example, he regularly listed those evolutionists who had preceded Darwin as forerunners. "Of all the representatives of the theory of development from pre-Darwinian times," he wrote of Lamarck in the second volume of the *History*, "no one has as powerful a claim to the honor of being a complete and all-around forerunner in relation to Darwin and Haeckel as this scholar of the French Revolution and the First Empire."[118] In the *History* Zöckler listed no fewer than twenty forerunners since Erasmus Darwin, to say nothing of the five he named from earlier in the eighteenth century. As late as 1892 he identified August Weismann's critique of inheritance of acquired characteristics as a sign of "defection and division within the camp of those who believe in descent."[119] This understanding of Darwinism as descent helps make clear why Zöckler took as his central task the defense of a foreshortened age of life on earth. He may have been just as little able to accommodate his Christian faith to Darwinism as Charles Hodge, but his opposition was not as much directed at what Hodge took to be the fundamentally anti-teleological implications of natural selection as they were against what Zöckler understood as the anti-biblical implications of descent.

When Zöckler did deal with natural selection, he acknowledged that it was "a purely mechanical explanation," but he did not think Darwin had been the first to have established it.

> Considering the unrestricted natural laws of inheritance, the tendency to variation and differentiation, overproduction and the unavoidable result of the demise of a considerable portion of surplus individuals, and finally the survival of the fittest and most favored, there is reason to follow new and more basic paths to the goal of a purely mechanical explanation of the development of organic nature. The last law named, to whose great significance Spencer first called attention, can be designated as "natural selection," a shorter and less involved phrase than survival of the fittest.[120]

Zöckler posed the obvious question himself: Why was it that Darwin's name had been attached to the modern theory of descent? His answer reveals again that Zöckler viewed natural science as an enterprise which carried with it more authority than other intellectual pursuits. Darwin, "the man who established descent and transmutation on firm pillars," supported his theory with observations, combining a vast store of knowledge with his experience as a natural scientist. Herbert Spencer, Büchner, Hudson Tuttle, and others all based their conclusions on philosophical explanations alone.[121]

In places like these, where Zöckler was either praising or simply summarizing Darwin's achievements, one is tempted to entertain the possibility that Zöckler wished to accept transmutation, provided it could be incorporated into the short time available to him. Although he might have accomplished this resolution through his claim that change happens far more rapidly during the earliest stages of development than during later ones, nowhere did he appear willing to apply this pattern to organic laws of descent. His attitude seems to have been that somehow Darwin's empirical labors, which were valuable and acceptable, could be cleanly separated from the undesirable parts of Darwin's theory, most of which were not new.

Zöckler's critique of Darwin in the 1870s confirms this impression. In 1871 Zöckler distinguished between Darwin himself and those scientists who ran roughshod over Darwin's doubts about deriving all life from one first cell. He denied that Darwin's reference to the divine creation of a small number of progenitors of plants and animals was just a clever concession to orthodox Christians; "rather," he wrote, Darwin "holds fast to the foundations of the religious and moral world order with total seriousness."[122]

Two years later, after the appearance of Darwin's *Descent of Man*, Zöckler adjusted his view of the English naturalist substantially. He confessed he had overestimated the difference between Darwin and some of his extremist followers. He even went so far as to observe that the common assumption that "most so-called Darwinists are more Darwinian than Darwin himself" had been shown to be pure illusion.[123] In the *Descent of Man* Darwin had astonished some of his friends by his open agreement with Vogt, Büchner, and Haeckel about the "literally brutal" origin of mental and moral human capacities, without any reference to a free and independent act of the creator. But Zöckler revealed that in spite of his disappointment with Darwin, he was not yet ready to equate him with the radical Darwinians. Granted, Darwin had made it difficult for a theologian like Zöckler to defend him, for although he did continue to refer to God, "there was no place left for a real living God." All that Zöckler could find outside the forces of matter was an abstraction, "a powerless straw man in the sense of the half-belief of deism and rationalism."[124] But for all that, Zöckler nowhere equated Darwinism with an atheistic anti-teleological position.

By the time he completed the second volume of his *History* some six years later, Zöckler's disappointment with Darwin had softened somewhat. No longer did Zöckler associate Darwin with Haeckel and the radical Darwinians. Once again the emphasis was on Darwin's difference from those who wished to banish teleology from nature. Darwin was now portrayed as a deist who saw grandeur in God, who directly created the four or five progenitors of both plants and animals and instituted secondary causes that would produce a foreordained pattern of transmutation.[125] He criticized pangenesis, the assumption of extended time, and Darwin's belief that missing evidence would someday be found as examples of

Darwin's piling up of hypotheses. The descent of humans, for example, was described as a deduction made possible only on the assumption that the general theory of descent was true.[126] Zöckler wrote as if Darwin would be acceptable if he would only abandon these unnecessary assumptions.

While in Zöckler's eyes Darwin himself retained a respect for religious belief, it now became his concept of natural selection that provided the means by which Darwin could be distinguished from the radical Darwinians. Natural selection, said Zöckler, was being misunderstood and misapplied.[127] In reality, Zöckler argued, natural selection was just a magic formula Darwin used to explain everything; it was a wishing wand by means of which he conjured up missing proofs. But since nature was selecting, it could be thought of as proceeding "with consideration and according to a definite plan."[128] In his survey of the four camps into which German scientists divided over Darwinism, Zöckler identified the radical group as that which misinterpreted Darwin's natural selection by removing all pretense of teleology from it.[129]

In a special section on monism—which Zöckler called Haeckelism, ultra-Darwinism, and an infringement on and denigration of Darwinian speculation—Zöckler distinguished between Darwin himself, who, he said, was anti-theistic but not yet atheistic, and those who wanted "an exclusively mechanical explanation of nature." He rehearsed the standard objections to the assumption that matter alone could serve as the ultimate explanation of everything, including the ability of mechanical explanation to account for mysteries like spontaneous generation. The monist message was hardly new, as the English materialist John Tyndall himself admitted by tracing its heritage to Lucretius in ancient Greece. Others saw its precursor in the pantheism of Schelling, Hegel, and Feuerbach. What was new was the way in which the position was established and justified through natural philosophy. Zöckler marked the birth of this modern version with the discovery of the conservation of energy, for it was the application of conservation to the world of mind that permitted this ultimate form of reductionism.[130]

Beyond his denunciation of monism as an illegitimate extension of Darwin's ideas, Zöckler associated its appeal with undesirable social and political forces that wanted to grant humans prerogatives that had once belonged exclusively to God. Clearly Zöckler believed that the cause of the appearance and popularity of such radical natural philosophy was external to both the scientific and religious communities. It was a political ploy of the left.

> The real triumph of this modern natural wisdom, that which gives it its major appeal in the eyes of the masses and which has brought about and continues to bring about its rapid spread especially in the political and radically religious circles of liberals, progressives, and socialists, is the clever trick of setting aside the Creator and teleology, a stroke it teaches or at least seems to teach. Only because it contains an instruction to have done with the riddles of organic life without the help of a personal creator do the masses succumb to it.[131]

THE IMPOSSIBILITY OF RECONCILIATION

When all was said and done and the technical objections to Darwinism had been stated, Zöckler remained concerned that his contemporaries still might not comprehend the fundamental threat to religion that Darwin had precipitated. In the *History* he discussed over thirty attempts to reconcile Darwinism and religion that had cropped up in one fashion or another, a phenomenon he saw to be especially popular in England and North America.[132] He required more than three times as much space to refute these systems than he had taken to set them forth. Much of this refutation, however, consisted of summary statements of points he had made elsewhere in print.

One deficiency, Zöckler alleged, was common to most all the attempts to reconcile Darwin and religion; namely, they moved directly to accommodation before determining what the factual content of Darwin's system actually was. Must the biblical account be interpreted in Darwinian terms? Might not the twenty-year-old descendance theory yet come into disrepute?[133] Zöckler reminded his readers of Descartes's vortex theory, Newton's emission theory of light, William Herschel's theory of the sun, William Harvey's preformation theory, and Georg Ernst Stahl's phlogiston theory of combustion, all of which, he said, had faded after their first victory. Darwin's theory had so many obvious problems, "that its unsuitableness to be used in any given attempt at reconciliation, or even to admit of one in general, must be conceded."[134]

Zöckler repeated his conviction that there was nothing yet proven in Darwinism that required abandoning the Christian teaching of direct creation by God. Darwin did not deal at all with the creation of inorganic nature. As for the origin of life, even if the issue of a personal creator was left out of the discussion, there was still plenty of controversy remaining. Since Louis Pasteur had sounded the death knell for spontaneous generation, anyone who resorted to this explanation for life's beginning on earth did so in contradiction of the inductive principles of scientific research. Of the other explanations sometimes provided—that seeds of life came to earth on meteorites, that life always existed, or that an animated earth gave birth to organic life—none presented itself as a compelling choice. Darwin's own preference for God's initial direct creative activity seemed by far the best.[135]

But what about descent? Zöckler freely admitted that the idea was hardly anti-religious in itself. Nor was one compelled to assume that the classifications of modern botany and zoology represented original acts of creation. "The question," wrote Zöckler, "is of course how far the scope of rational and scientific admissibility reaches, which primal relationships or lines of descent may one concede have been made probable through analogies from the circle of facts known to us, and which are to be condemned as deceptive assumptions?"[136]

To begin with, Zöckler ruled out all scientific justification for going back beyond Darwin's own four or five animal progenitors and a few less primary forms for plants. He then asserted that Darwin's theory had undergone so many qualifications and modifications of fundamental significance that his projection of even a minimum number of progenitors had been weakened beyond repair. For example, as result of criticisms Darwin's principle of natural selection had been modified at so many points, especially with regard to the formation of morphological differences, that it had been as good as surrendered.[137] Modifications and additions to the theory, such as sexual selection or Haeckel's biogenetic law, might have helped, observed Zöckler, were it not for their lack of paleontological cogency. Zöckler reviewed his oft-stated objections in this area, including those directed toward the use of descent to explain the origin and age of humans.[138]

As he brought his magisterial historical study to a close, Zöckler addressed some of the general implications of a Darwinian view that severely threatened his Christian understanding. From his perspective these undesirable conclusions emerged from the intimate linkage between human beings and animals promoted by the principles of natural selection and descendance.

From a theoretical point of view the heart of Darwin's theory, natural selection, was frequently admitted to be incompatible with Christian morality. Since it was alleged by Darwin to act blindly, it could not be forced to conform to any ethical standard. When carried over to the social realm, natural selection had already been used to justify some very unchristian sentiments. Did not Spencer entertain the possibility of allowing unnecessary children to suffer a painless death, and did not Haeckel complain about keeping incurable prisoners alive?[139]

When it came to the principle of descendance there was a similar challenge to humanity's special theological status. It was not surprising to Zöckler that some Darwinists had begun to evaluate human behavior in terms of animal behavior. When this was done, noted Zöckler, human ethics was relativized, sin excused as a temporary regression or atavism, war ennobled, and the foundation of the legal system undermined.[140] Even religiosity was explained as a social instinct, an artificial confusion of the human spirit. And what replaced religious ethical ideals? The answer in Zöckler's view was self-condemning: egoistic utilitarianism.[141]

An ironic product of this egocentricity, which Zöckler took delight in exposing, was the replacement of a respect for animals with their exploitation. Zöckler explained that some theologians among those who conceded humanity's animal past tried to reestablish human uniqueness by emphasizing the present differences between humans and animals. Since humans had allegedly evolved far beyond other species, they had no reason to treat animals with respect. These theologians urged people to set themselves in opposition to other organic life, justifying human exploitation of animals with the Genesis mandate: be fruitful and multiply. Zöckler condemned this viewpoint and use of Scripture as distortions born of an improper scientific presumption.[142]

The tendency to locate personal meaning solely in humans

themselves, ignoring God, providence, and a future world, struck Zöckler as equivalent to a loss of hope. How any theologian could still see the possibility of a reconciliation between Christianity and Darwinian morality was beyond him. The naiveté of such people was most evident, he said, in the acknowledgment that natural selection was irreconcilable with a Christian moral standpoint, but that descendance somehow was not inconsistent with Christianity and informed the church at a deeper level.[143] In 1894 he reviewed the latest attempt by a German to build a bridge between descendance and religious thought by viewing the world portrayed in modern evolutionary terms as the fulfillment of Christianity. The author, wrote Zöckler, accomplished just the opposite, for what resulted was not the fulfillment but the destruction of the Christian heritage.[144]

In the end Zöckler remained bold and even optimistic. All the attacks, all the tearing down of the Christian doctrine of creation, ultimately would only strengthen the biblical view. Christianity would survive while the distortions introduced by Darwinism would fall by the wayside. Zöckler's final assessment carried overtones of confidence for the future and resignation about the present. The final victory would be his.

> Darwinism should be understood above all from a pathological point of view if one properly wants to determine whether it has to be credited with a certain legitimacy in the area of ethical aspirations and interests. Darwinism is a great and shining phenomenon of our time, but, as the wealth of errors, one-sidedness, and contradictions locked up in it show, it is nevertheless a phenomenon of disease. As with every disease, it has to run its course through the required crisis to healing . . . We do not fear that our Christian world culture will die of this disease. We confidently expect that humanity will be healed from this religio-moral and scientific epidemic. But it seems to us that we will have to prepare ourselves for a long time of evil.[145]

At the end of his life in 1906, no doubt because of the so-called eclipse of Darwinism around 1900, Zöckler felt even more confident. He sensed that Darwinism was declining.[146] His life had been spent in an attempt to confirm the harmony between the truth of nature and the truth of Christianity. Truth was his utmost concern.

"To establish and defend Christian truth" stood as the goal of the apologetical journal he had helped to found.[147] In his eulogy of Zöckler at the funeral, Victor Schultze summed up Zöckler's life with the words: "Everywhere before him stood the truth as a guiding star—the search for truth for truth's sake. He measured everything by it."[148] For Zöckler truth was unambiguous because it depended on God, not on humans. It existed independent of us, constraining us by its eternal and unchanging power. Time, he believed, would prove him right.

5

Rudolf Schmid and the Reconciliation of Science and Religion

THE MANY GERMANS who took comfort from Otto Zöckler's scholarly defense of fundamental orthodox beliefs about God's relationship to nature felt that Darwin's vision of the past simply had to be false. For religious conservatives everywhere the issue was simple and straightforward: if Darwin was right, then, as an American conservative put it in 1865, the Bible was "an unbearable fiction" and Christians had been duped by "a monstrous lie" for nearly two thousand years.[1]

At the other end of the spectrum were those who either secretly or openly admired what they saw to be David Friedrich Strauss's courage in facing up to the hard truth about the universe. True, Strauss forced the issue in an unpleasant manner, but in light of the popularity of scientific materialism in general it was difficult to argue that the grand multitude Strauss claimed to speak for did not exist.

If the orthodox and Hegelian theological positions on the relationship between natural science and religion represented opposite ends of the spectrum, there were many for whom neither option was attractive. Not content to join the all-or-nothing stance of right- and left-wing theology, these theologians were convinced that some intermediate position was still possible. What joined them to the right and left, however, was the common assumption that, although it might be more subtle and complex than the theologians on the extreme acknowledged, there was but one truth about God and nature, and it was accessible to human reason.

Mediating theologians had always been marked by their appreciation of conflicting theological perspectives. They rejected the extreme positions of both Zöckler and Strauss, but in their own writings they revealed sympathy for the convictions that had motivated each. Where the reaction to Darwin was concerned, this meant that they respected the need both for scientists to employ naturalistic explanations, avoiding reference to a supernatural agency, and for theologians to insist that natural science did not eliminate divine purpose from the universe. How these two concerns could be combined in the second half of the nineteenth century is seen in the life and work of Rudolf Schmid.

A Circuitous Route to the Ministry

Rudolf Schmid was born in 1828, the oldest of five children and the only son of Karl and Auguste Schmid. Growing up in Marbach, Schmid soon found that his parents' desire that he become a pastor, the calling of his father, was exactly what he wished for himself. Indeed, it was not only in the choice of his career that he resembled his father but also in his entire disposition. The warm recollections of family life and the great respect Schmid held for his father are imitated to a remarkable degree in the description Rudolf's own son later gave of him. The picture that emerges both from Schmid's unfinished autobiography and from his son's completion of that account is one of a kind and generous man, calm, even-tempered, and moderate in all things. If happiness and contentment with one's life are measures of success, then Rudolf Schmid clearly knew how to live. Near the end of his eighty years Schmid could say that he had been happy to live and he was happy to die.[2]

As a child Rudolf would frequently accompany his father on pastoral visits to area schools and to parishioners. These trips through the Black Forest region of the southwestern German state of Württemberg provided young Schmid a chance to develop a fascination with nature. With his father's help he became especially enamored of rock collecting,[3] an interest that showed itself through his lifelong attraction to the study of geology.

At the tender age of eight years he left home to live with his maternal grandfather in Marbach, where he could attend a Latin school and begin preparation for the eventual state examination leading to a lower seminary, university, and the ministry.[4] At the time it was understood by everyone, including Schmid, that he would someday become a pastor like his father. Once he later began studying natural science at school, however, the possibility of dedicating himself to natural science arose in his mind. "But the humble assets of my parents and my awareness of their unspoken wish regarding my choice of career made this idea seem to me as but a temptation that should be rebuffed."[5]

Although his grandfather was by profession a civil servant in Marbach, he harbored a lively interest in things intellectual. Young Rudolf listened to animated conversations between his grandfather and a young clerical assistant about the controversies surrounding Strauss's *Life of Jesus* and *Doctrine of Belief*. What he learned from these experiences had little to do with what Strauss had written in his books—of that he understood "as good as nothing." The lesson, rather, was that "these were important questions, and that being a pastor was something fraught with responsibility."[6] Rudolf applied himself earnestly in his preparation, making particular use of the Ersch and Gruber Encyclopedia and the many other books from his grandfather's library. Goethe and especially Schiller, who had been born in Marbach, made a strong impact on Rudolf's young mind. It is no wonder that Schmid looked back to these years as those which had guided him into the idealistic direction his life took.

There were two examinations that students who wished to enter the lower seminary at Blaubeuren had to take. On the first, which twelve-year-old Schmid took in 1840, the aspiring adolescent came in second. One year later Schmid passed the second test with higher marks than all others in his school. Entrance to the seminary which Strauss had come to some twenty years earlier was guaranteed for Schmid. The path that he would follow, however, would vary immensely from that of the irreverent Strauss.

For the next four years Schmid pursued the prescribed regimen of studies for seminary students. It was here that his love of natural science was cultivated in earnest. He investigated some upper layers of Jurassic rock accessible to him, and while they revealed little of interest in fossil remains, the same was not true of the exposed strata in the stone quarries he found. He was also drawn to the study of comparative anatomy. For Lorenz Oken's treatment of natural history he felt an instinctive sympathy, although the overtones of nature philosophy were too abstruse for him to grasp. In these endeavors, as in his many dissections of animals, he was alone. The other students did not share his natural curiosity about scientific phenomena.[7]

From 1845 to 1849 Schmid continued along the road to the ministry in Tübingen. During the first two years of preparation in philosophy he heard I. H. Fichte, whose condescending teaching style he did not appreciate. Other instructors made better impressions, but it was more the recent philosophical classics themselves than the teachers that he later recalled. Although Hegel's thought was still popular in the 1840s, Schmid confessed that he could never grasp the system as a whole. Much more to his liking were the writings of Kant, the elder Fichte, and Schelling. From the natural sciences most students took astronomy, but Schmid, following his earlier inclinations, enrolled in the geology course of Friedrich Quenstedt, whose geological excursions proved to be unforgettable.

In the middle of the second year at Tübingen theological study was added to the philosophical training. Schmid found that his exposure to philosophy had brought him to "a kind of Fichtean and Schellingian pantheism, colored with the outlook of Schleiermacher."[8] The tumultuous intellectual upheaval generated by Strauss, Feuerbach, and others in the Young Hegelian movement of the 1840s brought the young seminarian to a crossroads. Looking back to 1847 from his vantage point in the twentieth century, Schmid observed:

> At that time it was not yet the custom, as it is today, that a young man for whom church doctrine had fallen into ruin would style himself a reformer and therefore [continue to] covet an ecclesiastical teaching post in spite of his dissent. I thought simply, "With the worldview you have now you cannot be a pastor," although a clerical teaching position still seemed to me to be the most desirable career.[9]

This was the time at Tübingen when Ferdinand Christian Baur, the leader of the school of higher criticism, and his opposite, Johann Tobias Beck, the biblicist, stood at the height of their fame. Schmid's reaction to the crisis was unlike that Strauss had displayed at the same point in his Tübingen years. Rather than succumbing to the appeal of pantheism and becoming its ardent proponent, Schmid simply lost interest in Baur, Beck, and theology itself. Not until mid-century would he work through his dilemma. Until then his attention, like that of virtually everyone else, was diverted by the political events of 1848.

Already in the *Hungerjahr* of 1847 the student body in Tübingen was drawn into political activity. When the poorer classes in Tübingen threatened to destroy a nearby mill, the city officials concluded they would not be able to retain control of the situation without the assistance from the students. In collaboration with the university administrators, the students were armed with sticks and sent quickly to protect the mill. Although they rose to the cause of defending law and order, the students did donate their bread rations for several days to demonstrate their sympathy for the plight of the poor. The confrontation came to nothing, but the adventure resulted in the students continuing to organize themselves in military fashion.[10]

With the news of the uprising in Paris at the end of February 1848 an already tense atmosphere encouraged rumors of all sorts. On March 24, for example, word came to Tübingen of an invasion of Württemberg by the French, possibly by as many as forty thousand men. Again the students united to confront the danger, this time arming themselves with rifles and scythes. They rallied to the exhortations of one of their professors, who spoke to them from the balcony of a university building and urged them to defend the homeland. Throughout the night the rashly assembled army, Schmid among them, prepared to meet the invading hordes. Morning brought word, however, that the whole matter had been a false alarm.[11]

The unrest of the times was undeniable nonetheless, and it infected and incited the students. Freedom was the watchword. To the students it became an occasion to make demands on their superiors. Numerous petitions began to circulate in Tübingen, prompting university officials to establish a commission whose task it was to summarize all the demands in one grand petition. Schmid was named to the commission, which, in its enthusiasm, produced a document so offensive in tone that it was rejected by the Tübingen administration.[12]

In the wider political context of 1848 the German states assembled representatives in Frankfurt to create a new German nation. The degeneration of the so-called revolution of the intellectuals put Schmid into a difficult situation. The students were totally unable to accept the fact that the refusal of the new national crown by Frederick Wilhelm IV of Prussia in April of 1849, and the resignation of Archduke Johann as the titular head of the Frankfurt Assembly shortly thereafter, meant that their hopes had come to naught. Nor could others in the southern states of Germany simply surrender their allegiance to the new constitution. Oaths of loyalty to the National Constitution and to the regency which had assumed control following the resignation of the archduke were sworn by many in the Tübingen area. Schmid, while mocking these oaths as worth little or nothing in light of the circumstances, still acknowledged idealistically that the regency was the sole national entity left. When, therefore, the regency issued a directive that everyone who recognized its authority join in support of the uprising in Baden, Schmid found himself caught between his idealistic principles, which demanded that he join those of his comrades who were leaving to fight, and his common sense, which told him that the whole business in Baden was doomed. In spite of knowing that participation in such a venture would bring his parents great pain, Schmid reluctantly resolved to follow his conscience. On June 19, 1849, at 7:00 P.M. he and some forty other volunteers left their quiet university town to do battle for the national cause.[13]

As he suspected, the effort proved to be a fiasco. For his part, Schmid ended up marching around pointlessly with a detachment that could not decide what course of action to take. Prussia was rapidly crushing the uprising, though knowledge of the military losses around them eluded the ragtag company. On the morning of June 26, as the volunteers assembled before a tavern in the little village of Weissenbach, Schmid's father suddenly appeared on the street. He had come to bring his son home. His action was prudent, for within a month Prussia had quashed the rebellion and soon thereafter executed the major who had commanded Schmid's detachment.

His father consulted with the Tübingen officials about Rudolf's situation. The student was directed first to go home so that his mother could see him, then to return to Tübingen where he was put into the student prison for four days and assessed a small fine. That he continued to feel that his action had been principled was evident from his adamant refusal to pay any fines. Eventually, after they had accumulated for almost eight years, the assessed fees were dropped.

With seminary behind him and political order restored, the twenty-one-year-old Schmid was in need of employment. He accepted a position as house tutor for three children of a city official in Heiligkreuztal. The situation, however, was not good. The father, who lived apart from his wife, was demanding and unreasonable. Schmid found that in the real world the pantheistic conceptions he had arrived at in seminary were just not relevant to the practical needs of people. He was struck with a realization that had somehow escaped him to this point in his life, namely, that he had overlooked the reality of evil in the world. He came ever more decisively to the conviction that sin was "not only the absence of good, but a real power that was matched by just as real a power in redemption through Christ."[14]

Without doubt his unhappiness in Heiligkreuztal prompted him to some degree to reject the extreme doctrines he had picked up during his seminary training. But whatever the full explanation, there was no longer doubt in his mind that he could become a pastor. He registered for the first of the examinations necessary for the ministry, which he eventually took in August of 1850. After seeing his oldest charge through her confirmation in May of 1851, Schmid made himself available for what amounted to an ecclesiastical apprenticeship. In short order he was called to serve a religious dean in the picturesque Württemberg town of Calw.

In his new apprenticeship Schmid was directed by a man from the old supernaturalistic school. In addition to the regular Friday evening discussions of the upcoming Sunday sermon with his mentor, Schmid was assigned a specific number of the sick to visit. But more formative for his career than either of these duties were the educational responsibilities he shared. By himself Schmid taught religion in the upper class of the girls' school in the area, and he also filled in for his superior during a period of illness in the instruction of confirmants. His dean took the instruction of his charge more seriously than most, since he brought Schmid with him on his numerous school visitations. In the two summers Schmid was in Calw he traveled to no less than seventy different schools.[15]

Not long before he left Calw Schmid experienced the misery and tenuousness of life among the region's factory workers. In June of 1852 the Nagold River overflowed its banks following an unusual late spring storm. In addition to the loss of life and the destruction of homes, an epidemic swept through the area. Schmid realized to his anger and frustration that no one was concerned about the conditions under which the people lived. In spite of his pleading with the factory owners to improve the welfare of their workers, the youthful cleric was repeatedly brushed aside with the retort that he did not understand the matter and that, when it was a question of competition, one could deviate from what the Christian ethic seemed to call for.[16]

From Calw too came a brief glimpse of Schmid's future mediating theological sense, for it was here that he came into contact with Emma Gärtner, a liberally minded daughter of the botanist Carl Friedrich von Gärtner. Emma had absorbed the Straussian religious perspective from Christian Märklin, friend and university confrere of Strauss, who had earlier apprenticed in Calw. Although she was nine years Schmid's senior, he found that neither her age nor her Straussian views prevented them from a rewarding mutual intellectual understanding that continued to last until Emma's premature death.

Perhaps it was his duties as a teacher that helped Schmid decide the next step after his apprenticeship in Calw was over. Whatever the cause, he let it be known to the ecclesiastical authorities that he would gladly serve as a teacher in a lower seminary like the one he had once attended in Blaubeuren. Soon thereafter he was called to the school in Maulbronn, where he remained for just over a year. Far more decisive for his career than his teaching during this time was an event that had nothing to do with Maulbronn or even with Schmid himself. It was, rather, the marriage of his sister Pauline to Julius Köstlin, later professor of theology in Göttingen, Breslau, and Halle.[17]

Returning from a vacation in Switzerland, Italy, and Tirol in the fall of 1853, Schmid stayed overnight in Stuttgart at the home of his new brother-in-law. Julius Köstlin had just returned from his own trip to Scotland and had not yet gone back to Tübingen to resume his tutorial post there. Naturally the talk was of travel, whereupon young Köstlin showed Schmid the album from his Scottish journey. Encountering a picture of Inverary Castle in the western highlands, Julius asked Rudolf innocently if he had any desire to visit such a place. Schmid's nonchalant assent was challenged by Julius's father, who asked what his answer would be if the question were meant seriously. From this sequence of unlikely events it resulted that Schmid received an invitation to become the house tutor in the family of the duke of Argyll. For the next four years Schmid lived with the family of the duke and duchess, assuming instructional responsibilities for three of their sons, aged eleven, ten, and eight.

Why had this family determined to have a German instructor for their children? On seeking advice from their Scottish prelate, the parents of the children were informed that they would do well to obtain a recent graduate of the Tübinger Stift, the reputation of which was widespread even in Scotland. When Schmid arrived, he had against him his paltry knowledge of English. But in his favor were the generally amicable relations that existed between Great Britain and the German states. Prince Albert of Coburg was the consort of Victoria at a time when Englishmen knew themselves to be the industrial leaders of the world. There was in 1854 less of the hostility that would mark the competitive relations between Britain and the new German empire at the end of the century. Indeed, Schmid and the duke of Argyll worked against this growing antagonism throughout the rest of their lives, and their correspondence indicates their increasing concern as relations between the two countries worsened.[18]

Schmid possessed a love of young people which, when combined with the natural pedagogical inclination that had already expressed itself, made him a success with his Scottish charges. So deep and permanent were Schmid's friendships with his students that he confessed thereafter to have led a double life, a British one and a German one.[19]

The time spent in Scotland in the company of such highly placed people left a lasting impression on Schmid. Not only did he come into social contact with well-known individuals from the British intellectual and political scene, but on occasion he was also able to spend valuable time in discussions with them. Among the guests of the duke were scientists such as Charles Lyell and Richard Owen. There were also theologians like Samuel Wilberforce, whose famous encounter with Thomas Huxley lay, unbeknown to him, in the very near future. And there were others. The writers Thomas Macaulay, Alfred Lord Tennyson, the American Harriet Beecher Stowe, the explorer David Livingstone, and various political colleagues of Schmid's statesman host, including William Gladstone and Lords Russell, Palmerston, and Glanville, all crossed Schmid's path during his stay in the British Isles. Schmid even met Queen Victoria when, in the company of the family on a visit to the villa of the duke of Sutherland, the queen engaged him in conversation. Of all of these encounters Schmid was most affected by his extended discussions with Tennyson, Harriet Beecher Stowe, and Richard Owen.[20]

Schmid returned from Scotland in 1858 at the age of thirty. His intention was to prepare for the second theological examination, which had been postponed by his sojourn to Scotland. His plan was altered when he was requested to serve in some interim capacities by the ecclesiastical authorities. In the midst of these duties, in the summer of 1860, Schmid finally put behind him the last of the examinations necessary for him to assume a permanent pastorate. One year later he became pastor of St. Kilians in Heilbronn, where he remained for seven years. At thirty-four Schmid married Thusnelde Köstlin, sister of his brother-in-law Julius Köstlin, and settled into a life's work that carried him from the ministry into the religious education of the young.

THE ACADEMIC CLERIC AS MEDIATOR

The practical demands of his work in the last few years had removed any hesitation Schmid might once have had about God's providential role in the world. When he temporarily lost his voice because of a disease in 1867, then lost his young daughter and his mother soon thereafter, he was determined to retain his conviction that God's will was nevertheless being done. It was in this context that he received a call from King Karl of Württemberg to assume a post as pastor in Friedrichshafen on the Bodensee, the summer residence of King Karl and Queen Olga.

His duties in Friedrichshafen placed much more time at his disposal than he had enjoyed in Heilbronn, largely because the region was three-quarters Catholic. During the summer he did have special clerical responsibilities to the royal family, but normally he was able to reserve his mornings for something he had always wanted to do—conduct an academic study of natural science, especially as it related to theology and religion.

It was during the ten-year period in Friedrichshafen that his views solidified into a publishable form. First to appear was an essay entitled "The Developmental Issue Raised by Darwin: Its Present Status and Its Stance Regarding Theology," followed quickly by his book *The Darwinian Theories and Their Relation to Philosophy, Religion, and Morality* of 1876.[21] One piece, a brief account of the scientific and religious thought of Julius Robert Mayer, came out just after he left Friedrichshafen in 1878.[22]

In spite of the pleasantness of the arrangement on the Bodensee, which had even permitted him a return visit to his beloved Scotland in 1871, Schmid concluded that if he were ever to serve a wider audience than the small flock he served in Friedrichshafen, he had better attempt it before he moved too far beyond his fiftieth year. His wish to try something different was granted when he was named dean in the Franken town of Schwäbish-Hall in 1878. Here he began to acquire administrative experience in the ecclesiastical hierarchy, and within two years he had become the district school inspector. In another two years he felt sufficiently ready to take up a new position as head of the lower seminary in Schöntal. The six years in Schöntal Schmid counted as the best of his life. He loved being able to occupy himself exclusively with questions of theological study and education. In addition to overseeing the training of the young according to his own philosophy of education, he himself could get into the classroom.

Other than the works he wrote at Friedrichshafen and Schöntal we have but three publications of note, two from the year after he left the lower seminary. Of the two studies from 1889, one dealt with the teaching of the Old Testament to adolescents, and the other was an interpretation of the "days" of the Genesis account of creation. The third publication was a book from his retirement entitled *The Natural Scientific Confession of Faith of a Theologian*.[23] In the book Schmid claimed that he had broadened his scope from the relationship between the Darwinian theories and religion to that between religion and natural science in general. In fact the same heavy emphasis on questions of creation and biological science that marked the earlier study remained. The similarity in the stances of the two books and Schmid's other writings permits them to be examined together.

There is no doubt that Schmid represented the mediating theological tradition as it survived into the latter decades of the nineteenth century. As a youth in Tübingen he was powerfully influenced by Schleiermacher,[24] and throughout his life he expressed his distaste for theological parties. "My whole personality strives against the wretched participation in party conquests," he wrote. "Whoever wants to be a party man can be. I do not like it. Within this framework I have learned much from men of all parties and all conceivable positions apart from the battle cry of the party watchword."[25] His deeds must have matched his words, for his son testified to his open heart and his ability to see and value what was good and noble in those who believed differently from him. He also noted the strong positive reaction that his father's open-mindedness and his cheerful nature elicited from such people.[26]

As it was to many minds in the late nineteenth century, the opposition between the real and the ideal made a great deal of sense to Schmid. In 1889 he wrote that he recognized humanity was in a period which emphasized the real. But in such times it was especially necessary to nurture ideal factors, particularly for the benefit of the young, who were at an age when they needed inspiration.[27] But beyond his uncompromising insistence on two matters—the centrality of the doctrine of salvation and the preservation of a teleological worldview—Schmid showed a remarkably flexible capacity for compromise. Indeed, Schmid provoked considerable criticism of himself by his attitude of openness. Yet his rejection of an unbridgeable opposition between Christianity and natural science was clear to all from his writings. His attitude remained the same throughout his life: "Demand complete freedom for natural science and hold fast to the positions of Christianity in their full extent."[28]

If Schmid believed there was, as he said in his *Scientific*

Confession, an "absolute peace" between natural science and religion,[29] that did not mean that the believer had no compromises or concessions to make. Belief in the historical reality of the Genesis story and the doctrine that death came to the animal world first through the Fall of humankind, for example, were "distortions [*Verdunklungen*] of biblical doctrine."[30] These distortions hurt religion rather than helped it. We have become so used to treating questions like the origin of new species in religious terms, Schmid wrote, that any attempt to throw scientific light on them, any attempt to uncover intervening agencies, was seen as an attack on religion. There was a real danger here, noted Schmid in the very first piece he wrote on the subject. When people who thought this way became the majority, it could lead to an unfortunate condemnation of science by the church, as occurred with Galileo in the seventeenth century.[31]

Although he was willing to concede that extremists such as Ernst Haeckel had also contributed to the polarization of the scientific and religious communities, Schmid called for the abandoning of the assumption that Darwinism automatically meant opposition to the Christian conception of things, and for a tone that treated Darwinians in a more worthy fashion.[32]

For such sentiments Schmid was criticized for being a new believer *(ein Neuglaübiger),* though explicitly spared the opprobrious label of "unbeliever." "If by a new believer," he answered, "I am one who tries to learn from theologians now living, particularly concerning the historical events about which the Holy Scriptures inform us, then I am willingly a new believer."[33] To those who wished to retain and defend what he identified as "a devout and positive Christianity," Schmid later offered a warning. When theologians dealt with a new interpretation of something that up to then had been judged to be part of the fundamentals of Christian faith but that also belonged to the realm of nature or history, they should evaluate not only the validity of the reasons given for the new view but also whether the previous perspective now under attack really belonged to the fundamentals. Schmid cited the relationship between Christianity and the Copernican worldview, suggesting that the traditional explanation of creation in terms of six twenty-four-hour days had gone the way of Ptolemy's outmoded system of the heavens and that in so doing scientists had helped to clarify the insignificance of our ideas of space for determining the religious meaning of heaven.[34] It was irrelevant to the central question: "Whether the world as it exists today was created by God in periods measured by millions of years or by days . . . has absolutely nothing to do with the salvation which Christ has brought to us."[35]

In Schmid's eyes it was not inherently threatening to Christianity to ask questions concerning the "intervening agencies" that might be involved in natural phenomena. Darwin's question about how the first individuals of each species came into existence was not in opposition to the believer's conviction that God was responsible for the process. A closer acquaintance with the details did not disturb the faith of a believer; rather, it added to his knowledge about the method of God's operation.[36] According to Schmid there was nothing wrong with asking "How did God do it?" even where human origins were concerned.

> One can allow natural scientists quietly to investigate . . . and see if they can find out something about the ways God has called the human race into existence. One will not see one's Christianity, one's belief in creation and one's belief in the Bible disturbed if the natural scientists find that the dust of the earth, of which the body of humans consists, was an inorganic mass at the creation of humans, or that it was an animal organism, or if they find nothing at all.[37]

Of course Schmid was aware that there were those, including natural scientists, who claimed that the results of scientific investigations *were* harmful to a religious view of the world. His own view was, as will become evident below, that such an assertion exceeded the limits of scientific inference. One drawback to the theologian's entry into the debate was the lack of consensus among scientists about Darwin's theory. Schmid felt that it made his task more awkward, since he had to deal with a theory that rested on such pure hypotheses at its foundations. "It would be more appropriate and fruitful," he wrote in his earliest treatment of the subject, "to delay the discussion until the solution of the question, as far as it is attainable, has been more clarified in the realm of natural history . . . than is possible at present."[38] Nevertheless, the "cry of victory" over Christianity and religion some were sounding meant that the theologian could no longer sit idly by.

THE DARWINIAN THEORIES

In Schmid's book, as in that of Charles Hodge written two years earlier in America, it was deemed important to specify how Darwin's ideas compared to various existing theories, some of which were assumed to have been around for some time.[39] As one reviewer of the book commented, Schmid demarcated three separate claims being made in the literature in place of the usual two theories of Lamarck and Darwin, a novelty the reviewer did not find particularly helpful.[40] Schmid delineated a theory of descendance, a theory of development, and a theory of selection. His concern was not only to explain what each of the theories entailed but also to indicate how the appearance of Darwin's work had affected them. In sum, Schmid wanted to evaluate the Darwinian versions of the theories.

By a theory of descendance *(Descendenz)* Schmid did not mean what one might have expected.[41] The theory as he defined it did not concern itself explicitly with transmutation. It did assume that new species had arisen since the original appearance of life, and that some explanation for the origin of these new species was demanded. But what Schmid called the theory of descendance merely asserted that when new species appeared, they did so by emerging from previously existing life as opposed to arising directly from inorganic matter. In other words, Schmid here was deliberately opposing any theory of spontaneous generation from inorganic matter, although he did not speak of an *Urzeugung* by name. *How* life descended from earlier life constituted the theory of development.

When he described the theory of descendance Schmid noted that it was clearly the governing principle of scientific investigation into the origin of new species, but Schmid felt he could also welcome it in the name of religion.[42] For one thing it did not in principle exclude other possible forms of an original generation of organisms; for another it was admittedly a hypothesis whose truth was at best merely probable.[43] His own view was that it was "extremely plausible," given evidence from geology, plant and animal geography, and comparative anatomy. Linking new species to species previously living qualified as a "real" preparation of the new species for Schmid in a manner that generation directly from inorganic matter did not. Schmid was particularly impressed that it would be hard to account for phenomena such as homologous and rudimentary organs without assuming descendance from preexisting life.[44] He was totally unimpressed with claims that the Creator, or whatever one wanted to argue was responsible for generating new species, produced them directly from inorganic materials, particularly where the higher species were concerned. That would simply require too great a jump for Schmid, and it would fly in the face of the extensive taxonomic information that had been uncovered by Darwin and others.[45] Clearly the notion that existing life was best explained in terms of preexisting life was not a novel notion after mid-century. Darwin's special merit, Schmid asserted, came from having the courage to trace the idea of the descent of species in a scientific manner.[46]

Schmid turned next to what he termed the theory of development *(Entwicklung)*. This theory differed from that of descendance in that, unlike the latter, it purported to explain how new species developed from existing ones. Most of the time Schmid equated the theory of development in general with what he described as the Darwinian theory of development in particular. Darwinian development asserted that new species arose from previously existing species through a series of imperceptibly small transitional states. Perhaps because of the growing controversy within the scientific community over the size of the variations required for transmutation to occur, Schmid occasionally referred to an alternative theory of development in which the new species resulted from a fundamental restructuring of the "germ" of an older species, producing a more dramatic leap between the old and the new than was permitted in the Darwinian version. He left this alternative largely undefined, however, focusing his efforts on the gradual development described in Darwin's great book.

The theory of development, he wrote, was more uncertain than descendance; in fact, Schmid refused to grant that development of any sort had been established. Ernst Haeckel might declare that his studies in embryology proved the reality of an origin of species through development, but in so doing, according to Schmid, Haeckel was overstepping his limits. He conceded that Haeckel's work established the *possibility* of real development, but more than that he disputed.[47] The case was similar to the testimony of geology, for even though the work of geologists had revealed many transitional forms, one could not assume that it would do so everywhere. In the lower classes of animals, where the forms were most numerous, the record also contained examples of forms whose appearance on the scene was sudden.[48]

Turning to the highest form of life, the human race, Schmid acknowledged plenty of evidence for development over time. The strongest support for human development came, according to Schmid, from comparative philology, which established developmental patterns of language. Less powerful, but not inconsistent with a theory of origin via development, was the evidence drawn from comparative ethology and archaeology.

While none of the recent finds of human remains—Neanderthal, Cro-Magnon, and Java man—had produced a consensus regarding the bridge from animals to humans, Schmid felt that it was probable that humans and apes possessed a common ancestor. But the difference between humans and animals was nevertheless absolute owing to the self-consciousness of humans. Exactly when the human race came into existence remained an open question. Schmid insisted, however, that one could meaningfully speak of the first humans who had made their appearance at a definite point in the past.[49]

The evidence pointed to the development of an already originated human species, not an *origin* of humankind through development. All three sciences, philology, ethology, and archaeology, led back to starting points where humans already existed with the essential and distinguishing attributes in place. A key question here for Schmid was whether the agency responsible for inciting the development came from outside or inside the organism undergoing the development. Schmid felt that the evidence favored a thoroughly internal development. Still, no one yet had successfully identified the inciting agency.

When all was taken into consideration the theory of development was seen to be a hypothesis, like the theory of descendance.

It was more problematical than descendance, wrote Schmid, and it could go beyond its hypothetical status only if we succeeded in uncovering the formative forces present in the developmental process that were responsible for producing the new species.[50]

In spite of the more problematical status of the theory of development, Schmid declared that it, like descendance, was not incompatible with theism.[51] Nor did evolutionary development through an innumerable series of generations mean that new species had no proper beginning. Humans, for example, would have originated in those individuals in which self-consciousness, the identifying characteristic of humans, appeared for the first time.[52]

The theory most closely associated with Darwin's name, however, was the theory of selection. Schmid carefully explained the theory, noting that variations in the characteristics of an organism were ever present, that individuals possessing characteristics more favorable to their preservation would more likely survive the struggle for existence than would others, and that the decisive characteristics were transmitted to the next generation, which would then contain individuals that enhanced the desired characteristic to an even greater degree or would add to the inherited characteristics new traits favorable in another direction.[53]

Schmid did not hesitate to point out the specific strengths selection theory offered; for example, "in the realm of observed facts" the mimicry that was encountered in nature was easily explained by selection theory. Butterflies and insects that looked like the leaves of plants possessed an obvious advantage over those that did not. Natural selection was particularly apt for explaining why they obtained just such advantages.[54]

In choosing mimicry as evidence of the power of natural selection Schmid was indeed providing a strong argument, for neither the special creationist nor the Lamarckian could handle the case as persuasively. To argue that God gave the butterfly this advantage seemed to require a further explanation: why was it given only to some, and why did God find it necessary to rely on deceiving predators? As for Lamarck, he had not suggested that it was within an animal's capacity to regulate its color or other characteristics used in mimicry, hence he also offered no effective explanation.

Natural selection might have been superior in this instance, but that was no reason in Schmid's mind to rely on it exclusively. Like Zöckler, Schmid exonerated Darwin himself from the most radical conclusions. He noted that Darwin innocently employed both natural selection and Lamarckian use and disuse when they were in fact contradictory. Because of the contradiction

> there arose the group of neo-Darwinians [*Neu-Darwinianer*], who, more Darwinian than Darwin himself, declared natural selection to be the exclusive principle of the development of the species, and the group of neo-Lamarckians [*Neu-Lamarckianer*], who, to be sure, did not oppose a contribution from natural selection, but who found the causes for the higher development of organisms more in Lamarck's principles than in Darwin's.[55]

The head of the neo-Darwinian school, according to Schmid, was August Weismann, who championed the claim that acquired characteristics were not inheritable. Others too deserved to be described as "more Darwinian than their master," including Ernst Haeckel, who employed "pure selection theory," and Alfred Russell Wallace. The latter, Schmid noted, qualified his explanation of the development of physical characteristics of living organisms by exempting mental characteristics from the process.[56]

Schmid did not concede the noninheritability of acquired characteristics. He cited the work of the Viennese botanist Richard von Wettstein, who accounted for the gradual adaptation of water plants to land by the inheritance of acquired characteristics, as evidence that the explanation of some new forms required more than natural selection.[57]

Nor was selection theory without its difficulties. Unlike theories of descent and development, selection theory could not, according to Schmid, explain why species had remained permanent for thousands of years. In descent and development one could say that the generation of new species had slowed to a standstill, but the agencies specified in selection theory were still very much active. Schmid rejected the reply that thousands of years was not enough time to permit observable effects of selection, since, he wrote, the conditions of existence had undergone tremendous changes within human history alone. Selection should have had an effect. Furthermore, the natural tendency among organisms was to reduce, not increase individual differences; even artificially induced changes were known to revert when left in the wild.[58]

In his initial description of the theory he did not emphasize the random nature of the variations produced, but eventually he did get around to what he sensed to be the vulnerable aspect of Darwin's theory, the origin of variations. Darwin did not explain the origin of useful characteristics; rather, he simply assumed that they would somehow appear in new generations. But how could he know this, since their appearance was a matter of chance? Not only that, but Darwin assumed a second series of chances in the next generation that somehow coincided with the first. Over numerous generations, argued Schmid, the probability of obtaining ever more adaptive variations was infinitely small.[59]

Not only in this instance but also later in the book Schmid misunderstood the role of chance in Darwin's theory. By overlooking the capacity to accumulate and retain favorable variations Schmid miscalculated the odds which natural selection involved. He saw the chances of accumulating favorable variations becoming geometrically less probable with each generation instead of being enhanced.[60] In making this error he joined the ranks of innumerable others who got off the track, much to the annoyance of more careful Darwinists.[61]

Of course Schmid also judged selection with respect to its relationship to religion. The number of voices from the religious community declaring that no harm would come to a theistic view or to a Christian conception of God and creation from the idea of an origin of species through descent *(Abstammung)* was, Schmid acknowledged, on the increase. But this concession was directed at the theory of descendance, not at "real Darwinism," by which Schmid meant selection.[62] Schmid confessed he was disinclined to selection himself, but he indicated that he would examine it to assess whether it was in any way compatible with theism.

In the theory of selection the transmutation that occurred was not produced from within the organism as in the theory of development. In selection the change in organism was directed from outside. For Schmid selection raised an obvious question: since either the development happened by chance or it did not, which one was it? In his treatment of the issue it became clear that Schmid simply could not seriously entertain the prospect that selection theory was to include chance at its very heart. "No one," he wrote, "who makes a claim to be contributing a serious word worthy of attention will say by chance; rather [they will say] by necessity."[63]

Schmid's tone was one of reluctant concession. Having ruled out the possibility that anyone would seriously assert that development happened by random chance, he proceeded to admit grudgingly that it was at least logically possible to conceive of a process of selection operating totally outside the individual organisms in the species. For selection theory so envisioned, the agency would no longer be singular—it would be the amalgam of all forces and conditions acting on the organism from outside.

But what could possibly be gained from such a perspective? Schmid found it incomprehensible that the first of the two explanatory causes in Darwin's theory, the inclination to produce variations, could be "absolutely indifferent" to the systematic idea of the organism itself and to any progressive element in development. In characteristic nineteenth century German fashion, he could not conceive how natural selection, acting by virtue of the amalgam of forces and conditions affecting the organic world, could escape being captured by rational law. Open-ended development, in which the future could not be anticipated even theoretically, amounted to irrationality for him, and he spoke of this defect in Darwin's theory as a problem. Even selection, the second explanatory cause in Darwin's scheme, was applied "purely from outside, like individual variability."[64] Darwin might insist on extremely gradual variations as a means of minimizing the obvious logical problem involved, but no matter how gradual the variations, that was not a sufficient account of why even the first advantageous variation was selected.

Thus, if the amalgam of forces and conditions that constituted the external agency of change in selection theory was subject to rational law, then the overall system could be associated with a highest intelligence. Such an arrangement had the unfortunate effect of removing God a bit from the process of development itself, but at least selection theory so understood was compatible with theism.[65] "When the materialist uses the word chance," Schmid observed, "he merely hides or avoids the necessity of having to put [the words] plan and disposition in its place."[66] In this regard he saw the matter as did his friend, the duke of Argyll, with whom he must have discussed Darwinism many times. In the latter's book, *The Reign of Law*, the duke also attacked Darwin's use of the word chance as a cover-up of his ignorance of the causes of variation. Because natural selection could not even suggest the law under which new forms were introduced, the duke concluded that it could originate nothing in spite of Darwin's misleading reference to natural selection's producing of modifications. It could "only pick out and choose among the things which are originated by some other law."[67]

Schmid had set himself the task of examining dispassionately each of the "Darwinian theories" both for scientific plausibility and for their suitability to a theistic view. With regard to the second issue his conclusion was clear: if these theories were understood as rational scientific analyses, then theologians had no right to restrict scientists in their investigations regardless of which of the theories was involved.[68] Schmid's injunction was general because the one case he was willing to rule out, in which change was brought about through truly random variations, did not meet the criteria of rational scientific analysis.

Mechanism, Monism, and Materialism

Of course Schmid was neither the first nor the last writer in the nineteenth century to understand the theory of evolution by natural selection as he did. In his mind the problem introduced by chance variations, which even the materialist Ludwig Büchner had called the "great weakness and inconsistency in Darwin,"[69] was the likely reason why Darwin had confessed in *The Descent of Man* that he had probably attributed too much to natural selection. Schmid praised the love of truth evident in Darwin's admission, which, he noted, stood in marked contrast to the absolute refusal to admit any doubts about selection on the part of Haeckel and his followers. Darwin's concession, according to Schmid, also allowed for the possibility of change due to the action of as yet unknown agencies inducing deviations of structure.[70]

Schmid differentiated Darwin's claims from those of Ernst

Haeckel on yet another front. Whereas Darwin made no attempt to explain the origin of life, a "mechanical explanation of life" was put forth in its most systematic and logical form by the naturalist from Jena. Schmid was clearly more sympathetic to Darwin's reference to the Creator's breathing of the first life into one or a few original forms than he was to Haeckel's assertion that all matter, form, and motion associated with living organisms merely resulted from the increased complexity of matter and motion. Haeckel's view, said Schmid, was nothing but a way of hiding our ignorance.[71] Even if we could produce life artificially, a prospect Schmid found highly improbable, it would not have explained the origin of life.[72]

The general problem, as Schmid saw it, lay in the misuse of mechanistic explanation. Mechanical interaction of matter involved the assumption that there was a uniformity of law in the occurrence of events. Schmid had no objection to this. He did, however, insist that the causal principle associated with this uniformity was different in the physical world from that of the psychical world. There was causality in both realms; but the causal relationship that obtained between two psychical phenomena or between a psychical and a physical phenomenon was not the same as that which existed among physical phenomena alone. Among physical phenomena alone the relationship was mechanistic, and the uniformity it reflected could be expressed mathematically. The causal relationship operating between psychical and physical phenomena was not, all our experience told us, mechanistic in this same sense. There could be no denying the mutual causal influence, but, Schmid implied, one could not relate physical and psychical phenomena in the same manner that physical phenomena are related via mathematical equations. The realm where causality reigned in the form of mechanism, he concluded, could act as a support, foundation, and instrument for the realm where causality reigned but mechanism ceased.[73]

It was the monists, of course, who obliterated this distinction. Schmid named Haeckel, Strauss, Büchner, and others as the leaders of those who urged that all states and processes in the world, spiritual and ethical as well as physical, could be explained by means of the pure mechanistic interaction of atoms.[74] He could grant that this view possessed some logic, Schmid wrote, if it claimed merely that our observation of order led to a view in which the world, because of the presence of scientific law, came about necessarily. Such a version of monism would not require, but would not rule out, a personal author of the world. Schmid could readily conceive of this possibility because he believed the acknowledgment or denial of God resulted not from scientific investigation or logic but from a moral act of the individual. "The assumption of a Christian or an anti-Christian worldview is by no means the result of our scientific investigations, but an act of our personal choice."[75]

To no one's surprise Schmid saw a problem in monism where morality was concerned. Although he had defended Darwin from the extreme positions that were taken in his name, Schmid still had to acknowledge that what one called Darwinism "in the real sense" was pure selection theory. And when one conceded the most radical consequences of selection theory, it was hard not to object on moral grounds. Not only had theologians noted that for humans love was replaced by egoism, but moral action itself was made impossible if mechanistic explanation was extended into a general determinism.[76]

The advocates of monism did not shy away from the elimination of purpose from nature. Schmid quoted well-known natural scientists to say what Darwin's accomplishment signified. The Englishman's theory showed, according to Hermann von Helmholtz, that the formation of organisms could occur through the blind administration of natural law, without any intermingling of an intelligence. Ernst Haeckel, whose *Riddle of the Universe* stood as the high-water mark of the polemic against theism, concluded even more boldly that if one were logical in one's analysis, then Darwin's theory had to lead to a mechanical conception of the universe. After the turn of the century Hugo de Vries identified the use of natural causes, without recourse to any teleological element, as both the supreme value of Darwin's theory and the reason for the universal acceptance of the idea of descent.[77] Schmid deemed it important to note that in making claims like these Haeckel and others had left natural science and passed over into philosophy. Schmid did not object to such a step; indeed, it was necessitated by the scientist's need for a coherent view.[78] But by the same token philosophical assertions should not be taken as scientific conclusions simply because they were made by a scientist.

Even theologians in the mediating tradition had their limits, and in the bold claims of Haeckel and Helmholtz, Schmid found one of his. What he could not tolerate was Haeckel's determination of the necessity of a mechanical view of the universe in which God played no part. He was even willing to set aside what he deemed to be the fundamental incompatibility between the claims of radical materialism and the existence of human moral conscience for the sake of argument. If one did that, then what monists like Haeckel had asserted, that a teleological view of nature was not necessary, was true. Denying the *necessity* of a purposeful world order, however, was a far cry from establishing the conclusion that the world was not in fact purposeful or that God did not in fact exist.[79]

Monists like Haeckel and Strauss held that the general reign of causality invalidated teleology. But Schmid denied that the two were mutually contradictory and exclusive. Again for the sake of argument, he set aside his own assumption that there was but one manner in which human knowledge corresponded to the real world. How would we know, he asked, even if causality and teleology *were* contradictory, that causality held preference over teleology. Both approaches involved what he identified as anthropomorphic forms of reasoning, yet both were necessary and so valuable that we should not want to have to sacrifice either. But if one insisted on asking which was to be given preference on the assumption that they were contradictory forms of reason, only one of which corresponded to the real world, then who was to say that final cause was inferior to efficient cause?[80]

Schmid acknowledged that one expected a theologian like himself to defend a teleological worldview; hence he cited philosophers and natural scientists who also argued that teleology could not simply be banished as the monists alleged. Among the numerous philosophical thinkers in the idealist tradition whom Schmid listed, he was clearly pleased to be able to include Friedrich Vischer, the friend of David Friedrich Strauss. But perhaps it was the natural scientists, even Darwinians, from whom he hoped to draw his most convincing support.

Oswald Heer, Albert von Kölliker, and Alexander Braun were some of the scientists he cited. His strongest cases were Karl Ernst von Baer in Germany and, because of his thoughts about human development, Alfred Russell Wallace in England.[81] Even Darwin himself did not come out *against* design, wrote Schmid, except for one passage in *The Expression of Emotions in Man and Animals*. Citing the German translation of 1872, Schmid quoted Darwin's observation that those who believed in design would find it difficult to account for blushing. But, he continued, nowhere in the *Origin of Species* or the *Descent of Man* did Darwin reject the teleological view.[82]

Schmid, of course, maintained that mechanism and purpose were not mutually exclusive categories.

> The knowledge according to which something exists does not exclude the question why it exists, nor does it replace that question. If I am to answer both questions satisfactorily, then I may and must first ask . . . if that which I call cause and effect in the language of causality also falls under the category of finality, where that very cause is simultaneously a means and the effect simultaneously a purpose.[83]

Clearly Schmid's conviction was that cause and effect *could* be simultaneously means and end. Once again his emphasis was consistent with that of his friend, the duke of Argyll. "All things in nature may either be regarded as means or as ends," wrote the duke, "for they are always both."[84] In the English edition of Schmid's book he provided an introduction, in which he noted that everyone was aware that Darwin's theory, which had given rise to strange and aberrant developments, was received more widely and with more unbridled enthusiasm in Germany than in England.[85]

Schmid's treatment of mechanism versus teleology depended on the German tradition of teleomechanics, though it was not worked out as thoroughly here as it was in the works of other German scientists and philosophers of the time.[86] Schmid's observation that both mechanism and teleology were anthropomorphic was vaguely reminiscent of Kant; certainly his willingness to leave open the question of which should be given preference, teleology or mechanics, intersected with the neo-Kantian theological trend of his own day, represented in the Ritschl school.

What must be conceded, however, is that Schmid never entertained the possibility of abandoning a correspondence theory of truth. It never seriously occurred to him that both mechanical and teleological accounts might be deficient in the sense that neither was necessary. Schmid assumed, along with the majority of the intellectuals of his time, that because there was only one truth, scientific and religious accounts of the world aimed to depict existence "as it really was." He thus viewed mechanical and teleological explanations as inevitable. Not only might every phenomenon in the world be viewed from these perspectives, but "it is in fact regarded from either one or the other standpoint."[87] Yet each was complementary to the other without being a contradictory means of representing one and the same reality. In words reminiscent of Zöckler, Schmid denied that one could simultaneously be a Darwinist and a Christian. "Something cannot be true from a natural scientific viewpoint and false from a religious one, nor, conversely, [can it be] true from a religious viewpoint and false from a natural scientific one. Further, I cannot, to appease myself, be a Christian with my heart and an atheist with my intellect [*Verstand*]."[88]

The two other options examined in Chapters 3 and 4 above, though each was also consistent with a correspondence theory of truth and its accompanying realistic interpretation of natural science, left much to be desired in Schmid's view. Otto Zöckler, who assumed that in a religious vision numerous widely accepted scientific conclusions must be regarded as erroneous results that did not correspond to reality, and David Friedrich Strauss, who conceded that the results were in fact true, and that the scientific view of the world based on them would eventually replace at least the Christian worldview, represented extreme positions to Schmid. As a mediator, however, he was able to recognize kernels of truth in both of these perspectives because their authors, like Schmid himself, spoke the language of correspondence theory.

There was another possibility as well, but it involved abandoning both realism and truth as correspondence, and it was as unacceptable to Schmid as it was to Strauss and Zöckler. One could conclude that religion could and should simply cut itself free from natural science because truth, whether encountered in religion or in natural science, did not involve knowledge that corresponded to the world "as it really was." The truth of religion was qualitatively different from the practical knowledge of natural science. This stance was represented, as we shall see in the next two chapters, by the Ritschl school.

Those who like Schmid acknowledged the need to investigate the competing claims of scientific and religious knowledge were, in so doing, rejecting the call to view science and religion as wholly distinct and mutually antagonistic spheres of thought that must be kept separate. A sympathetic reviewer of Schmid's second book praised him for not giving in to "the favorite means at the moment of minimizing the cause of irritation between natural science and theology by confining the Christian revelation of God to the person and work of Christ." The reviewer noted that Schmid knew well that even such a restricted area was not safe from attack, and that the impoverishment of revelation by excluding nature from its realm was unacceptable.[89] By means of such a view, noted the duke of Argyll, scientists who were afraid of being regarded as hostile to religion, and theologians afraid of being thought of as fearful of science could live in peace, each going his own way as if there were no problem. The duke was clear about what he thought of the doctrine of the strict separation of science and religion. It was intellectual cowardice: "There can be no such treaty dividing the domain of Truth. Every one Truth is connected with every other Truth in this great Universe of God . . . To these, and to all who are troubled to reconcile what they have been taught to believe with what they have come to know, this doctrine affords a natural and convenient escape."[90]

As Schmid took stock of the general relationship between natural science and religion in light of Darwin's work, he judged that much had been learned as a result of the issues that were crystallized in the debates over evolution. For the most part Schmid acknowledged openly that developments in natural science in the nineteenth century had forced concessions from the traditional religious community. In Germany, however, there was a tendency toward polarization, in contrast to the growing trend toward reconciliation found outside Germany. Without naming him Schmid echoed Zöckler's observation of the lack of a tradition of natural theology in Germany. Unlike England, there were fewer in Germany who wished to unite a religious interest with natural or historical questions.[91] If Schmid, like Zöckler, appreciated the need for work on the history of science and religion, his respect for the exegesis of Scripture also linked him to those claiming to hold a biblical perspective.

Nevertheless, he could not accept the pragmatic judgment of belief in individual immortality, the efficacy of prayer, miracle, and a personal God that Gustav Jäger had put forth in a book of 1869. Jäger had argued that those who had retained beliefs were at an advantage in the struggle for existence over those without them. Schmid could admit that Christianity imparted strength to the believer, but limiting its truth to its usefulness was not acceptable.[92]

With those in Germany who had taken up the theme of the reform of religion in light of Darwin's theories, Schmid found no one with whom he agreed completely. He appreciated some of the aspects of Strauss's theology, for example, the aversion to miracles, but he clearly went along with the criticism of Strauss by Heinrich Lang, the spiritual leader of Switzerland's reform theology and the friend of Strauss who rejected the latter's bold elimination of design from the universe. But Schmid could not bring himself to accept Lang's acknowledgment that special providence was no longer possible, that prayer was merely of subjective and psychological significance, and that the moral order showed itself only to the extent that guilt and punishment stood in a natural causal connection with one another.[93]

Schmid wanted to believe that any concessions that had to be made because of natural science not only would not harm the essential message of religion but also would enhance it. The religious view of the world recognized the work of an almighty creator and ruler in the universe. It was fitting that such a view should contain knowledge of the Creator's work; indeed, every correction and enrichment of our scientific knowledge was only a correction and enrichment of our knowledge of the constitution of divine creation.[94] Schmid defined what he called the religiously inclined individual to be one "who sees in everything nature offers him . . . an educational school of divine wisdom and love."[95]

The Responsibilities of a Teacher

The practical implications of Schmid's position became evident quickly, when he had to instruct young students. Schmid did not hesitate before the task. Over the years Schmid had observed what was successful in teaching and what was not. The head of his Latin school, for example, had lost the respect of the students through his overly friendly interaction with them. His history teacher there had been too pedantic, while the Greek classics instructor had no idea how to make his subject come alive. The favorite in those days had been Karl Köstlin, later professor of aesthetics at Tübingen. It was the thoroughness of his knowledge and his teaching that made Schmid single him out. Köstlin knew how to motivate his students.

Equally important were Schmid's views on classroom discipline; specifically, he rejected corporal punishment as an effective incentive to learn. At the Marbach school one of the most caring and committed of his teachers had learned over the course of his career to abandon the use of the stick as a punishment for mistakes. It was he who was able to boast of more successes on the state examinations than anyone in the land. From this instructor, and from his own father, Schmid inherited his decided dislike of the so-called cudgel system of learning.[96]

As for the goal of the seminary instructor, Schmid saw the challenge to be in building a bridge between the learning of childhood and that of the university, and in doing so while leaving the student's piety, child-like faith, and sense of truth intact. He held that education should underpin, not undermine religious faith. He further believed that a teacher should provide a model for the students. His own willingness to take time for his charges generated fond and lasting memories of him. At grave side one of his former pupils eulogized him with words that few teachers could earn: "He made our youth and school days golden."[97]

In the Old Testament instruction of the *Obergymnasien,* given to students in their seventh and eighth years, there were naturally a number of challenges to be faced, not the least of which was the Genesis account of creation. It would be a mistake, wrote Schmid in his booklet on the contemporary problems of teaching religion, to lead the student directly into biblical criticism itself. Young people did not have the wherewithal to wrestle with the cases where the Old Testament criticism clashed with historic confessional writings. Even if the teacher used the traditional understanding of Genesis as a leading thread, which was then adjusted in accordance with the criticism, the student would not know how to handle those points on which it was necessary to cut the thread.[98] So it fell to the instructor to guide the student through the mine field of criticism. But which route was the teacher to take?

Schmid's answer was dictated by his conviction that the central issue was truth and that truth was one, that it was attainable, and that scientific and religious expositions of truth were complementary. First of all, a teacher of religion must possess a religious account of truth, and for Schmid that meant that he must revere Scripture. But where it was a question of facts, wrote Schmid, we cannot avoid facing them head on. Nothing hindered the teacher from making use of the results of criticism in instruction, nor, as in all cases, from following the conviction of truth conscientiously won. However difficult it might appear, Schmid decided, the instructor had no alternative. In retrospect he did not regret his choice.

> I know I have served my dear young friends more in this way, I have better provided for their future and, to the extent that it is possible for us humans, more safely protected them from someday experiencing a shaking of their foundations than if I had handled things differently. The results of their own thinking might work out to this view or that—that is beside the point, provided [they] are really the results of conscientious research and that they always retain the conviction that both religious and scientific drives for truth have one source.[99]

Since it was always the truth alone that made us free and gave us light, such a use of the treasure of knowledge held no doubtful results, only blessed ones. If, for example, the theory of descent, which was at present merely probable, were raised to the level of certainty through further research, "we would say of this knowledge exactly what we have to say of every progress in our knowledge of nature: this knowledge would give us a new and deeper look into the nature and manner of divine creative work, and such an insight would not be a disturbance, but an enrichment of our religiosity."[100]

It would seem from these words that Schmid felt the ascertainment of scientific truth to be an unproblematical process. In fact, Schmid did assume that scientific knowledge could be final and certain. In the piece on Robert Mayer, whom Schmid honored because of his connection to Mayer's beloved Heilbronn, Schmid treated the subject explicitly. The purpose of the essay on Mayer was to demonstrate that the founder of one of the certainties of natural science opposed not only those who used his discovery to support materialism but also those who unambiguously defended his results as wholly consistent with a religious worldview.[101]

Schmid cited approvingly Mayer's own declaration in his work on the mechanics of heat that the essence of natural law and the touchstone for the correctness of theories was the quality of general validity arrived at through inductive inference.[102] Development by natural selection, he observed, might be losing its scientific significance, but Mayer's discovery of the mechanical equivalence of heat was another matter. Mayer's discovery was "not a hypothesis." Rather, "it forces itself upon the mind as truth with all the concentrated power of proof from experiment and observation, from logical and mathematical reasoning, and from philosophical intuition. In spite of the relatively short time it has existed, it stands forever as firm as any of the most certain achievements of human knowledge."[103] Mayer's achievement was clearly out of the ordinary, however. Most of the time we wallow in the realm of the merely probable. Nevertheless, from his discussion of hypothesis in science it was clear that certainty was not easily achieved.

Schmid defended science against those who rejected its theories on the grounds that they were only hypotheses, and as such were not certain. Of course those scientists should be reproached who treated unproven hypotheses as if they were already proven or who presented results as laws or facts before it was really known whether they were laws of immutable validity or facts at all. But a general rejection of hypotheses was hardly justifiable; indeed, hypotheses were "frankly indispensable" to research in every branch of science and especially in natural science where unsolved problems abounded more than in most sciences.[104] Look at the crucial role of the hypotheses of the ether, the atom, and the molecule. None of these could be perceived empirically, but some of the noblest triumphs of science would have been impossible without them.[105]

Schmid saw no need for concern because of the use of unproven hypotheses in science. History had shown how one hypothesis was replaced by another when the second withstood critiques better than the first. Copernicus had superseded Ptolemy, the organic nature of fossils had replaced the notion that fossils were freaks of nature. Ultimately the use of hypotheses represented the confidence we have in the correspondence between nature and our minds. Hypotheses for Schmid were attempts to get things right. "Humans must abide by the belief that the incomprehensible is comprehensible, or else they would cease to investigate it," Schmid quoted from Goethe.[106] A citation from Robert Mayer made his reliance on the correspondence between thought and things even more explicit: "What is subjectively thought correctly is also objectively true. Without the eternal harmony pre-established by God between the subjective and objective world all our thinking would be fruitless."[107] Although Schmid wrote in such a manner that the attainment of certainty was reserved as a real possibility, he was not at all quick to assert that a particular result of natural science had in fact achieved certainty. The Darwinian theories of descent, development, and selection, for example, were still "mere hypotheses" in his last book, though by then at least descent was acknowledged to be a firmly established hypothesis. Development, while apparently the view of most, was still to him an open question.[108] The distinction between fact and theory, while unproblematical in Schmid's mind, rarely produced an uncontested consensus; hence Schmid's bold confidence resulted more in an ostensible spirit of open-mindedness than in the actual surrender of a specific doctrine in the face of scientific conclusions.

Nevertheless Schmid's stance was courageous. The only way a teacher of religion could proceed, he proposed, was first to form convictions for himself about the most probable course of history based on conscientious biblical study in light of modern research, and then simply to draw the most consistent conclusions possible. To the objection that the results of criticism were not yet certain Schmid replied that the correctness of the traditional religious view was even less certain.[109]

On some issues Schmid's mediating position did not sit well with more orthodox minds. Schmid's concern was to acknowledge that the foundation of Christianity had to be based on historical fact and therefore to subject Christian doctrine to scientific scrutiny.

Only in this way, he thought, could the indispensable and transitory trappings of the faith be jettisoned without violating the integrity of religion or science.[110]

Regarding the virgin birth of Christ, for example, Schmid concluded that historical scholarship had produced no coherent biblical tradition concerning the circumstances of Jesus' birth. It was, of course, true that one could not disprove a virgin birth on the basis of historical science any more than one could prove it. Because those who doubted it and those who believed it could both be considered intellectually respectable, however, Schmid concluded that the uniqueness of Christ's divine sonship did not require a virgin birth.[111]

The resurrection of Jesus represented another difficult issue. Schmid recognized that here too there were those who assented to it and those who did not. He was personally inclined to accept the resurrection because of the eye witness accounts given in the New Testament. Schmid also argued that the resurrection did not necessarily violate laws of nature. What happened to the unique individual of Jesus might one day happen to the whole of creation. Nevertheless Schmid did review the various naturalistic explanations of the New Testament event, pointing out their strengths and weaknesses. In addition he observed that many earnest Christians opposed Christ's bodily resurrection. His own choice was to view it as the foundation of hope.[112]

In the more focused context of the relationship between Genesis and geology there was a specific challenge that highlighted the dilemma of the religious instructor. How was the teacher to find time for such individual searching when he was not a professional theologian? Here Schmid offered no solutions. Since there was no good text on the subject, Schmid could only recommend that teachers do what he had done, namely, consult the professional Old Testament scholars they trusted for advice on the best literature available. He directed the reader to the work of Julius Wellhausen and others as writers whom he had found helpful, defending Wellhausen in particular against charges of disrespect to Scripture and of forcing the divine factor out of the history of Israel.[113]

In an article in the *Jahrbücher für protestantische Theologie,* published in the same year as the study of religious instruction, Schmid presented his own interpretation of the "days" of creation mentioned in Genesis. Noting the formula ending each of the first six days, "The evening and the morning were the first day," Schmid asked why the intervening night was never alluded to. The answer, he decided, was that there was no night yet. These were God's days, different from those of humans. Further, the seventh day had no end since the formula was not used with it.[114] Schmid denied he was merely trying to smuggle in the old view that the six creation days were equal to six periods of the world or of the earth. Such an identification was not at all part of the mentality of the Hebrews. Schmid added that his interpretation depended on a willingness to follow Wellhausen in locating the priestly codex (Genesis, Exodus, Leviticus, Numbers) at a more recent date than was traditionally done. One could not interpret a document from the earliest times, laden with myth as it would have been, to have intended such precision.[115]

As for other concessions, Schmid went beyond agreeing that the earth was far older than was usually allowed and beyond supporting the likely physical evolution of life. Cherished doctrines of Christianity such as the providence of God, miracles, and the efficacy of prayer, though defended as essential components of the Christian worldview, were reinterpreted to conform with a nature constrained by mechanism. The infinitely complex causal chains of natural mechanical interaction were not inconsistent with these doctrines once it was realized that God relied on the inviolate conformity to law to accomplish divine ends. Miracle, for example, did not necessarily run counter to the laws of nature; rather, miracle was the emergence of something new, which, while drawing attention to God's sovereignty, stood in harmony with the whole.[116]

To explain how providence could be reconciled with causal law, Schmid appealed to an analogy with human self-consciousness. Humans constantly interfered with nature without violating causal control, hence God might do likewise. But Schmid was careful to note the difference between the two cases. It may be that in voluntary motion of the body we exhibit an interaction between a freely acting will and a causal system, but our own ability to carry out this action did not eliminate the mystery present there.[117] If humans could make nature serve their ends without comprehending the manner of the interaction, surely the Creator could be guiding nature without our understanding it as well. Clearly God did not violate our freedom by compelling us to recognize himself in nature; on the contrary, as the presence of evil in the world showed, God's interaction with nature represented a far greater mystery than did human interaction. But the presence of mystery did not eliminate God's capacity to interact with mechanical nature according to some purpose any more than it did in the case of humans. Not only providence, but the efficacy of prayer could be justified in this manner.[118]

What was absolutely necessary for a responsible treatment of the whole subject, repeated Schmid, was a recognition of design in nature. His analysis of Darwin's theory of natural selection revealed to him that intelligent design was not only reconcilable with causal mechanism but actually postulated by a scientific contemplation of nature. When something was explained as the result of natural law, prerequisite to the explanation was an assumption of the order represented by the law. To Schmid this order was an expression of God's presence. He specifically rejected an arrangement in which the order existed apart from God, an order which God appropriated to accomplish his ends. That might be the way human intelligence related to nature's causal reality, but it did not transfer into the divine realm. The laws themselves revealed a teleological acting creator.[119]

The rejection of design in nature was therefore equivalent to the destruction of both religion and natural science for Schmid. The two endeavors thrived together in harmony or they perished together. Failure to appreciate their inextricable mutual dependency led to much misunderstanding and error. Throughout history, for example, failure to find a causal explanation for some natural phenomenon led religiously minded people to attribute it directly to God, thereby cutting short the task of science. The temptation to resort directly to creation instead of seeking secondary causes was a fundamental flaw, in Schmid's view, since creation applied only to the origin of matter, force, and laws themselves.[120] Beyond these, religion was obliged to accommodate the results of scientific inquiry, especially where disproof was concerned. When something was shown to be erroneous through scientific investigation, then religion had to remove it as a constituent part of its worldview.[121]

Schmid, then, did not like the selection theory, even the "rational" selection theory he understood to be in principle compatible with theism. He preferred that the formative principle for new species be seen to reside within the species instead of being located outside it. He would in no way concede that selection theory by itself was sufficient to account for development. The attempt to explain a species' origin exclusively by means of the selection theory he regarded as a patent failure.[122] Scientists who based religious and philosophical inferences on any of the Darwinian theories were drawing premature conclusions. They had to show theologians the same respect he had demanded from his theological colleagues toward scientific investigators.[123] True to his stance within the mediating tradition, Schmid's position asked each side to meet the other half way.

Because Schmid had always felt that education should support rather than erode religious faith, and because he took seriously the expectation that the teacher should be a model for the student, he took time and great concern with the youths in his care. "The ages 14 to 16 seem more modest than 16 to 18," he once said. "But the imprint that the spirit quietly and unobtrusively makes then usually endures for life. To be permitted to reach out a hand here is a noble work that I happily do daily."[124] Such an attitude, which could not be hidden from the students, produced a lasting and reverent respect for him.

The personal qualities that made Schmid so effective with people and the successes he enjoyed because of them could hardly go unnoticed by Schmid's superiors. After merely six years at Schöntal Schmid was offered the first in what amounted to a series of appointments in the ecclesiastical hierarchy. By decision of the king he was named prelate and general superintendent of Heilbronn, the city he had left more than twenty years earlier unable to deliver a farewell sermon because of problems with his voice. In just over a year he was brought to Stuttgart, the capital of Württemberg, as court preacher and member of the Protestant Consistory. In addition he was given directorship of a commission for educational institutions, and he eventually assumed duties for the administration of several royal girls' schools. Two years before he retired he became field provost of the Württemberg Army Corps, in which capacity he was responsible for Württemberg's Protestant chaplains.

Of all these tasks and more, the office that brought him greatest personal esteem was his service as the king's pastor. He did not have long to function in this capacity, however, for in 1891 Schmid had to cut short a trip to Britain to officiate at King Karl's funeral. His unquestioned loyalty was as much appreciated by the new King Wilhelm as it had been by his predecessor; indeed, King Wilhelm and Queen Charlotte later paid great homage to Schmid by arranging a special train for themselves and others to make the three-hour journey to visit Schmid's family the day after his funeral.[125]

As Schmid neared seventy he found that preaching had become more and more a burden. After losing his wife in 1896, Schmid took a trip to Scotland for what was to be the final reunion with his old friend, the duke of Argyll. The following year he was venerated by the theological faculty of Tübingen University with the awarding of an honorary doctorate. In his letter of thanks Schmid revealed just how much the academic recognition meant to him: "A greater honor has never in my life been given me."[126]

His retirement in October of 1898 brought a welcome relief from most of the diverse responsibilities that his rise through the ecclesiastical ranks had brought to him. While he retained a few advisory roles, Schmid made good use of his greater freedom. With his daughter he made yet another trip to Scotland, this time to visit the oldest of his former pupils, now the new duke of Argyll. He also devoted his time and energy to social work with the disabled and the retarded, efforts in which he never before had time to participate. To the many distinctions that had come his way he was able to add in 1906 the Karl-Olga Silver Medallion for his altruistic endeavors.[127] Finally, of course, he was able to complete his *Natural Scientific Confession of Faith,* composed, he said, out of gratitude for the honorary doctorate he had received from Tübingen.

Rudolf Schmid died content with his life. As he neared his end, his son noted, "he looked back on his life with deep gratitude in increasing measure."[128] Although restrained by the natural infirmities of old age, he declared on his eightieth birthday that he never felt better in his life. Less than two months later he experienced a light stroke, from which he recovered. But in July of 1907, while escaping the summer heat in Obersontheim, he was afflicted with several light strokes and then, on the last day of the month, a severe stroke left him helpless and unable to finish the *Lebenserinnerungen* he had carried through 1858. On August 7, 1907, surrounded by his children, he breathed his last. His son recorded the final scene with an observation befitting of one whose life had been so meaningful: "He lay there, a picture of peace."[129]

Obituaries of Schmid celebrated the personal qualities that had made it possible for him to live triumphantly through a century of contention between natural science and religion. "His sunny, childish sense was a rare and treasured gift of the specially favored. His trust in people never left him; it enabled him to be an exceptional leader and teacher of youth."[130] Others made it clear that his success had been due not to his oratory, which was not his gift, but to his sincerity, warmth, and an unfailing ability to carry the undying power of youth with him into his old age.[131]

PART III

Nature Lost

The material of our ideas and the object of our faith are completely incommensurable.

Wilhelm Herrmann

6

Wilhelm Herrmann's Encounter with the Theology of Albrecht Ritschl

THE THREE THEOLOGIANS examined thus far all shared a desire for their views about the relationship between natural science and religion to be understood by the public. Strauss, who periodically surfaced in public controversy throughout his career, styled his *Old Faith and the New* as a confession, as did Schmid in his *Natural Scientific Confession of Faith of a Theologian*. Zöckler's articles appeared as often in popular journals like *Daheim* or his own *Beweis des Glaubens* as they did in theological organs, and his *Primal History of the Earth and of Humans* as well as the later *Doctrine of the Origin of Humankind* both derived from a series of public lectures delivered in Hamburg in the spring of 1868.[1] The theology of these individuals was seen by them and by the reading public to be embedded in the real concerns of German cultural life.

Apart from Ritschl himself, whose textbook *Instruction in the Christian Religion,* though written for use in German high schools, turned out to be too difficult to use there,[2] the neo-Kantian theologians of the late nineteenth century did not come to the public's attention like Strauss, Zöckler, or Schmid had. This was especially true of the young Wilhelm Herrmann, whose 1876 work on *Metaphysics in Theology* foreshadowed the coming of a Ritschl school.[3]

Although the esoteric concerns of the theologians who began to identify with the sentiments of Albrecht Ritschl would exert an impact on the future of academic German theology, few of those who took their cue from Ritschl expressed their difference from mainstream speculative, orthodox, and mediating theological approaches by way of the problem of religion and natural science.

Because, as noted above, Ritschl's theology could be understood as a turning away from the scientific worldview so visible in the early Imperial Period,[4] most Ritschlians, including Ritschl himself, revealed their antipathy to the methods of natural science by ignoring them in favor of alternative approaches they chose to pursue in detail.

The young Wilhelm Herrmann is the exception. In the early phase of his career Herrmann took upon himself the task of articulating theology's right to refuse to grant natural science the deference which men like Strauss, Schmid, and even Zöckler thought it was due. To do so Herrmann sought to provide the kind of careful philosophical examination of what he called "religion in relation to knowledge of the world" that was missing from the work of Ritschl but that, in light of the aggressiveness of scientific naturalism in the closing decades of the century, seemed to be so desperately called for.[5]

It should be made clear, however, that Herrmann's defense of theology in an age of science was intended neither to confirm nor to deny the existence of the warfare between science and theology so frequently debated among historians of the post-Darwinian era. Herrmann understood the two endeavors to be so radically separate that they could not intersect, let alone do battle. For this reason Herrmann could and did repeatedly demand that scientists be afforded complete freedom in their work. By the same token Herrmann made no attempt to reconcile science and religion in order to demonstrate that there was no enmity between them. Because public opinion tended to separate into two camps—those who believed a battle had to be fought and those who maintained that reconciliation was possible—Herrmann's position was like the tie sticking out of a suitcase. His message simply did not fit into the usual nineteenth century baggage.

As Herrmann saw it, scientists and theologians normally attempted to discuss science and religion on the assumption that they shared a rational ground. In true Kantian fashion he argued that this hasty and often unconscious presumption resulted from not thinking through one's own scientific presuppositions. In his work he undertook to show that the employment of reason itself revealed how those very scientific presuppositions colored the results obtained from them. Unless one realized this, one would continue to pursue scientific truth as if it were something possible to achieve, when in fact all the knowledge of science constituted a self-enclosed framework surrounded by a limitless night. Herrmann rejected above all the notion that theology involved a religious knowledge that "rounds out the scientific explanation of the world, accompanying [it] as a stopgap measure."[6]

Since natural science had progressed to a self-consciousness of its methods sufficient to identify the limits of its knowledge, Herrmann felt that one might have expected theology to have responded accordingly. But orthodox and liberal theologians alike remained caught by the hope that they could continue to negotiate a harmonious solution to their conflict with scientists.[7] Such a hope was pure delusion to Herrmann. Far better that theologians should recognize that the impossibility of arriving at truth with the tools of science held great significance for religion. Herrmann boldly declared that a cosmos ruled by natural law must be acknowledged as a place alien and hostile to the essence of human existence, which was based on moral purpose. While humans *qua* humans could never be reconciled to the purposeless realm of cause and effect, the inability of acquiring final truth through causal analysis meant that they were free to claim a cosmos of purpose if they chose to do so. Since the theologians and scientists of his day did not understand either the relativism of scientific explanations or the real nature of Christianity's viability in a modern world, Herrmann took up the task of persuading them that "religion, in comparison to other intellectual forces of culture, must try to prove itself to be a power stronger than all of them."[8]

Preparation of a Future Ritschlian

In the third volume of his massive study of justification and reconciliation Ritschl had denied that the probabilities which natural science uncovered in nature could ever be consolidated into what he called "real knowledge."[9] Such a blatant but at the same time facile dismissal of the relevance of our knowledge of the world to our spiritual concerns aggravated virtually everyone. It proved unacceptable to natural scientists, who must have found Ritschl's words extremely presumptuous in light of the scientist's image as the unbiased seeker of truth. It was equally distasteful to theologians of the three schools already discussed. Strauss certainly could not agree, since his theology amounted to an apotheosis of a scientific worldview. Zöckler appreciated Ritschl's focus on the centrality of justification, but neither he nor Schmid could agree that the conclusions drawn by natural scientists on questions like the origin of the human race were unimportant to theology. No one who held that truth was to be understood as an accurate correspondence between human thought and reality could accept Ritschl's radical separation of religion from natural science.

There were those, however, for whom a message of the immunity of religion from the heartless implications of a scientific worldview was more than enticing. Would that there was an escape from the awesome prospect that humans were merely the improbable by-product of a mindless and totally indifferent mechanical universe, that the uniquely spiritual characteristics humans felt themselves to possess were produced in such fashion, and that consequently the only meaning of human life was that which the individual was capable of manufacturing in the present. To some nineteenth century minds a Darwinian universe was no place for human beings to be.

Such a person was the young theology student Wilhelm Herrmann in Halle. Unlike the other theologians in this study, there is little biographical information available about Herrmann. Martin Rade, later a colleague of Herrmann, observed of his friend's career, "[it is] not as if there were a particularly exciting course of events in this scholarly life to report. On the contrary, there is not much to tell."[10] Like Zöckler, Schmid, and Ritschl, Herrmann was born into a pastor's home. Wilhelm's father, Johann Wilhelm Herrmann, was himself a pastor's son and the brother of two more pastors. The penchant for the ministry had begun with Wilhelm's grandfather, who had become a cleric in Bibra against the wishes of his peasant father.[11]

Wilhelm was born in Melkow in the Altmark on December 6, 1846. True to the pattern already encountered in the lives of Schmid and Zöckler, his father oversaw his early education, supplementing the instruction he received in the village school. Johann Herrmann was intrigued with Schleiermacher, whose collected works he had managed to acquire and whose plaster bust he kept on his desk. Wilhelm's mother came from a more orthodox background in which the preference was for Hengstenberg's theology, but over time she came to appreciate the less rigid approach taken by her husband and later by her son.[12]

When young Wilhelm went off to the gymnasium in Stendal, he did not distinguish himself as a student. But he was a solid pupil whose interest in philosophy was clear from his industrious reading of Kant's *Critique of Pure Reason*. When he not only passed his final examinations but successfully prepared a weaker colleague for them as well, he received a trip to Switzerland as a thank-you gift from his friend's family. Nothing quite like that opportunity had ever come his way before. During this first foray beyond his modest surroundings in Melkow, Herrmann developed a fascination for mountain climbing that he retained his entire life.[13]

When Wilhelm went off to the university in Halle, two of his older brothers were already studying there.[14] Wilhelm had six brothers and sisters, so financing his university education strained his father's resources. Wilhelm was able to contribute to his own expenses by repeatedly winning the philosophy prize question even though he was matriculated in Halle's theology faculty. The philosopher Herrmann Ulrici tried to get him to switch to philosophy, but one of his theology professors, F. A. Tholuck, was impressed enough to make him his amanuensis and to offer him housing for over two years. Tholuck, an influential figure in neo-pietism, had been Ritschl's teacher over twenty years earlier. How much of an influence Tholuck was on Herrmann has been disputed,[15] but there is no doubt that it was in Tholuck's house where Herrmann first met Ritschl.

Herrmann's university study was interrupted by Prussia's war with France, which drew into its sphere all the German states except Austria. Herrmann came back from service in France with a distaste for military drills and a stomach illness that plagued him long after the war.[16] He was able to resume his studies in June of 1871. After completing his first theological examination, he became a house tutor and was eventually able to combine his duties as tutor with those of a teacher of religion in Halle's gymnasium. By January of

1875 he had completed his habilitation thesis, a work on Gregory of Nyssa, and assumed the post of *Privatdozent* in Halle.

It had been one year earlier, however, that Herrmann met Ritschl when Ritschl visited Tholuck's home. When Tholuck, unsolicited, recommended the aspiring theologian to Ritschl, the latter, in characteristic style, replied curtly with a Latin phrase which meant that no one was obligated to do more than he was able.[17] Embarrassed, Herrmann felt unjustifiably rebuked. Once he had completed his work on Gregory of Nyssa the following January, he sent a copy to Ritschl in Göttingen with a letter explaining that he had not asked Tholuck to recommend him, and that he had come to Ritschl's work not via Tholuck but via Max Besser in Halle. Besser had recommended Ritschl's works "as a means of freeing me from the spell of the education I have acquired partly because of and partly in opposition to Halle's stimulation." He added that he had in fact been much impressed by Ritschl's work and that he wished to study it further.[18]

Ritschl not only read Herrmann's work on Gregory of Nyssa and wrote a notice of it for the *Jahrbücher für deutsche Theologie,* he also softened a bit his earlier rebuff. He noted that he had not intended to reject Tholuck's request; he merely wished to place limitations on Herrmann's expectations. Ritschl conceded that his behavior could have been seen as rude, but he went on to note that he really had little to offer a young theologian, since all parties—right, left, and center—opposed him. Associating oneself with him, cautioned Ritschl, might do more harm than good. Clearly at this point there was not yet a group of followers who constituted anything like a Ritschl school. In fact, Ritschl concluded his reply to Herrmann with the observation that he had been stung by others who had approached him, adding: "Incidentally, the people who seem rude are not the worst ones."[19] To Tholuck he reported that he had written a notice of Herrmann's work and would look for other ways to help Herrmann advance his career.

This interchange occurred at the end of January 1875, well prior to the attacks on Ritschl that were yet to come. The first volume of the *Justification and Reconciliation* had been out for almost five years, but the really controversial sentiments were contained in the third volume, which had made its appearance less than a year earlier.

According to Rade, Ritschl had a few students, as did virtually any university professor, and the volumes of his study of justification came to the attention of the scholarly world without fanfare.[20] Zöckler's review in *Beweis des Glaubens,* for example, was complimentary. In spite of numerous specific objections to heterodox ingredients rooted in Ritschl's subjectivity, it would be unjust, wrote Zöckler, "to count him without further ado among the representatives of theological radicalism and to deny the various valuable things in his efforts to contribute to the positive furtherance of Protestant theology."[21]

What, then, had Ritschl said in his third volume that had made young theologians like Herrmann sit up and take notice? First and foremost Herrmann heard Ritschl attempting to refer the scientific treatment of Christianity to the *overall* enterprise of systematic theology as opposed to the rational examination of individual doctrine. Although systematic theology depended on the proper delimitation of religious representations *(Vorstellungen),* its definitions could be assigned only with respect to their place within the whole system, "because the proper fixing of individual [components] already presupposes insight into the coherence [*Zusammenhang*] of all parts." This made it necessary to take into account the teleological relations between individual components and the whole. Logical accuracy, wrote Ritschl, did not guarantee truth. "Individual conditions . . . must be shown to be necessary through their relationship to the whole. For real scientific proof can only be undertaken for the whole system."[22]

Ritschl introduced here a historical observation which Herrmann would develop further in his own work. Medieval theology had proceeded in just the opposite fashion. It had declared the whole to be unprovable on the assumption that Christianity taken *in toto* was suprarational *(übervernünftig).* Nevertheless, it carried out rational proofs for each individual doctrine, "because Christianity did have a certain analogy to human intellectual endowment."[23] But the measure of the truth of Christianity was not to be found in the rational demonstrations of natural religion which had resulted from the fatal application of a "Platonistic monotheistic cosmology and the Stoic assumption of the law-giving power of conscience" to Christianity. What ensued was what Ritschl called "the Catholic epoch," in which the mediation of creation through the Logos became a central feature of the work of the Redeemer, and the moral dimension was subsumed into natural knowledge. In spite of the Reformation, theologians from the Middle Ages to the appearance of deism had presumed that theology should not contradict natural reason but use it as a support.[24]

The cultural development which challenged this perspective was the onset of a Protestant era with the appearance of Kant's philosophy. Here the knowledge of the world was subjected to the standard of "the self-responsibility of human beings and the determination of universal morality." Now the idea that Christ was the purpose and middle ground of creation meant that creation should be understood in the context of Christ's role as the lord of the moral kingdom of God. Cosmology had become dependent on ethics.[25]

Ritschl did not think, however, that theology after Kant's achievement had yet escaped from the Greek assumptions that resulted in its use of categories infected with the characteristics of nature. Not even Schleiermacher, with his "half-Spinozan cosmology," had laid hold of Kant's teleological worldview. As the young Herrmann finished the Introduction to Ritschl's third volume, he knew something quite different was ahead.[26]

What appealed to Herrmann as a student was the boldness of Ritschl's challenge to "Platonistic cosmology." Ritschl's alternative perspective began with the assumption, no doubt gleaned from Hermann Lotze, that fundamental to all human intellectual and spiritual activity was the capacity to make a judgment of value. In religion one ascribed a characteristic value to the judgment that the universe was established and directed by a personal God. In science value resided in the utilization of the laws of cognition for the purpose of mastering nature, while in morality one endeavored to master oneself. Because of their common derivation from interest and value, all three—religion, science, and morality—rested on egoism.[27]

Drawing on the third volume of Lotze's *Microcosm,* Ritschl maintained that the peculiar feature of a religious worldview was that it formulated an idea of a whole or completed reality. To Ritschl this was expressed in many religions, both monotheistic and polytheistic, by the idea of God. Furthermore, because human beings saw themselves as possessing kinship with God the ruler and sustainer of the world, "the idea of God and the worldview ascertained according to it means that everywhere human beings are helped over the contrast between their natural situation and their spiritual feeling of self, and ensured an elevation or freedom over the world and their usual interaction with it."[28]

When Ritschl turned to the specific question of the relation between religious and scientific knowledge, he reinforced the difference in their interests. In setting out to master things in the world, human beings could in no way guarantee that the acquisition of scientific knowledge would end in a highest law of being from which one could derive the world as a whole. To do so would mean that one had "stepped out of the application of the exact method of knowledge, thereby proving oneself to be an object of an intuiting fantasy like the religious idea of God and the world."[29]

Yet the smuggling of the religious imagination into philosophy was precisely what the Greeks had done, according to Ritschl. In so doing they were able to guarantee a comprehensive worldview through the unity of the Divine Idea. But in the process they had "surrendered" (in later editions the word was "neutralized") the personal nature of God, and with it the possibility of God's active influence on the world. As a result human beings had no option but to "view themselves as a passing emission of the world soul or as a link in the spiritual development of humanity, the progress of which subdued and degraded them into dependence."[30] Ritschl's response to this situation was unequivocal, and it announced the perspective that would be adopted not only by Herrmann but by an entire generation of burgeoning theologians.

> If one thinks that this worldview should be preferred to that of the Christian it is of course because the official representatives of Christianity are just as little aware as the theological dilettantes among their opponents of the basic principle of the Christian's judgment of self—that the individual person is worth more than the whole world, and that he tests this by believing in God as his father and by serving in the kingdom of God. For the Christian worldview shows itself to be perfect religion through the disclosure of the spiritual and moral collective purpose of the world, which is the real purpose of God himself.[31]

Herrmann would assume as one of his responsibilities a detailed demonstration of the inadequacy of what Ritschl called theoretical knowledge and of what Herrmann would name the scientific knowledge of the world *(Welterkennen)* to replace the personal worldview defended in Ritschl's work. The germ of Herrmann's later argument, however, was to be found in his teacher's review of Kant's critiques. To interpret the world as a unified whole, even to decide whether the causal relations among forces in nature were created *(geschaffen)* or eternal, was not a chore that could be addressed in mechanical natural science without overstepping the limits of theoretical knowledge. And nothing, of course, could be inferred about teleological laws or a purposeful intelligence from within the bounds of experience.[32]

It was near the end of his study where Ritschl emphasized the misunderstanding which arose when one made laws that applied to the narrower context of natural science into world laws. He particularly objected to those who argued that his view, which interpreted the universe in personal, teleological, and miraculous terms, was impossible given the mechanical considerations of natural science. Miracles and purpose, wrote Ritschl, were an unavoidable part of any attempt to explain organisms or nature as a whole. If readers objected to the application of conscious will to nature because it was presumptuous, they ought to realize that the principle of causality on which natural science relied was also a presupposition of our thought, hence it too should not be applied to nature. Both applications were either rational steps or delusions.[33]

If the religious view of the world was suspect because it arose from a need of the human heart, one should remember that even the simplest study of nature, in proceeding according to the rule of causality, "emanates from a need of human reason, and its validity is subject to the same suspicion; namely, that humans observe something in their will and then find it behind the phenomena because they want to find it there."[34]

As he pondered the implications of Ritschl's bold but brief counterattack against the naturalistic worldview of the latter half of the nineteenth century, Herrmann became convinced that what was tossed off in passing by Ritschl should be driven home more aggressively to his age. Like Ritschl and many other contemporaries,

Herrmann accepted the depiction of natural scientific explanation given by Emil DuBois-Reymond in his lecture on the limits of natural knowledge in 1872—specifically, that natural scientific cognition was the reduction of changes in the material world to the mechanical motion of atoms.[35] But he was also convinced that not only natural scientists but theologians as well erred greviously and to everyone's detriment when they failed to understand the difference between humankind's desire to master or control the natural world and the need to arrive at a complete and final understanding of nature. The young Herrmann set out to instruct those of his day on the issue, and in the process he drew attention to the sentiments buried in volume 3 of Ritschl's *magnum opus*. "To Albrecht Ritschl," Rade writes, "it was allotted after but moderate successes in thirty years of academic work suddenly to see himself as head of a theological school that irresistibly attracted young talent. The first academic who broke the path and pledged [his] adherence to Ritschl was Wilhelm Herrmann."[36]

THE *METAPHYSICS IN THEOLOGY* OF 1876

Not long after Herrmann habilitated in Halle in January 1875 he began teaching as a *Privatdozent*. In April he wrote to his brother that he enjoyed his new status as a dispenser of wisdom to expectant students. But Herrmann was not particularly successful in Halle. He began with ten to twelve students in the spring of 1875, and in spite of his preference for speaking extemporaneously for at least part of his lectures, he never attracted more than twenty students during his four-year tenure.[37] Naturally he did not become wealthy from his earnings as a *Privatdozent*. He lectured on symbolics, the history of dogma, and the subject Ritschl had brought into German theology, the kingdom of God. But he was not close to any of the mediating theologians who constituted Halle's faculty, a position due more than anything else to his bold publication in 1876 of *Metaphysics in Theology,* followed a year later by a critical review of R. A. Lipsius's *Textbook of Evangelical-Protestant Dogmatics* in *Theologische Studien und Kritiken*.

At the outset of his 1876 polemical tract Herrmann embraced what he called "a psychological investigation of the religious mind," in contrast to the cognitive approach of classical theology. Later in his career Herrmann would identify Schleiermacher as the first one to provide a persuasive critique of rationalistic religion.[38] Here, however, he referred to the Friesian theologian W. M. L. De Wette's "Thoughts on the Spirit of Recent Protestant Theology" from 1828. De Wette indicated that the theology of his time was marked by the realization, with which De Wette clearly agreed, that the content of religious intuition was far superior to its momentary expression at any given time. Herrmann noted that theologians at the end of the nineteenth century were not comfortable with De Wette's position because De Wette had not insisted on associating metaphysical expressions of theological ideas with religious intuition; hence he had not required a conceptually articulated world- and life-view.[39]

Herrmann here introduced the theme that betrayed his association with Ritschl, the relationship between theology and metaphysics.[40] Were modern theologians reasonable in their demand that in order to possess the proper contents of Christianity, we must make sure that we acquire a unified view of the world? Herrmann interpreted this demand to mean that modern theologians required metaphysics in order to reach a final and definitive knowledge of the actual state of the world. In this they were like some scientists.

> [Modern theologians] have colleagues in their work in some enthusiastic representatives of natural science, who, unsatisfied with their intercourse with individuated existence, raise themselves from the basis of empirical investigation to the idea of a world whole and a unified ground of things. This cooperation permits theological speculation to hope that it will, with such help, be able to reach the rock bottom of reality and uncover the secret of the unified gigantic construct of the universe.[41]

Herrmann explained that it was not difficult to understand where the assumption that theology and natural science possessed a point of intersection arose. If in natural science we acknowledge that the train of thought pursued could never be brought to a conclusion within science, then theology, which spoke of a knowledge that did not have its origin in the scientific explanation of the world, seemed well suited to offer a solution to its riddle. For some theologians, then, it was natural to import metaphysics into theology as a welcome continuation of a science up against its limits. And for some scientists it was all too easy to forget the limits of science in the face of the impressive mastery over phenomena that had been achieved. When this occurred one began to believe that one could solve nature's riddle, that agreement of experiment with theory brought with it the right of locating human purposes among the series of unalterable events.[42] In the long run, however, mastery of the world was not what humans were seeking.

Herrmann in no way opposed scientists' right and responsibility to master their environment; nor did he think that the success scientists enjoyed could be ignored. The unlimited extension of the mechanical interpretation of the world over that which appeared in space had to be presumed in any "proper theory of knowledge." Mechanical mastery of nature was fine as long as it was kept in perspective. Herrmann illustrated his position using the notion of the atom. The arbitrary limitation natural science imposed upon the divisibility of matter with the concept of the atom was, he wrote, free of contradictions as long as it is only a question of an explanation of nature for the purpose of the mechanical mastery of nature. "But on the question . . . whether the concept of the atom is suited to designate the real independent elements of the world nothing at all is decided."[43]

Beside the need to dominate the world humans also possessed another need, "a drive for knowledge [*Wissenschaftstrieb*] that does not slacken because of practical successes in the mastering of nature." Herrmann identified this drive as the natural human requirement for a unified understanding of the world as opposed to the need to control it. Here one acknowledged that arbitrary notions of natural science like atoms were not without contradictions, and one was led beyond such unsatisfactory concepts to metaphysics.[44]

Herrmann noted that natural science could remain untouched by the results of metaphysics. If, for example, natural scientists were found by metaphysicians to be employing concepts that were self-contradictory where claims of actual existence were concerned, then the scientist simply called them "fictions of undoubted practical success" and continued to use them. Scientists were therefore freed from having to concern themselves with the fundamental riddles surrounding their area of investigation since these riddles about the actual existence or essence of things had nothing in common with their work.[45]

Thus far Herrmann had associated metaphysics with what he called a "compulsion to know [*Wissenstrieb*]," although he implied that the results of metaphysical pursuits, while they might help clarify how we are to think about what exists, could not yield certainty of any kind. Metaphysical thought involved presuppositions which sought to go beyond the conceptually contradictory categories of science to a unified explanation of the world that was free of contradictions. But a metaphysical system free of contradictions did not guarantee, according to Herrmann, that a system corresponded to the world as it really was.

Herrmann's real interest, of course, was not the relationship between natural science and metaphysics but the link between metaphysics and theology. His examination of natural science, however, enabled him to ask whether metaphysics was any more suited to serve the theologian than it was the scientist. Did metaphysics function merely to reveal the limits within which the theologian worked like it did for the natural scientist, or could the theologian actually make use of metaphysical concepts?

Systematic theology, according to Herrmann, involved the representation and grounding of a religious view of the world. With this claim Herrmann introduced a new consideration that had not been part of his treatment to this point, the dimension of ethics. "Every religious worldview is an answer to the question: how is the world to be judged if there really is to be a highest good?"[46] A religious worldview, then, was different from a metaphysical worldview. In the latter one asked about the general forms or presuppositions under which all being and all events could be thought without contradiction. The correctness of the forms had no relation to the goals of one's will. This was not the case, however, in a religious worldview. There one *was* indifferent to the kind of consistency present in a metaphysical system, but one was vitally interested in one's capacity to exercise the will and even to interpret one's existence in the world in light of it. Herrmann depicted religious individuals as those who knew themselves to stand in a constant struggle with nature but also as those who valued the human moral capacity over the ability to construct a contradiction-free account of being. The religious individual introduced order into the chaotic diversity of experience not by measuring reality against presuppositions regarding its material or mental constitution (which was the method of metaphysics), but by viewing experience as subject to the highest good.

When he referred to the presentiment *(Ahnung),* which the religious individual could not dispel, that the world beheld in moral terms could be juxtaposed along side that seen by the metaphysician, Herrmann made use of a category that had been employed by his neo-Kantian forerunner Jakob Fries three quarters of a century earlier.[47] But he quickly added that neither the right nor the occasion for the harmony which was felt to exist between the two came from Christianity. Not the right, since the certainty the Christian possessed came not at all from insight into the structure of the world but from the power of the highest good over the life of the mind. Not the occasion, since there was nothing in the religious perspective that would make the Christian willing to subordinate the world to the highest good except the overpowering feeling for the value of the kingdom of God.[48] It was a total misunderstanding of the nature of religious faith to attempt to gain insight into the physical world from the religious impulse. Herrmann's bold position, as laid out in the *Metaphysics in Theology,* ran counter to the speculative theology of his day and he knew it. But he was neither apologetic nor timid. His conclusion was that metaphysics had no place in theology because knowledge of the world, which imparted no certainty, was automatically subordinated to the indubitable demand of value in the religious perspective. Nature became the servant, the Christian the master. "Our belief in God as our Father in Jesus Christ enables us to judge the world no longer to be something contradicting our personality, filled as it is with such a definite moral content, but something that can be subsumed as a servile means under our moral destiny, the highest good."[49]

In this understanding physicists were not the bearers of bad news because the universe was not an alien place in which human existence was, so to speak, out of place. But Herrmann's confidence that the world did not really negate the realm of personality did not mean that the lack of negation could be demonstrated cognitively.

In fact that could not be done any more than could the gospel message be made accessible to speculative reason. Between the knowledge of facts *(die Erkenntnis des Tatsächlichen)*, which was always incomplete, and the feeling for the value of the good *(das Gefühl für den Werth des Guten)* there was an inextinguishable difference. To make the latter in any way dependent on the former was tantamount for Herrmann to surrendering the otherworldliness of the Christian idea of God.[50] On the contrary, the religious world-and-life-view of Christianity, in which the world of nature served merely as means, was only accessible through the feeling for value. "It is only possible and only makes sense to a moral community dominated by one moral ideal." The nonethical domain of mastering the world stood beneath the moral as the common field of activity for those engaged in the pursuit of the scientific explanation of the world.[51]

Herrmann knew that his downplaying of metaphysics would not set well with any of the existing schools of theology, all of which assumed that truth consisted in a correspondence between thought and reality. To meet the objection that he knew would be voiced, namely, that his position made metaphysics indifferent and ancillary to Christianity, Herrmann went directly to the heart of the matter. What, after all, was "the truly real [*das wahrhaft Reale*]?" For Herrmann the truly real was the moral spirit, "in whose reality all else shared to the extent it could legitimate itself through its positive relation of purpose to moral spirit."[52] Clearly this was not the case for the metaphysician.

We must admit, Herrmann pointed out, that in Christianity we mean something totally different by what is truly real from what we intend in metaphysics. In the latter what is meant is "the real that produces [*das hervorbringende Reale*], by which we make comprehensible to ourselves the possibility of all being and all events in general." In Christianity, by contrast, one develops certainty of the truly real not through concepts but "in interaction with the nontranslatable experience of the value of the Christlike good."[53] We ignore this fundamental difference, wrote Herrmann, to the detriment of both. If one attempted to become acquainted with the operation of the moral will through notions of mechanism, one did not have purely ethical religion in view. Likewise, mixing in the moral dimension where it did not belong led to a surrender of the freedom of scientific investigation.[54]

So separate was religion from metaphysics that for Herrmann a Christian was free to embrace either an idealistic or a materialistic metaphysics. Herrmann based this conclusion both on the independence of the religion of morality from metaphysics and on the ultimate relativity of a conceptual mastery of the world. As science (*Wissenschaft*) in the general sense, metaphysics was confined to the mind's "purely knowing function, which cannot, but also should not, give an answer to the question posed by the moral spirit about the truly real." And in its own sphere metaphysics unpacked the conceptual apparatus it used "to get close to things factually by making clear the immanent presuppositions necessary to it." As a science independent of the moral will, then, it belonged, like all "mechanical" attempts to master the world, to the realm of the facts. For our religious judgment it was nothing more than one fact among others, together with which it built the world that Christ had overcome for us. Where making the religious task harder or easier was concerned, it made no difference at all whether the dogmatic metaphysics which the Christian follows was materialistically or idealistically directed.[55]

Of course Herrmann was writing as a systematic theologian, and systematic theology was itself a science. What, then, were the implications of the separation of religion and metaphysics for theology? Herrmann was quick to point out that whatever related to confirming the actual state of the world was not to be part of theology's task. However, theology could turn to metaphysics for concepts that permitted the articulation of the distinction between a morally determined spiritual realm of purpose and a nonmorally determined world of means. "We must demand of metaphysics that it provide us concepts that correspond to those relations."[56] But Herrmann knew that he had to prescribe limits here, for if metaphysics were to make use of concepts whose meaning was restricted to an ethical context, then theology would have to unmask what amounted to pseudo-metaphysics as a surrogate of religion.

The overall task of theology, according to Herrmann, was "to demonstrate that the problems of the moral spirit were solved when that spirit, having acquired Christian goodness, participated in the religious worldview of Christianity." Confusing this responsibility was a number of what he called "pseudo-problems" that had no meaning in a proper theological perspective. One such problem was the contrasting of consciously volitional relations among humans to the mindless causality apparent in the world of events. In every exercise of the will except the moral, it did no harm to view the will as subject to causality. When, however, we experience the value of the good which surpassed all else, we judge ourselves to be free in a sense which would not allow our will to be treated among the causal events captured by "the play of the understanding."[57]

Emboldened by his youthful confidence, Herrmann hardly paused before he pronounced his almost ascetic conclusion: human beings could never be satisfied with the experience of the particular and could find full satisfaction only independent of everything that endured in the flux of empirical existence. Once one had recognized that moral purpose fashioned a determining ground for our will, one judged oneself to be free and saw oneself as possessing a form of will superior in value to everything otherwise conceivable. This was why human beings could never surrender their ultimate responsibility by appealing to circumstances as an excuse.[58]

Building on the absolute distinction between natural and moral realms that had its origin in Kant, Herrmann forced his reader to choose between them. Near the end of his work Herrmann justified the fundamentally dualistic nature of his perspective. Christianity did not demand continuity, he wrote, between the reality confirmed in religious faith and the scientifically explained worldview. The worldview articulated through concepts could be used for "contemplative gratification" through the impression of harmony it might provide. But any unity between nature and religion would remain "uneasily foreign" to the Christian, of whom it was demanded that the world as scientifically explained be subordinated to the higher religious reality.[59] In the middle of his treatise Herrmann explained that while the Christian conquered the world by virtue of God's power, it was not God's power over the physical world that he had in mind. The power of God became clear when the Christian saw how Jesus had conquered the world spiritually. "To speak of a physical power of God seems to me to contradict a healthy supernaturalism."[60]

It was Ritschl, according to Herrmann, who introduced the notion of the kingdom of God not only as the context in which humans were conquerors of the world but also as one in which they assumed a moral onus. The Christian faith in the reality of the highest good was never present in its inner richness, explained Herrmann, unless it was seen in its realization, and that could only occur when it was related to the means by which the faith was realized—the world. In this way and through this need the moral task of the religious community in the world was born, as Ritschl so successfully had pointed out.[61] In the end, then, the challenge to overcome the world lay with the individual, and whether it would be met depended not on the possession of a unified worldview but on each person's valuation of the human capacity for moral choice.

The Beginnings of a Ritschl School

There is no denying that Herrmann's early works contained a sharp and aggressive tone. Herrmann had become convinced that theology as it was being pursued exhibited a fundamental flaw which had to be corrected. Ritschl, whose interpretation of basic Christian doctrines had not only avoided the error of mixing metaphysics into theology but had also explained the history behind the development of the tendency, had not made a point of critiquing contemporary theology for its failure so much as he had made clear his understanding of the relation between religion and morality. It was Herrmann who made explicit the dissatisfaction with the assumptions of speculative, orthodox, and mediating theologians alike implicit in Ritschl's work. Because of his youth and lack of recognition, Herrmann's attack simply could not go unchallenged.

The response to the *Metaphysics in Theology* was noteworthy because the vehement quarrels it precipitated began the coalescence of a group which took its cue from Ritschl's theology. It was not that the work received widespread attention. The author, after all, was an unknown licentiate in theology, and the work with which he greeted the theological public was so succinctly written that it cried out for more elaboration and clarification. From his allusions to the kingdom of God, his treatment of Melanchthon, and his general position on metaphysics, Herrmann had associated himself with a well-respected Göttingen professor, but Ritschl's stature in 1876 was not sufficient to guarantee that every young scholar who invoked similar views would be heeded.

The most important response to Herrmann appeared in the liberal theological journal *Protestantische Kirchenzeitung*. Three contributions to this publication during 1876 and 1877 dealt with Herrmann's position in one way or another. First came a brief review by Richard Adalbert Lipsius, a well-respected senior theologian in Jena and an acquaintance of Ritschl. When Herrmann replied to Lipsius in the form of a critical review of the latter's newly published textbook on dogmatic theology, it provoked a defense of Lipsius from Georg Graue in the *Protestantische Kirchenzeitung*. In the same issue in which Graue's diatribe against Herrmann appeared, there was another treatment of Herrmann by a young theologian named Otto Pfleiderer, Herrmann's senior by seven years. Together these responses to Herrmann went a long way toward solidifying the notion of a Ritschlian theology in the mind of the public.

The relationship between Lipsius and Ritschl was an important factor in Herrmann's emergence onto the theological scene of the 1870s. The two senior theologians routinely sent copies of new publications to each other as a gesture of their mutual respect. In 1871, however, Ritschl commented in a letter to Ludwig Diestel that he was not overly impressed with Lipsius's *Faith and Doctrine*, which he had just received. Ritschl definitely approved of Lipsius's recent call to Jena, but he viewed him as "more a traditionalist than he thinks," by which Ritschl meant that Lipsius stood in the line of Schleiermacher and the Lutherans, who felt that cosmology was an integral part of theology. Ritschl noted to Diestel that he was not close enough to Lipsius to share his criticisms, since their practice was not to send accompanying letters when transmitting publications.[62]

In light of the criticisms Lipsius published of Ritschl's *Justification and Reconciliation* in 1874, it is possible that Ritschl's view reached Lipsius through Diestel. But in early July of 1876 Lipsius attempted to smooth Ritschl's obviously ruffled feathers with a copy of his recently published *Textbook of Protestant Dogmatics*, sent this time with a note in which Lipsius indicated that he was happy to have been able to agree with Ritschl "in a series of essential points."[63] Ritschl, who could be sensitive and difficult, was not assuaged. He replied sarcastically that if he were more practiced in the latest criticism he might try to prove Lipsius's letter was not genuine, since its respectful tone did not square with Lipsius's earlier negative evaluation of his work. Nevertheless, he would take the gift as a sign of Lipsius's good intent.[64]

Lipsius did not reply to Ritschl, as their common friend Diestel was sure he would. In the meantime Herrmann had written Ritschl, noting that Lipsius had drawn on Ritschl in his textbook without acknowledgment. One week later Lipsius's critical remarks about Herrmann's *Metaphysics in Theology* appeared in the *Protestantische Kirchenzeitung* as part of an introduction of his new textbook to the journal's readers.[65] Within a month Ritschl learned from Herrmann, no doubt with satisfaction, that the latter had been given a free hand to review Lipsius's text in *Theologische Studien und Kritiken*.[66]

Herrmann subsequently regretted the tone he adopted in his long review of Lipsius, though Martin Rade, later Herrmann's colleague at Marburg, called the piece "the most interesting thing Herrmann wrote during his Halle period."[67] Had Herrmann been older the review would most likely have been read as a critical appraisal by a neo-Kantian theologian who, while identifying with Kant, had strayed from the fold. Because Herrmann was young and unknown, however, it became clear very quickly that he had overstepped the limits of propriety when he dared to disparage so thoroughly the work of an established member of the theological community.

In Herrmann's eyes Lipsius was trying to please everyone at the same time, including Ritschl, whose discontent with Lutheran theology had been clearly registered. Lipsius apparently agreed, for example, with Ritschl, whom Herrmann now openly named, about the *exclusively* religious character of the need to represent human experience of the world, appearing to admit that the scientific explanation of the world was completely different from the religious. Herrmann was happy to agree with Lipsius that neither of these two kinds of explanations of the world could leave its sphere and enter into the other. But how then, asked Herrmann, could Lipsius insist that dogmatics must exclude what was not consistent with our knowledge of the world?[68]

Herrmann was unwilling to concede Lipsius's acceptance of a

theoretical correspondence between human thought and reality. Lipsius demanded that the statements of dogmatics be seen as approximations of objectively valid propositions, even though as approximations they were figurative and could not be taken at face value. But Lipsius believed that the results of one era could be "more reasonable" than those of an earlier period, that a future time would reject present forms of expression in favor of a more basic purification of the idea of religion from sensual elements. To this Herrmann replied in remarkably modern language: "If generally speaking the material of our ideas and the object of our faith are completely incommensurable [*incommensurabel*], then it is an empty game to select from our concepts those which might fit best as the designation of this object."[69] Lipsius, in other words, did not really agree that the circles of science and religion were totally separate; rather, there was a higher unity joining religion and science which, though hidden to the mind, existed as the presupposition of dogmatics.[70]

Herrmann also took Lipsius to task for his claim that by exhibiting the lawful nature and the necessity of religious ideas within the human intellect one justified the religious view of life. In Herrmann's view Lipsius's reliance here on empirical psychology expected something of it that it could not deliver. "Nothing is explained, only described," Herrmann wrote. The important question, which had to do with the purposeful essence of human beings, belonged to a realm not accessible to empirical psychology. Herrmann was not about to concede that true religious experience could be understood through deterministic law, and he would spend the rest of his career resisting what he believed to be a basic error. Much later Herrmann humbly confessed that he had spent thirty years trying to show that religion sees the truth of its ideas to be grounded solely in the experiencing *(erleben)* of its content.[71] Here at the beginning of his career he was far more cocky. In decidedly presumptuous fashion for a young theologian, Herrmann recommended that the elder Lipsius consult the philosophers Friedrich Albert Lange and Wilhelm Windelband and the psychologist Wilhelm Wundt, all of whom, he condescendingly suggested, were worth reading by theologians who wanted to make their dogmatics "scientific" through psychology.[72]

Herrmann's attack on the very foundations of nineteenth century intellectual culture was simply too much for some to endure. Georg Graue's protest, entitled "In Defense," exemplified the outrage felt by many. Graue noted that he was no doubt also setting himself up for mistreatment by Herrmann, but such a risk was necessary in this case.

Graue's critique of Herrmann's review of Lipsius revealed the great gulf separating Herrmann from the rest of the theological community. The perspectives of Graue and Herrmann were so dissimilar that communication was virtually impossible. Graue complained that Herrmann had misrepresented Lipsius, since he knew very well that the latter had repeatedly emphasized in his textbook that "the objective reality of the religious condition as well as the unified worldview it built on the basis of religious experience . . . cannot be proven strictly scientifically."[73] But Graue saw no contradiction between this admission and Lipsius's demand that dogmatics reject whatever was inconsistent with what Graue called the "scientifically insured collective intellectual property of the present." Striving to achieve a unified worldview was not inconsistent with the religious drive, wrote Graue. Again it came back to the question of truth; what united religion and science was the one truth lying behind them both. "For it is the essence of religion to awaken and to spur on the drive for knowledge of the human spirit that is directed to the whole and the general and, instead of being satisfied with the knowledge of individual truths, to seek the truth, that is, the truth which embraces and interprets all particulars in a unified whole."[74]

Writing from a perspective in which the one truth, independent of us, dictated the conditions under which we must try to match our ideas against it, Graue characterized Herrmann's misunderstanding as an insistence that the propositions of dogmatics were of the same kind as those of science. To Graue and Lipsius science and religion *were* two separate spheres, but both were embedded in a higher unity which could and must be approached conceptually. The propositions of dogmatics and those of science—though unlike—were therefore, to draw on Herrmann's categories, commensurable. To Herrmann, however, the two domains were incommensurable, and consequently they did not share in a higher unity whose single truth, by constraining our attempt to articulate it conceptually, made theoretically possible what Lipsius and Graue understood to be the objective validity which our knowledge approximated better and better with each passing generation.

Graue rejected Herrmann's characterization of Lipsius's dogmatics as a continuation or completion of a scientific view of the world because to him propositions of dogmatics and of science were not homogeneous *(gleichartig)*. But, as we have seen, Herrmann was not impressed with Lipsius's claim that science and religion were separate realms. If he had really meant it, according to Herrmann, then Lipsius would not be concerned to give science the upper hand by rejecting from dogmatics anything that contradicted the present-day explanation of the world. To Herrmann any dominance that was to be exerted should be the other way around. In a letter to Ritschl Herrmann said of Lipsius: "He has . . . not grasped that it is the purpose of religion to regulate [*regeln*] the relation of humans to the world."[75] Understanding among Graue, Lipsius, and Herrmann, to say nothing of agreement, was simply impossible.

Graue went on in his extended critique to defend Lipsius's use of psychology as a means of giving scientific grounding to religious ideas, and to defend Lipsius's position on the ever improving nature of dogmatics as an expression of the truth. Near the end of his defense of Lipsius he could hold back no longer. Lipsius, Graue noted, was a highly respected theologian with a long list of accomplishments. What had Herrmann achieved? That Herrmann had been allowed to discuss a book like Lipsius's in what Graue took to be a tone of audacity and presumption offended him. He castigated Herrmann as "but a copy of Ritschl" who was so firmly entangled in the thought patterns of his master that he could not understand any other terminology. Herrmann had behaved like a mediocre school teacher who became angry at a gifted student whose essay he could not understand.[76]

In conclusion Graue let his reader in on the secret that at least made sense of why Herrmann had written in such a preposterous manner about Lipsius. He reminded them of Lipsius's remarks about Herrmann which had appeared a year earlier in the *Protestantische Kirchenzeitung*. An acquaintance of Herrmann in Jena, explained Graue, had received a letter from Herrmann in which he expressed his anger at Lipsius and invited the acquaintance to show Lipsius his complaints. Graue attributed Herrmann's treatment of Lipsius to the pettiness of Herrmann's wounded pride, and he rebuked the *Theologische Studien und Kritiken* for allowing Herrmann to defame the work of a revered colleague.[77] Graue could understand Herrmann in no other way.

The final critique of Herrmann to appear in the *Protestantische Kirchenzeitung* was penned by Otto Pfleiderer, a young representative of the disintegrating mediating school. Speculative, orthodox, and mediating theologians might harbor fundamental differences, but they were at least joined together by the assumption that knowledge of the world constituted an essential component of theological truth and that it was at least theoretically attainable. Both the speculative and orthodox schools were so confident of their respective no-nonsense depictions of the true nature of things that their impatience with their detractors exposed the dogmatic flavor of their views. Mediating theologians, however, prided themselves on the carefulness of their reason in their examination of the complex relationship between thought and things. They were the ones, therefore, who were most threatened by Herrmann's rejection of their right to assume that reason necessarily corresponded to reality.

Throughout the *Metaphysics in Theology* Herrmann had used Pfleiderer, and to a much lesser extent Isaak Dorner, as the kind of theologians he opposed.[78] As a young theologian in a prominent post, Pfleiderer was the perfect choice to represent the mediating theology Herrmann attacked. There was no doubt about the polemical tenor of Herrmann's style, and he must instinctively have avoided directing his critique to those closer by. Of course Herrmann had to interact personally with the mediating theologians at Halle, but, we are told, "at the same time [he] despised them scientifically." For their part his Halle colleagues returned the dislike, especially after the publication of *Metaphysics in Theology*. He was even advised by them to go into the pastorate for a while to cure himself of his one-sidedness.[79]

The subsequent round of the ensuing controversy between mediating theology and the Ritschlians did not betray the eventual outcome. Pfleiderer's critique of Herrmann was an impressive denunciation of what he dubbed "the Ritschl school." But Ritschl's theology, based as it was on assumptions that sounded at once both traditional and conservative, was on the rise while the theology of mediation was on the decline. The carefully reasoned position of the middle ground, designed to accommodate all sides, in fact held attraction for fewer and fewer among the young. Pfleiderer would make the Ritschl school a specific target of his criticism throughout his career, but it was the rise to dominance of this very group which has been cited as the reason why Pfleiderer never achieved a stature in Germany commensurate with that he enjoyed outside his native land.[80]

Pfleiderer's critique of Herrmann's work came at the end of a two-part overview of contemporary theology entitled "Silhouettes from the Religious Science of the Present." Pfleiderer defined the beginning of the present as the resurgence of "modern realistic empiricism," whose impact on the science of religion had been to deprive it of any truth of its own in favor of empirical studies of the religions of the world.[81] Although he began with Feuerbach, Pfleiderer's major interest was the implication of the rise of neo-Kantian thought on contemporary theology.

From a theoretical point of view, according to Pfleiderer, Feuerbach and the neo-Kantians had a great deal in common. Feuerbach rejected the approach of Hegel, who, "because religion is truth . . . had the task of mediating the truth of religion in its full objectivity via reason." Feuerbach made religion a subjective matter, opening the road to a skepticism whose danger was "hidden by the rose bushes of beautiful feelings and experiences of the heart." For neo-Kantians, also, religion belonged to the domain of the practical, having nothing to do with the objective world of the actual. The resulting dualism was different from pure materialism only because neo-Kantians tried to remain aware that the mechanical worldview they did embrace existed solely on the level of phenomena.[82] They believed they could cling to "the rock of duty," which Friedrich Albert Lange had declared was protruding "from the sea of doubt and despair." But, cautioned Pfleiderer, "it is only a proper rock if it rests on the firm ground of objective being." If it was but an abstract "Ought," completely loose from this ground and standing in no thinkable relation to it, then it was not a rock, but "a castle in the air and a mirage."[83]

Pfleiderer evaluated the general impact neo-Kantianism was having by citing Eduard von Hartmann, who described the dilemma facing neo-Kantians as they confronted the contradiction between the understanding and the heart. Either the understanding was right, in which case the protest of the heart originated in the dispositions of feeling from an earlier period in the culture and would disappear with time, or the heart was right, because it had grasped a higher form of truth with the unconscious reason of instinct than the understanding did with its abstract discursive reflection. In the latter case the understanding would ultimately recognize the ideals of the heart as truth; hence in neither situation was the contradiction permanent. Pfleiderer's citation of Hartmann was close to a conscious rejection of all alternatives to a correspondence theory of truth, for Hartmann denied that the contradiction between heart and understanding "follows with necessity from our mental organization." It represented rather what he called the conflicts of a transitional phase.[84]

When Pfleiderer turned specifically to neo-Kantian theologians Herrmann received the great bulk of his attention. It was above all due to Ritschl's theology, Pfleiderer began, that modern neo-Kantian thought was experiencing favor in theological circles. He quickly set a polemical tone of his own to match the one he had sensed in Herrmann's exposition. The *Metaphysics in Theology* had been written especially against Dorner and himself because they were Swabians, for Swabian theologians were known to possess a speculative bent. Moreover, Herrmann's tortured German must have been designed to hide his ideas, because it first had to be translated into honest German before it could be understood.[85]

When he came to the substance of the work Pfleiderer observed that from its very beginning it was the same old neo-Kantian song about metaphysics having to do only with a knowledge of what was actual and religion only with the moral so that the two had nothing in common. That Herrmann and Pfleiderer were speaking different languages became evident when Pfleiderer assumed that Herrmann's characterization of the task of metaphysics, which he quoted in Herrmann's own words, exposed the fallacy that laid at its foundation. There was no hope of mutual persuasion of either author by the other, so different were their respective starting points. Pfleiderer could only react with horror to Herrmann's contention that it was all the same to religious individuals whichever dogmatic metaphysics or philosophical perspectives they consulted. He merely quoted long sections from Herrmann's work as if they by themselves clearly condemned their author. Pfleiderer confessed to being "truly astounded" at the degree of self-contradiction Herrmann's approach contained: "Can philosophical concepts be used like empty boxes into which one can stuff the most diverse contents equally well? How can anyone, with even the most superficial knowledge of recent philosophy, fall back into the medieval scholastic view of a purely *usus formalis* of philosophy in theology?"[86]

The complete inability of Pfleiderer's nineteenth century theological perspective to grasp the twentieth century mentality Herrmann was helping to introduce became explicit when Pfleiderer addressed Herrmann's conception of truth. To him Herrmann's view represented a return to the old medieval doctrine of the totally separate "double truth" of philosophy and religion. Ever since the Middle Ages there had been nothing but "the skepticism of undermining nominalism (empiricism)"[87] behind this doctrine. For a while the inconsistency of this teaching with the true faith of the church lay hidden, explained Pfleiderer, but with the radicalism of the Renaissance its mask had been torn away, exposing its paganism to all. Now neo-Kantian thinkers had made the ultimate meaning of the dualism plain and clear: "the simple negation of all objective religious truth." Pfleiderer struggled to understand where Herrmann could possibly be coming from, but his conception of truth as correspondence with the objectively real made success impossible.

> It is really difficult to grasp how one can prize "doubled truth" . . . as the saving anchor of religion and theology. "We mean by the truly real in Christianity something totally different from [what it means] in metaphysics"—what kind of genuinely reasonable response are we supposed to make to that? Is there something else outside the truly real than the not truly real? . . . As there is but one reason, so there is but *one* truth, but one world of the real, be it

sensually or intellectually real, which together constitutes the entire *one* reality, the object of our reasonable thought . . . Only the sophism of the two-fold, half-skeptical and half-believing neo-Kantians expects of reason the torture of uniting in one and the same conviction a doubled worldview that is completely indifferent and in part completely contradictory one to the other.[88]

In the *Metaphysics in Theology* Herrmann had attacked Pfleiderer's "rationalism" in the context of the latter's understanding of the person and work of Christ. Drawing on Pfleiderer's understanding of Jesus as the means but not the ground of salvation, Herrmann had asserted that mixing up the theological and metaphysical tasks in Christology was even more condemnable than it was where other doctrines were concerned.[89] Naturally Pfleiderer could not sit idly by in the face of what he heard to be an accusation of Pelagianism. He replied with a charge of heresy of his own, observing that Herrmann did not know what the historical concept of rationalism entailed, and adding that what Herrmann put in its place was "quite simply narrow-minded positivism, which necessarily results from the Ebionite incapacity to raise oneself above what is historical in the appearances to what is eternal in the metaphysical principle." But the positivism that came from allowing the Christian religion to be consumed in an ethical judgment of the world that might well stand in contradiction with our theoretical explanation, Pfleiderer continued, possessed no criterion of truth other than "trust in the superficial testimony of human authority," which to Pfleiderer amounted to dogmatism.[90] Pfleiderer, true to his age, wanted to be constrained by the single truth of reality, not master over it.

Had it not been for Herrmann's extension of the antimetaphysical stance Ritschl had adopted in his interpretation of justification and reconciliation into a thoroughgoing attack on the assumptions common to all the theological traditions of the day, it is highly doubtful that Lipsius, Graue, and Pfleiderer would have begun referring to a "Ritschl school" as they did. Others were being drawn to Ritschl's focus on the ethical nature of religious experience. Although none merely adopted Ritschl's position *in toto*, eventually the ranks of the Ritschlians would include the theologians Julius Kaftan (Berlin), Hermann Schultz (Göttingen), Johannes Gottschick and Theodor Häring (Tübingen), Max Reischle and Ferdinand Kattenbusch (Halle), Hans Hinrich Wendt (Jena), and Otto Ritschl (Bonn).[91] The historians of religion Adolf Harnack and Friedrich Loofs were also influenced by Ritschl and are frequently identified with the Ritschl school.

But it was Herrmann who first exposed the implications of Ritschl's approach for religion in its relation to science. If it required impetuousness, enthusiasm, and a certain recklessness to take the offensive against the theological establishment, the youthful Herrmann was just the person needed. With his next contribution he made it crystal clear that the denunciations of his efforts had hardly been the last word.

7

The Existential Critique of Science and Theology

IF HERRMANN WAS to pursue seriously what he had begun, he could not do so as a mere *Privatdozent*. All of his critics had been quick to point out his inexperience and lower rank as a licentiate in theology. Herrmann changed that with the publication in 1879 of his lengthy study *Religion in Relation to Knowledge of the World and to Morality,* which gained him the status of full professor and resulted in a call to the theology faculty in Marburg. Later in his career Herrmann left behind these early critical analyses of religion and science in favor of a more positive concern with religious community; hence relatively little attention has been paid to the 1879 work.[1] Because of its focus on cosmology, however, it addresses precisely the concerns that are relevant to an investigation of the loss of nature to theology at the end of the nineteenth century. In addition, Herrmann's 1879 work revealed how much more capable he was than Ritschl of understanding and articulating a fundamentally new foundation for theology.

THE PROVISIONAL NATURE OF SCIENTIFIC KNOWLEDGE

Herrmann's first concern was to assure the reader that just because he believed what was valuable to the Christian community lost its original sense if made into an object of worldly knowledge, that did not mean he rejected all scientific grounding of Christian faith. On the contrary, it was the task of the theologian to justify the generality of belief. Herrmann reiterated, however, that in his view this could not be done when the position of faith was examined as the raw material of metaphysics.[2]

In a fascinating historical account of how theology had come to be dominated by metaphysical concerns, Herrmann laid blame at the feet of the medieval church. The attractive and persuasive power of the church had always been its capacity to present to humanity's moral spirit a world it could recognize as its own. Like the existential thinkers to follow him, Herrmann portrayed the natural world as a hostile place for humans, one in which the characteristics that made them uniquely human also estranged them from the world. For Herrmann it was the human moral capacity that contradicted and opposed what Camus would later depict as the benign indifference of the universe. But religion's ground in morality was lost during the Middle Ages.

How and why had this come about? The medieval church, wrote Herrmann, "stamped out the character of an institution of salvation so energetically in its doctrine and disposition that the real religious task of the church, to represent the reconciled community, only came into play surreptitiously."[3] Herrmann had been influenced by Ritschl, whose similar concern with the here and now was everywhere evident in his *magnum opus*. Ritschl had once written to Diestel that it was clear to him "that in dogmatics the idea of redemption is only properly interpreted as a means for the highest purpose of the kingdom of God."[4] Throughout his career Herrmann made the kingdom of God, understood in this manner, the foundation for "regulating," as he had put it to Ritschl, "the relation of humans to the world."[5]

Although in the Reformation Luther's emphasis on justification by faith opened the door to a proper valuation of the moral basis of religion, the opportunity remained unexploited. As a result, according to Herrmann, "moral ideas, which originally were the property of the religious community, were placed in opposition to the church under the title of free humanity."[6] Herrmann faulted theologians for this secularization of religion, for they had failed to seek the foundation of religious truth in the life of the moral individual. Instead, they preferred to base their claim of the universal validity of religion on another spiritual power which they thought was not subject to the variations of subjective experience: the philosophical explanation of the world.

Although Herrmann identified the promise of objectivity as the attraction philosophical explanation held for theologians, that factor did not account satisfactorily for the historical development of church doctrine. *Why* was the prospect of an objective explanation of the world so alluring to the medieval church? Herrmann answered this question in social terms: the church wished to exercise control, and it could not do so if the basis of religion was the moral individual. If the gospel message as Herrmann understood it regulated all aspects of human knowledge,

> then one has to renounce seeking the validity of religious truth in its agreement with the knowledge of the world in natural science and metaphysics. The idea that this agreement also designates the limits of what is possible for the objects of faith was tolerable only as long as the church dominated the world and the knowledge directed to it through the arm of the state. For the domination of the world that the church exercised included the authority to prescribe its ways to our [knowing] of the world. [When] science is led from a purely religious point of view [it] expresses church doctrine much more as a limitation of that domination.[7]

Herrmann had already described the negative impact he believed metaphysics had on theology in his earlier work. Here he noted that as long as the church possessed external authority in society, it had been able to control by force the consequences of metaphysical propositions when they clashed with the interests of the religious community. To continue to insist that the religious community required legitimation by demonstrating the commensurability between Christian truth and secular intellectual culture, which Lipsius and others demanded, was thoroughly misguided according to Herrmann. Only a church that still wanted to rule the world interested itself in identifying the power of knowledge, which was "religiously indifferent," with religious faith. Such a stance was, however, inherently suspicious, since it drew the state into the religious community.

In the Introduction to his work Herrmann presented the general conclusions which he would explain more fully in the text. But he did not want there to be any doubt about his assessment of the situation facing the church in Bismarck's Germany. The unity of science and religion sought for and promised by so many had never been achieved.

> The theological proof for the general validity of the Christian religion produced [by this means] is only tolerable under the presupposition that the church rules and must rule as an external power in the world. In the Protestant church, where this presupposition is no longer valid, it has become completely pointless. It seems therefore to be justified to undertake the proof that Christianity is the universal religion by other means.[8]

More than any other member of the emerging Ritschl school Herrmann took upon himself the task of demonstrating to his contemporaries that the conclusions of natural science were necessarily provisional. While the effect of his philosophical analysis of science was more sharply critical than that which scientists like Ernst Mach and philosophers like Hans Vaihinger had carried out, their common dissatisfaction with what they deemed to be the widespread misunderstanding of the nature and scope of mechanical explanation unwittingly united them in expressing an unease with the naive overconfidence of their scientific age.

Herrmann's approach followed a definite strategy. Leaning heavily on Kant's critique of reason, he began by making a fundamental distinction between pure knowing *(das reine Erkennen)* and scientific knowledge of nature *(das wissenschaftliche Naturerkennen)*.[9] By "pure knowing" Herrmann meant the capacity of the conscious mind merely to make a representation of the world. It was the mental activity that commenced as soon as consciousness came into existence, that is, prior to other dimensions of the human psyche such as emotion or will. Of course this knowing could not exist apart from an experience of things and of their relation to one another; indeed, one could know nothing of the laws governing this activity or of oneself as a subject with a unified consciousness were there no content providing an occasion for the activity to commence.[10] Herrmann credited Kant with the distinction. Not only had Kantian epistemology made clear that pure knowing was uninfluenced by feeling and the will, but Kant's achievement meant no less than winning for theology the possibility of existing as an independent science alongside philosophy.

> By isolating the functions of pure knowing [Kant] made clear the impossibility of developing or coming to religious conviction

through [philosophy's] help . . . The interpretation of the world for humans as knowing subjects is totally different from that as moral persons. This knowledge, gained by Kant, vindicates on the side of exact science [theology's] obligation [to be] independent . . . On the other hand theology is released into freedom by this Kantian achievement.[11]

Herrmann proceeded to explain how pure knowing worked by reviewing Kant's exposition of the role of spatial and temporal intuition and of the concepts of substance and causality in knowledge. Humans were not free in pure knowing to refuse to employ these categories; hence pure knowing possessed a dimension of objectivity. By the same token, when consciousness, using such categories, unified a plurality of sensations into an object, it was Kant's contention that the result was not a product of sensation but an act of the understanding. Herrmann explained that in Kantian epistemology we know nothing at all about the *origin* of this unifying function of consciousness; we know only the results of our conscious activity.[12]

Herrmann undertook this exposition in order to establish a foundation for his eventual thesis that scientific knowledge, like metaphysics, could never be more than provisional knowledge. Pure knowing, as already observed, required empirical sensations for us through our conscious activity to become aware of its existence. But in spite of the unified objects of perception that emerged in pure knowing, there was another vital sense in which the world of pure knowing was indefinite. Herrmann followed the anti-realist coherence theory of truth initiated by Kant, which holds, in the words of Ralph Walker, that "propositions that cohere within the system do *not* form a determinate totality."[13] Scientific knowledge of nature is of necessity forever in flux.

> Our ideas of objects are not completely resolved into the same forms by the connecting and relating activity of consciousness. There is always left over . . . a manifold of modifications of consciousness that are intuited in spatial order . . . When we call the intuitions ordered according to the laws of knowing Nature, this Nature is not a completed entity for consciousness which is determined on its own; rather [it is] a plurality growing into ambiguity of which consciousness must seek to become master through its

unifying functions. The possibility of discovering new characteristics of nature goes into infinity, just as [do] the combination of elements in space and the definition of the relations of relative dimensions in mathematics. Further, if the knowledge of nature is directed to determining objects and the changes of their states as completely as possible, no definite limits can be imagined for this activity. It lies in the nature of our concepts that our attempt to complete the representation of an object can never totally be drawn to a close.[14]

Herrmann went on to deny that the distinction between substance and accident could be maintained in any final form. Quoting Kant, he observed that in all cases there would always be a broader context to which the predicate of an existing thing could be referred.[15] Pure knowing, without which the individual would not be a unified consciousness, had as its task the establishing of lawful relationships in the discharge of representations, none of which could be considered final. The ordering activity of pure knowing must be ready to begin anew at any time; that is, it must treat its product as having only "hypothetical validity."[16] Pure knowing, then, was the vehicle by means of which humans achieved the unity of consciousness, but by itself it did nothing more. It had no direction, so left to itself it would simply respond to experience of the world by unceasingly uniting sensations into ever new combinatorial products, not even being aware of the limitless nature of the possibilities.

Natural science occupied an important place in Herrmann's overall epistemological schema, although his discussion of it can easily leave the first-time reader confused. One does not realize that Herrmann wished to rely on a distinction between what he sometimes called "pure natural knowing" or "natural science remaining purely in its mode" and a more practical or applied understanding of the scientific treatment of nature. Like Fries, Herrmann had read Kant's endorsement of the notion that as in mathematics the possibility of new combinations continued forever, "in like manner the discovery of new features of nature, new forces and laws [proceeds] through continued experience and amalgamation of them by reason."[17] Kant's emphasis in this connection, which Herrmann repeated, was that natural scientific knowledge was restricted to phenomena and had neither a point of contact with metaphysics or morals nor needed one for its explanations. Herrmann commented on these sentiments from Kant that here "the independence of natural science from that which does not belong to the production of the object from the diverse [contents] of intuition is declared with the greatest decisiveness possible."[18]

But if Herrmann separated pure natural science from metaphysics so clearly in this passage, he did not make the same qualitative distinction when he later spoke of what might be called "practical" as opposed to "pure" natural science. For it was also natural science, according to Herrmann, which first made us aware that the products of pure knowing would reproduce in different guises forever. Here natural science involved a different kind of knowing, an intentional knowing that gave direction to our pure cognitive capacity. In the ordering of representations in pure knowing, wrote Herrmann, there was no requirement that one result "confirm" or go together with another. But practical natural science did presuppose a homogeneity of representations. "For only if this occurs do we experience the regularity of events that the tabulation of particular natural laws, which is the solution to the task of natural science, makes possible."[19]

Herrmann specified the new context in which he was now viewing natural science by identifying a difference in the way practical science commenced.

> When therefore natural science proceeds on the presupposition that nature according to its special laws represents not an incalculable multiplicity, but an order in keeping with our power of comprehension, then this presupposition of the coherent comprehensibility of nature is obviously constituted differently from the general judgments about things . . . which consciousness follows in representing things.[20]

The search for natural laws, according to Herrmann, required more than a unified consciousness. Herrmann drew on Lotze's characterization of the genuinely human subject when he declared that "the fundamental hypothesis of the scientific explanation of nature, the hypothesis of the comprehensibility of nature, is only possible for feeling and willing beings."[21] Much later he would write: "Where the courage of hypothesis is wanting, there is no will to know and consequently no possibility of knowledge."[22] Now the entire human psyche was involved, but in so doing it introduced what Herrmann had called a necessary prejudice *(Vorurteil)*. The prejudice, which was also the source of practical natural science itself, was present not because consciousness had to know nature to be unified but "because the feeling and willing person wants to be master of nature." It was the purposeful individual who, in imposing this requirement on nature, also placed limits on our knowledge of nature by making it inherently dependent on human purposes. In words which those of his time must not have understood, Herrmann wrote as if he were letting out a secret. "We hold nature so constituted to be real not because we know it [to be so], but because we want it [to be so]."[23]

Nineteenth century scientists were not used to thinking of the reality they were uncovering as something affected by their desires. For Herrmann such a situation was not only unavoidable, but it helped to define who humans were. Humans were persons only to the extent they exercised the capacity to realize value in feeling and willing; in fact, the idea of a coherent world was to him inseparable from personal existence, since without the conviction of its reality one's own existence was meaningless. Herrmann did not hesitate to draw a bold conclusion: we become certain that the world really exists as a coherent whole not because of concepts of the understanding but "because a solid bond between it and our individual existence is established by a value judgment." Herrmann's reliance on the earlier conclusions of Lotze and Ritschl was as clear as was his resonance with the categories of existential thinkers to come.[24] It was the power of the subjective impulse to which the representations of the mind ultimately owed their origin. Although that was no reason to discard them as imaginations *(Einbildungen)*, when we believe our own existence to be something which, because of the primary role of value, is closed to itself, "we treat the cognitive activity of the representing consciousness as a means [serving] the purpose of the person."[25]

Herrmann acknowledged that a problem arose for science when the needs of the feeling and willing person resulted in representations being treated as absolute when in fact they were by their nature relative entities that inevitably lost themselves again in the deeper context. He approved of Kant's solution to this difficulty; namely, the scientist must transform such representations into regulative ideas. "In this form they are nothing more than an expression of the confidence that the world which we claim as a means for [realizing] our purpose can be coherently explained."[26]

Herrmann's evaluation of natural science to this point resulted in the provisionality of its knowledge of the world. As for the question of truth, one could say that he argued against the claim that the propositions of theoretical or practical natural science corresponded to reality in any direct or necessary manner. Pure natural science was not constrained at all. Practical science, which did experience constraint, was directed not by nature's one truth but by the profoundly subjective influence of value, specifically the desire to control and master nature.

Herrmann was well aware, however, that his way of looking at scientific knowledge was not shared by his contemporaries. He realized that the limitations he was imposing on knowledge due to his neo-Kantian perspective were unacceptable to those for whom truth had to be accessible and could not be defined solely in pragmatic terms. To Herrmann their error lay in the failure to appreciate the pervasive role of value where the acquisition of knowledge was concerned. To persuade his opponents, he returned once again to a discussion of the significance of metaphysics.

The Certainty of Religious Truth

According to Herrmann, theologians like Strauss, Zöckler, and Schmid all took it for granted that there was a higher order of knowing in addition to the knowledge of objects. Herrmann thought he knew where this presumption had originated and, although he himself was no scientific realist, he explained it in terms remarkably similar to the objections of twentieth century scientific realists to an anti-realist interpretation of science. Must we, asked Herrmann, justify our conviction that nature is ultimately rational through "blind trust in happy coincidence?"[27] Herrmann understood that for many happy coincidence did not sufficiently confirm *(erhärten)* one's right to assume the possibility of a coherent explanation of nature. The alternative to which they appealed, however, was metaphysics.

Metaphysics, which to Herrmann involved the endeavor to determine the final essence of things, *would* increase our knowledge if the concepts it brought together were unchangeable. As he had already argued, however, perceptions were never final and complete, and the coherent propositions of science do not form a determinate totality. He pointed to the progress of empirical science in his own day as testimony to the necessity of modifying or possibly even setting aside concepts which up to then had served well to explain nature. Empirical research, he said, had a way of decomposing any system of concepts that aspired to finality; indeed, the vitality of scientific concepts depended on their hypothetical character.[28]

One sees here why Herrmann associated metaphysics with many of the propositions of nineteenth century natural science so readily. To him *any* manipulation of scientific concepts for other than strictly pragmatic reasons fell under the category of metaphysics because it presumed to provide knowledge "of a higher order" than that available through the understanding. Herrmann might sympathize with those who could not be satisfied simply to accept the mystery behind successful predictions based on scientific theory, but he was adamant that they not be mistaken about the hypothetical foundation on which any such account rested. Any effort to go beyond and, as he repeatedly put it, "to complete" knowledge of phenomenal objects was metaphysics. Since natural scientists in his day did not understand this, they also could not have understood why Herrmann viewed a great many of their claims as metaphysical.

Of course Herrmann was not opposed to metaphysics; he merely objected to the misunderstanding of its nature. What responsible metaphysics could do was to make clear how the presupposition that nature was ultimately rational imparted a "characteristic coloring [*eigenthümliche Färbung*]" to "the immediate conceptual apparatus of natural science." But metaphysics must not be taken as a real completion of the limited scientific knowledge of the world.[29] While Herrmann understood that metaphysics had its origin in the feeling and willing person, its attempt to go beyond pure reason by providing its own kind of practical explanation of the world was fundamentally misguided if it tried to furnish a cognitive account of the noumenal realm. The scientific explanation of nature was limited to the ultimately arbitrary manipulation of phenomena, and that was as far as cognition could extend.

Nowhere, however, had Herrmann replied adequately to his own question. Was nature susceptible to rational analysis solely because of "happy coincidence?" His strong anti-realism was too much for even some of the neo-Kantian philosophers of the time. In his review of Herrmann's book the philosopher Paul Natorp took him to task precisely on this point. Natorp complained that one could not justify the idea of eternal natural law, which did not arise out of experience and yet was something which the natural scientist could not simply take or leave, through reference solely to practical reason. Admittedly Natorp's understanding of natural law, unlike Herrmann's, was as a regulative idea that marked the goal "to which our knowledge approaches as it were asymptotically."[30] Natorp's dissatisfaction with Herrmann's anti-realism was similar to that of any realist. For Herrmann success in science must in the end be simply a miracle.

To reiterate, what inspired the investigations of natural science, according to Herrmann, was the desire to control nature, whereas the motivation for metaphysics was a dissatisfaction with the inconclusive nature of scientific explanation and with the insoluble mystery of the natural scientist's capacity for success. The fundamental role of the human will in both natural scientific and metaphysical undertakings was undeniable in Herrmann's schema. To him the irreducible foundation of the human spirit was its ability to employ the will in making value judgments on which all subsequent mental activity depended. In the last analysis cognition of the world was a means employed for the purposes of feeling and willing human beings.[31]

While it was true that both mastering the world by means of what Herrmann called the "mechanical" explanations of science and seeking to come to a complete or final scientific account in metaphysics ultimately depended on the directive function of human value, neither expressed for Herrmann the essence of being a person.[32] To accomplish this one had to guarantee the preservation of the person as such over and against the world. It was at this point that humankind employed a different means of going beyond pure reason to the realm of the practical. In religion, according to Herrmann, one was not content merely to demonstrate the subordination of the world as it is known through the natural science of a particular age, for example, the nineteenth century, to the human value of utility.

> On the contrary, where the preservation of the individual in general is concerned, the power involved in the presupposition about the essence of the world is meant to be one which subjects the world, however it might be, with hidden force to the *highest* purpose humans have. When the conviction of the reality of this power dominates the entire intellectual life of the individual, then that person has religion.[33]

Put succinctly, both natural science and religion were dependent on value, but they drew their lifeblood from very different values. Like Ritschl, Herrmann acknowledged that human beings existed in a physical environment to whose course they were subject. But as in the case of human cognitive powers, such physical powers were not of value in themselves and were important only "because of their relation to the specific purposes of persons."[34] The highest purpose a person could conceive was met in religion, for there humans declared their supreme value by forming the idea of a power dominating nature in their behalf, directing events toward their preservation whether or not they appeared definitely fixed. Later Herrmann explicitly contrasted his personalistic view of God and the universe to the impersonal portrait that had so moved David Friedrich Strauss.[35] Here he did not shrink from a bold conclusion, which announced what he thought about claims to truth. "From this it follows incontrovertibly that religion's practical explanation of the world must step forth with the claim of absolute truth. Since its judgment about the world expresses what is unconditionally valuable for humans as the power over the relations which rule him as a natural creature, religious conviction is unthinkable without that claim."[36]

In his own way Herrmann recognized what Kuhn has called "the incommensurability of differing paradigms," for Herrmann noted that what was absolutely persuasive to him would hold no court at all with those who were not grasped by the primal datum *(Urdatum)* of personal value. Although he suggested that doubt should be impossible to the religious mind since it was equivalent to doubting the certainty of one's self, he confessed that it was easier to view the world of cause and effect as the "real" world because its appearance derived from and was determined by the nature of the representing consciousness itself. By comparison, the world of purpose appeared fickle, wavering, and unpredictable, and one could not avoid wondering whether one had created the world of purpose because one so desperately wanted it to exist.

Herrmann resisted such doubt first by acknowledging that it was unavoidable if one treated the world of purpose like the world represented in science, and then by denying that one had to succumb to that temptation. The two worlds were incommensurable, regardless of how much metaphysics tried to connect them. Herrmann identified two false paths of metaphysics in theology, religious orthodoxy and religious liberalism, both of which undermined the persuasive power of religion by refusing to stand or fall with the "wavering" world of purpose alone. By insisting that scientific truth conform to its predetermined explanation of the world, orthodoxy not only mingled incommensurable worlds but denied freedom to science. Liberal theology continued the fruitless struggle to find the "essence" of religion that would forever elude the conceptual articulation of science. To Herrmann neither path realized that the gratification of science was momentary because in the last analysis it was utilitarian, and utility could not compare to the confirmation religion imparted to the personal soul.[37]

It was a huge mistake, according to Herrmann, to try to fight the widespread scientific materialism of his day on its own terms. In religion humans excluded the world of nature, which opposed their personal significance. Religion was not at all their product as creatures, nor did it result from their formal intellectual activities. True, both religion and metaphysics were "practical" explanations of the world because both sought to orient the individual to the larger context that lay beyond pure reason. But, Herrmann declared, it was easy to test whether our attempts to find unity amid diversity was a response to a religious or a metaphysical want. One need only check to see if the judgment made regarding the plurality of things emerged as a hypothesis, which occurred in metaphysics (the utilitarian hypothesis of the comprehensibility of the world), or as dogma, which revealed the presence of religion.[38] In a later essay Herrmann explicitly defended the right of dogma, which was based on the value of the preservation of the person, to supersede the approach of science in spite of the criticism and derision such language was bound to evoke from the scholarly community.[39]

If Herrmann reminded the theologians and natural scientists of his day of the papacy with his open defense of dogma, his detractors made the exact same impression on him. In their defense of the compatibility of science and religion they were not only unaware that the point of natural science was the practical mastery of the world, but also passed off the metaphysical worldview of science as if it were knowledge comparable to that obtainable on the level of phenomena. Herrmann referred to the response of a theologian who had been asked why the Protestant Union was not erecting a new confession of faith. It was because, the unnamed spokesman had said, natural science and philosophy had not yet brought their enterprise to a conclusion. To Herrmann this meant that one was being asked to suspend religious judgment until science and philosophy were able to finish their work,[40] a request that stood in complete contradiction to his anti-realist understanding of truth through coherence.

What irritated Herrmann as much as anything was the lack of humility many exhibited in the face of the human condition. Pfleiderer had mocked Herrmann's assertion that what was truly real in Christianity was wholly different from what it was in metaphysics. "Is there then beyond the truly real," Pfleiderer had written, "something other than the not truly real, i.e. the unreal, the purely imagined, the fantasy world of abstract ideals?"[41] In reply Herrmann could only shake his head in frustration at Pfleiderer's inability to recognize that others may see things differently from him. Pfleiderer and anyone else, Herrmann noted, who conceived of truth as something complete which one then sought to discover, *had* to latch on to such considerations. But to Herrmann they still amounted to presuppositions in the face of life's mystery. Natural science was to be left in complete freedom precisely because only then could it serve the purpose of the Christian community—the realization of the kingdom of God. Herrmann illustrated what he meant by the doctrine of creation *ex nihilo*. The point here was to put an end to all attempts to explain the world as it existed by reference to God. The ancient cosmogonists had understood what Pfleiderer could not grasp: the meaning of the Christian doctrine of creation was not to be understood as something qualitatively similar to the rest of our knowledge.[42] In the end Herrmann preferred the attitude of his mentor. "Kant used to say that one must not squander the name of God on the concept in which the scientific explanation of the world seeks to bring itself to conclusion. Many Christian theologians have no understanding of this reverential awe of the Holy."[43]

In the remainder of the book Herrmann changed directions. First he explained how the scholastics' dependence on Aristotle had abnormally "bracketed" Christianity in a manner that confined the ultimate ground of religious experience to an explanation of events in the world.[44] The bulk of his work, however, was devoted to developing the meaning of the moral foundation of religion, repeating the theme of the inaccessibility of a moral explanation of the world to a metaphysical one since the latter had to rely on concepts and their manipulation in science. Herrmann spent considerable time dealing with the historical unfolding of Christianity, including Kant's role in establishing the fundamental importance of the moral law for persons. Kant, however, had mistakenly identified the moral will as something totally independent of history, and as such Herrmann did not believe it could serve as the source of religious certainty. It was Schleiermacher who had corrected Kant's error by bringing the will back into direct relation with the historical person of Christ, thereby contradicting the notion that the moral will had no need of a revelation of God.[45] For the most part, however, post-Kantian theology had drawn on Kant's understanding of the moral will through reference to his emphasis on moral improvement, placing that alongside a continued employment of metaphysics as if both it and morality supplied similar kinds of support for faith. "That Kant's proof has as a presupposition the profound conviction of the uselessness of metaphysics, and that in [his proof] the value of religion is led beyond [his] moralizing purpose, to this no regard was paid."[46]

With the completion of *Religion in Relation to Knowledge of the*

World and to Morality Herrmann had communicated everything he had to say about the relationship between science and religion. Those who read his second work could not have missed the explicit rejection of the assumptions of all three of the major theological schools of the time. The book was impressive enough to earn Herrmann a call to Marburg, thereby freeing him from the stifling confines of Halle. But it would be the young generation of theology students who would come to Marburg in the 1880s to study with Herrmann that heeded his rejection of traditional theology.[47] Even then the students were drawn to Herrmann as a leading representative of Ritschl's theology. But it was only after Herrmann's two publications that the new theological tendency announced by Ritschl and articulated systematically by Herrmann emerged on the scene as a defined school.

THE REACTION TO HERRMANN'S EXTENSION OF RITSCHL'S THOUGHT

Herrmann's second work was reviewed less by professional theologians than it was by philosophers and others. Reference has already been made to the response of the neo-Kantian philosopher Paul Natorp. While complimenting Herrmann for his erudition, theological laymen were cautiously suspicious of his effort. The evaluation of Prince Ludwig zu Solms revealed that the radical nature of Herrmann's message had not come through to everyone.

In the last composition he wrote Solms pleaded for peace among contentious factions in the church. Citing "the great gain" of identifying and resisting unjustified metaphysics, which Ritschl had brought to theological science and which younger scholars were more and more recognizing, Solms argued that there was a justified as well as an unjustified use of metaphysics for theology. Although Solms concurred that overstepping the boundaries of knowledge through the power of the imagination landed one in unjustified metaphysics, he tried to justify a metaphysical activity in which one distinguished among cognitive results and made judgments regarding their value while remaining within the acknowledged limits of reason. In the end he failed to see what divided Herrmann and Lipsius. He rebuked Herrmann for his criticism of Lipsius, noting that their "complete agreement" on basic presuppositions was due to Herrmann's use of Kantian language in theology.[48] Like Lipsius, Solms assumed a correspondence theory of truth, arguing that the philosophical and theological approach would have to arrive at the same goal, "because truth can only be one."[49]

Others were more critical of Herrmann. From Tübingen came an endorsement of the need to purify dogmatics of foreign elements with a concern that Herrmann had gone too far. "What should one say if natural scientific principles and systems are attacked in the name of theology?" asked a reviewer in the *Theologische Literaturzeitung*. Herrmann's extremism gave theology the appearance of flight into an unhealthy dualism. Why was it important to separate nature and morality so exclusively? Just because our trust in God was personal, did that mean that God ruled the world without law? It did not matter to faith that nature did not conform to moral law; it was only important that nature served the highest good in accordance with God's will.[50]

Four to six years after the appearance of Herrmann's *Religion in Relation to Knowledge of the World* the book was still being reviewed in the journals. In 1883 the theologian Alfred Krauss published an open letter to Herrmann, who was by then well established in Marburg. Krauss was disturbed by Herrmann's division of reality into an unbridgeable opposition between nature and moral personhood. He resented Herrmann's restriction of nature to the status of a means sharing no common ground with morality, and he rejected Herrmann's characterization of moral personality as "the truly real" in comparison to nature. Personality to Krauss was a part of the world, part of the truly real. What contradicted the truly real was not nature but one of two possibilities: the truly unreal, nothing, or the untruly real, illusion.[51]

Since Krauss's perspective was built on the assumption that independent external reality forced our knowledge into correspondence with it, Herrmann's position was upsetting to him. The *conditio sine qua non,* wrote Krauss, was that religion also had to do with empirical data, that morality rested on objective relations independent of human formation of laws. Only the empirical reality of the relationship to God guaranteed that this relation was truth and not illusion. "As little as humans make natural laws, so little do they make moral laws." The moral order independent of us could only be investigated. Since both nature and morality were independent of human creation, their common subjection to law reinforced their ultimate unity in God rather than their essential opposition. Krauss's conclusion was simple: "No metaphysics in religion means—no religion."[52]

Krauss accused Herrmann of creating a new metaphysics in place of that which he wished to remove from theology. He conceded Herrmann's contention that our endeavor to know the world would never come to completion but, he maintained, we could approximate the truth. Krauss demanded that we use our approximation as a spur to further work, thereby guaranteeing "the unceasing progress of development and, at the same time, the preservation of piety and humility."[53]

This kind of opposition clearly brought greater attention to Herrmann's views. When in 1885 Lipsius published three long essays in which he attempted to distinguish his position as a neo-Kantian from that Herrmann had laid out in his 1879 work, it was apparent that Herrmann was emerging as the most articulate theological spokesman for a new outlook. Lipsius himself pointed out that Herrmann had progressed far beyond the status of a mere student of Ritschl.[54]

What bothered Lipsius more than anything else was the radical dualism he sensed was catching on because of neo-Kantian thinkers like Herrmann and the philosopher F. A. Lange. Although he agreed that from an epistemological point of view one could just as well interpret the universe in personal as mechanistic terms,[55] he could not approve of the complete and mutual independence of "the world of reality" and "the world of values" that Herrmann and others were urging. And yet much of what Lipsius himself had written could be mistaken for dualism because his neo-Kantian language sounded like Herrmann's. As already noted, many perceived Lipsius and Herrmann to be saying the same thing and simply could not grasp why Herrmann had attacked the Jena theologian in the first place. If Herrmann's critique of Lipsius had failed to communicate their difference, so too did Lipsius's belabored rejoinder ultimately miscarry.

Although this in-fighting among neo-Kantian theologians bewildered their contemporaries, the differences in their perspectives become easier to comprehend if one asks about their respective positions on nature's truth. Lipsius struggled, as had Kant himself, to balance the anti-realism implied by our limited capacity to know nature, which Walker has called a version of the coherence theory of truth, against the realism, unexpressible in concepts, entailed in the acceptance of things-in-themselves. In the end Lipsius chose to side with realism, emphasizing the element of Kant's perspective which stressed the constraining effect of that which is *given to* the subject. As Lipsius said of his efforts to clarify his relation to Herrmann: "The following exposition rests in essence on Kantian epistemology, but, [starting] from [Kant's] presuppositions, it attempts to reach an agreement with 'realism.' "[56] The sentiments of Isaiah Berlin at Oxford some fifty years hence would capture what Lipsius felt:

> True, we might never get to this condition of perfect knowledge—we may be too feeble-witted, or too weak or corrupt or sinful, to achieve this. The obstacles, both intellectual and those of external nature, may be too many . . . But even if we could not ourselves reach these true answers, or indeed, the final system that interweaves them all, the answers must exist—else the questions were not real. The answers must be known to someone: perhaps Adam in Paradise knew; perhaps we shall only reach them at the end of days; if men cannot know them, perhaps the angels know; and if not the angels, then God knows. These timeless truths must in principle be knowable.[57]

Lipsius did not deny Herrmann's claim that the universe, viewed in personal terms, brought with it an overpowering sense of certainty which could be "more firm" than that founded on scientific knowledge. The reality of moral and religious conviction must be experienced *(erlebt)* and cannot be proven *(erwiesen)*. Some theologians tried to deduce the former as dependent on thought in the same sense as was knowledge, assuming that only in this way one could attain "objective truth" or "objective reality." That, acknowledged Lipsius, was the fundamental error of speculative theologians. But he refused to conclude from this that the certainty found in the personal perspective corresponded to "a wholly different reality from the one which has to do with theoretical knowledge." This was the kind of thinking which gave rise to Herrmann's scandalous assertion that it made no difference to the religious task whether the dogmatic metaphysics adopted by the Christian was materialistic or idealistic.[58] Clearly there was to be no meeting of the minds between Herrmann and the neo-Kantian Lipsius. Where others were concerned the situation became even worse in the early 1880s, in part due to Ritschl's publication of his own *Theology and Metaphysics* in 1881.

For some time Ritschl allowed the work of Herrmann to stand as the supplement to his treatment of the positive fact of human freedom in the *Justification and Reconciliation,* with its implications for tempering the world of cause and effect. In 1878 Gustav Kreibig had written a book against Ritschl's understanding of justification and reconciliation, and a reviewer sympathetic to Pfleiderer's position had accused him of rationalism in the *Allgemeine evangelische-lutherische Kirchenzeitung.*[59] But in spite of these critiques and of the reception given Herrmann's works, it was not until the spring of 1881 that he could no longer leave it to Herrmann and others to lay out the new theological tendency his work had set in motion. In a letter to Harnack of April 9, he described how Herrmann Schultz's *Doctrine of the Divinity of Christ,* in which Schultz openly proclaimed Ritschl's influence, was not well received by pastors. In particular, an acquaintance of Ritschl had indicated his displeasure with it during a train ride to a conference in Hannover. "I tried to make clear to him," wrote Ritschl, "that only his Platonic realism [*platonischer Realismus*] was in jeopardy, that I was confident I could demonstrate a threat to Christianity in the continued acceptance of this wretched metaphysics in religion. Fourteen days ago I had already said in Herrmann's presence that it was necessary for us to make clear what it is we make use of against this metaphysics in our interpretation of the person of Christ."[60]

Herrmann, of course, had already taken up the challenge Ritschl had neglected to emphasize in his writing. Yet Ritschl's frustration over what he deemed continued misunderstanding now caused him to consider putting aside his ongoing work on the history of pietism to deal with the problem as he saw it. After brushing up on the appropriate sections of Lotze's *Microcosm,* he wrote to Herrmann in early April of his dilemma, alluding to the epistemology "which we follow" and the Platonism of the opponents "who constantly misunderstand us."

> It is the same theme of metaphysics in theology that I touched on perhaps six years ago and have left to you. You, however, have thought of the problem of cosmology; I meant the question of ontology, how one might have the thing to know. A whole series of circumstances has made clear to me the necessity of finally writing about this with direct exemplification of theological themes. I only wish I had the courage to take the matter on.[61]

By April 20 he had decided to do it. "Spener can wait," he wrote to Harnack. On June 6 Ritschl's own *Theology and Metaphysics* was finished, and in the fall it was published.[62]

The work, which has already been discussed above,[63] merely added flames to the fire. Ritschl was not the philosopher Herrmann was, and his foray into the field meant that he would eventually be taken to task for his misinterpretation of Kant and his confused epistemological claims. Pfleiderer delighted in attacking the master, for by exposing Ritschl's philosophical inadequacies he hoped to cast aspersions on the whole Ritschl school.[64] In his defense of Ritschl, Herrmann conceded that Ritschl's epistemology was not well worked out but noted that at least Ritschl had recognized what Pfleiderer had not—the immense significance of Kant's intention to make explicit the limits of science.[65] Pfleiderer ran roughshod over Kant's achievement, said Herrmann, in his arguments about the grounds of sensation and the grounds of religious feeling. Contrary to Pfleiderer, our certainty that the reality of the external world was grounded on our sensations did not rest on the same law of cognition as our certainty that the reality of God was the ground of our religious feeling of subjection. Pfleiderer must have become accustomed to imposing few requirements on the scientific knowledge of reality, wrote Herrmann, if he could see a sufficient proof for the reality of God in the perception sparked by such a feeling. To borrow categories from Thomas Kuhn once more, Herrmann appeared to recognize that he and Pfleiderer were talking across paradigms. "To me the indefatigable fury of his polemic against us is not comprehensible. We have nothing at all to say to one another. For the scientific means which put him in the position to attribute to his knowledge of God the same 'cognitive value' as his knowledge of the things in space are so obsolete that it would therefore hardly be possible for us to understand each other."[66]

Much of Ritschl's *Theology and Metaphysics* was directed against the Leipzig conservative theologian Christoph Luthardt, whose faithful representation of fundamental Christian doctrine in the language of traditional metaphysics had provided Ritschl with an ample supply of examples for critique. As a result of his treatment of Luthardt, the orthodox theologians and pastors—who up to that time had not been sure what they thought of the new theology—openly turned against Ritschl. After 1881 there appeared suddenly an upswing in the brochures directed against the alleged harm that Ritschlian ideas were having on the church.[67] The mounting criticism took its toll on Ritschl himself, who, due to his increasing reliance on tobacco, found himself unable to sleep until he gave up cigars and withdrew from the field of battle.[68]

What had confused orthodox thinkers about Ritschl and, for that matter, about Herrmann was the unqualified insistence of both on the centrality of a personal encounter with the historical Christ as the crux of the Christian religion and the irreplaceable content of divine revelation. Herrmann was uncompromising about the necessity of basing the content of Christianity on the historical Christ as opposed to locating the essence of religion in some transhistorical truth. In a lecture delivered in 1884, entitled "Why Does Our Faith Require Historical Facts?" Herrmann discussed Lessing's dilemma about the relationship between historical fact and religion. On the one hand he was unimpressed with the rationalistic Christianity of the philosophically educated theologians of his day, but on the other he did not recognize an eternal dimension in the facts of history either. Unable to cross "the detestable moat . . . no matter how often and earnestly I have tried to jump," Lessing proclaimed history and religion to be unjoinable.[69] Kant too had decided against the authority of history for religion. Objects of history were to him not matters of belief but facts, while religious objects lay beyond the realm of cognition.[70]

Herrmann acknowledged that many in his time accepted the alleged irrelevancy of historical facts to religion, but he insisted that Protestant Christianity must embrace them as necessary. True religion was not present when one simply declared, through the artificial authority of the church, that it was necessary to believe in the historical Christ. That, wrote Herrmann, was the solution of the Catholic church. Nor could true religion be captured by any mere scientific or artistic formulation of the meaning of the highest good. Only through an encounter with the unique life of the historical Christ could one come to understand the personal moral dimension at the base of a truly religious view of the world.[71] There could be no substitute for it.

This theme, which is in no way absent from Herrmann's earlier works, grew in dominance as he assumed prominence as a spokesman for the Ritschl school. The times were ripe for the coalescence of a theology like that Ritschl and Herrmann represented. What was required, according to Rade, was someone free of speculation and the mediocrity of mediation, someone who revered Scripture and history, someone who defended a practical message that could be preached and understood.[72] Although most parishioners could not follow Herrmann's philosophical analysis of the meaning of worldviews, they did understand that he made no apology for his insistence on what appeared to be the central teaching of Christianity.

Within theological ranks as well Ritschlians began to prosper. Their attitude toward higher criticism of the Bible put them in direct competition for chairs in Old Testament studies with the students of Julius Wellhausen, but in spite of that and the opposition of orthodox, mediating, and speculative theologians, Herrmann and others began to win what in 1881 Pfleiderer called "the overly gullible judgment . . . of especially the younger generation."[73] Three theological journals became organs for Ritschlian theology, the *Theologische Literaturzeitung* begun in 1876, the *Christliche Welt* in 1887, and the *Zeitschrift für Theologie und Kirche,* which brought Adolf Harnack and Julius Kaftan together with Herrmann as editors in 1891.[74] By the end of nineteenth century, as mediation theology waned and a new "liberal" theology replaced both it and the older Hegelian speculative outlook, the Ritschl school had grown dominant in appeal. At the beginning of his study *Kant, Lotze, and Ritschl,* Leonhard Stählin noted in 1889 that "no German theologian has a larger following than Albrecht Ritschl"; meanwhile in 1898 in America John Henry Wilbrandt Stuckenberg declared that Ritschl was "the most prominent name in German theology at the close of the nineteenth century."[75]

Natural Science in the Later Herrmann

Hasler has suggested that in conjunction with the transition from competitive to monopoly capitalism and imperialism around 1890 Herrmann began distancing himself from the theological system he had worked out in his first two major works.[76] What is clear is that Herrmann became less preoccupied after 1890 with the task of eliminating metaphysics from theology and more convinced of the need to represent the existential promise Christianity held out to his day.[77] But in the process of clarifying his position on specific doctrinal issues, Herrmann raised a number of questions in the minds of his contemporaries.

The absolute insistence on the power of a personal encounter with the historical Christ to open up the meaning of life with reference to the eternal seemed to the orthodox to make Herrmann one of their own. But his respect for historical facts turned out to be more complicated than theirs. Although for Herrmann there could be no compromise regarding the existence of facts,[78] clearly he understood by "facts" something more relative than they did. For example, the religious experience of a devout Israelite of biblical times was no longer fully understandable to people of modern times, "for the facts that affected him as revelations of God no longer have power over us."[79]

Herrmann's refusal to include miracles among facts was certainly unacceptable to conservative theologians. To Herrmann, miracle applied to the realm of faith and the personal view of the universe but not to the natural world. He opposed trying to explain miracle by saying that God broke through the natural order; even Jesus, he observed, had downplayed his power over nature in the New Testament by commanding witnesses to it to tell no one about it. The significance of the New Testament miracles was not that they were objects of faith but that they directed people to Christ so they could experience God's power for themselves. Herrmann even included the resurrection among the events whose supernatural aspects were not intended as objects of faith.[80] The natural order was the object of cognition and science; it was not to be confused with the realm of faith.

The lives of Jesus composed by Strauss and others in the nineteenth century did not, according to Herrmann, reach through to the historical Christ. Strauss's error was that he sought a basis for a religious worldview which possessed the certainty of a calculation. "Whoever demands that," Herrmann tersely added, "should go to the Catholic Church."[81] He revealed how much his emphasis had changed from the earlier works when he remarked that Strauss, like many others, held on to the illusion that Christianity was to be accepted because it was reasonable.[82]

For Herrmann history was not the past; rather, Jesus stood in the history to which we ourselves belong. Herrmann's colleague Martin Rade noted his friend's objections to the presumptions of historical work on Jesus, and quoted him to say that there was no scientifically defensible ground for our faith, and that the motivation for his entire literary activity was to defend that proposition.[83] Such an attitude was anathema to orthodox and liberal theologians alike. As Lipsius put it, Herrmann's reliance on the revelation of God in the historical Christ, exercised in isolation from metaphysics and from the search for the essence of religion from the history of all religions, was an insufficient protection of Christianity from doubt.[84]

There is of course a temptation to view the later Herrmann as a mystic, but Herrmann specifically rejected this label. He conceded that at least religious mystics recognized that revelation must find its grounding in the religious person's own life as opposed to the dictates of traditionalists or in knowledge and reason. But the God of the Christian faith was not one and the same with the mystics' sense of the Eternal, for the Eternal did not save us.[85] Only the encounter with the Christ of history could do that.

In his observations on the social circumstances of the last decades of the century Herrmann revealed further the "aggressive impulse" he called on Protestant theologians to embrace. The fact was that the German middle class and urban working class had joined the ranks of the unchurched. In light of this development Herrmann rejected the call of the *Deutsche evangelische Kirchenzeitung* to animate the middle class citizenry through a return to an orthodox Protestantism. Educated Germans were growing more and more alienated from the Christian faith, making an acceptance of traditional doctrine thoroughly impossible. Herrmann blamed the antagonism between educated Germans and Christians on the form in which the gospel was being proclaimed. The current theology of anxiety *(Angsttheologie)*, in striving to save itself before the world rather than wishing to conquer the world, hid the emancipating content of the salvation message at the very time when it must be made more understandable.[86]

But it was in his analysis of what to do regarding the working class that the full flavor of the later Herrmann emerged.[87] The occasion was the second Protestant Social Congress, and Herrmann addressed the group on the topic "Religion and Social Democracy." In his lecture, which was published in the opening volume of the new Ritschlian *Zeitschrift für Theologie und Kirche,* Herrmann remarked that in social democracy many German workers had hit upon a spiritual power which they understood and to which they enthusiastically responded. Since Christianity did not require a particular social order, a specific kind of state, or the protection of a historically developed form of private property, the church was indebted to modern socialism for "broadening its field of vision, deepening its cultivation of ideas, in short, of enriching its inner life."[88]

But while Christianity did not dictate a particular form of order, it could in no way sanction what German social democracy stood for. Herrmann did not have in mind the socialistic dimension of the party's political vision; rather he focused on its sentiments about religion as articulated by Marx, Engels, and others. It did not escape his attention that some in the party in 1891 had declared religion to be a private affair, that because the social democratic program was purely social and political, the individual had complete freedom to practice religion. Nevertheless, there could be no doubt about the real antagonism toward the church from social democrats. To them the propertied bourgeoisie needed the church as a means of controlling the threatening power of the workers, and in their eyes the church stood ready to be of service to capitalism.[89]

Herrmann reviewed the intent of the founders of social democracy in order to reject new forms of religion and to negate religion as such. The crux of Herrmann's argument against social democracy was cast in terms that were not only typical of his self-defined theological mission but that also revealed an emerging hostility toward natural science which marked his later career. German social democracy, unlike that in other lands, wanted to become a worldview.[90] Like natural scientists who became metaphysicians without even being aware that they were doing so, social democrats too had not understood the limits of human reason. In drawing out the contradiction between the social democratic and religious conceptions, Herrmann clearly associated social democracy and natural science. In their materialistic conception of history Marx and Engels had totally negated the historical life of the individual. As Hegelians they regarded the world as a web of mutually interacting processes whose laws of operation applied not to the forces which played at the surface but to the foundation beneath. Marx and Engels had gone beyond Hegel in seeking this foundation in nature and history.

> The unfruitful efforts of the philosopher have been outdone in the natural sphere through natural science. The discoveries of the law of conservation of energy and the reduction of organic life to mechanical and chemical processes made possible to us the idea of a natural order which wanted to be nothing more than the demonstrable interconnection of the real. Socialism achieves the same thing in the historical sphere. History originates not merely from the ends which people strive for consciously. To think that would be superficial. But not even ideas which people unconsciously serve shape history. To think that would be fantasy. The real course of all historical movement is to be seen in relations under whose pressure the individual acquires the most important motives of his will. The relations which ultimately determine the will of people most strongly come together in their economic situation. Consequently all struggles of history are at root economic struggles. The development of ideas in philosophy and religion can be traced back through numerous intermediate stages to those causes.[91]

Herrmann concluded with the admonition that social democracy had successfully influenced the thinking of German workers in two important ways. The first was to have convinced workers that the church was on the side of capitalism. The second was to have implanted the cognitive method of natural science and its application to the discernment of history so successfully in their minds that their understanding had escaped the bonds of reason and become a naturalistic worldview. "The worker assumes it is true that we [can] identify the real to whose recognition everyone can be forced by grasping its lawfulness."[92]

We are brought back to the fundamental difference between Herrmann's belief about the nature of reality and that of most of those around him. "We mean by the truly real in Christianity something totally different from what it means in metaphysics," he had written in his very first work. To him the truly real was the personal, the moral, the individual—that was why the intensely personal, supremely moral, and absolutely unique life of the historical Christ was the only rock of reality on which to base the Protestant religion. But, like natural science, social democracy made the universe and history into a machine and humans into animals. "For all those who cannot help but see humans [as something] different from animals the only rational thing is to take personal life, along with the ideas on whose truth it depends, as a fact. To be sure this fact cannot be understood or grasped by science with its cognitive means, but humans cannot let go of it."[93]

Herrmann's advice to his clerical colleagues about what to do in the face of the specter menacing the religious worldview left little doubt about the complete separation of natural science and religion he had come to, and it represents the culmination of the loss of nature from German Protestant theology. The situation as Herrmann saw it was indeed grave. The workers would not be rescued by a rationally argued demonstration that the worldview of social democracy and natural science did not represent an ultimate truth to which modern knowledge corresponded, that, like religion itself, it amounted to a *practical* solution to the problem of understanding the place of humans in the universe. The workers had to leave intellectual investigation to others more educated than themselves. But those on whom the workers relied could not meet their needs, for "the majority of the highly educated had capitulated to naturalism . . . It is not a question of drops poisoning the thinking of the people, but of a stream in which the people are swimming."[94]

How, then, was the stream to be resisted? "Defense," writes

Voelkel, "always led in Herrmann's case to counterattack."[95] Herrmann demanded that the church of his day not shrink from the conflict between natural science and religion but enter directly into it instead.

> We must be able to transplant ourselves into the conflict in which the people of our day, high and low, are losing faith in a God who hears prayers and punishes injustice. It must ring out loud and clear through the church: we also have the knowledge that is confusing you, but we do not fear it. Rather, we delight in it as in all truth. We must show that we also feel the conflict which brings distress to others, but that we are finished with it.[96]

Herrmann specifically denied that it was a question of eliminating or overcoming *(überwinden)* the opposition between scientific and religious views of the world—even to suggest a reconciliation would repudiate everything he had striven for his entire life. It was rather a matter of understanding the contradiction and then "becoming internally finished with it." He made no effort to deny that there was more than one way to accomplish this. For example, materialists and orthodox Christians each responded to the opposition in characteristic fashions. Herrmann described how he believed the church had to react: "We will become finished with the opposition by realizing that the Christian religion waxes stronger if it joins the knowledge of the lawfulness and endlessness of nature to belief in the living God, and by realizing that the Christian religion has foundations wholly different from the feeble attempts to dissolve the opposition."[97]

Although he did not say so in so many words, Herrmann's position amounted to claiming that since the discovery of knowledge corresponding to the truth of the world was *not* the real challenge confronting humans, then their only response to what existential writers would later describe as being "contingently thrown into an alien context"[98] had to be practical as opposed to cognitive. Herrmann had described two practical solutions, metaphysics and religion. Metaphysics was destined from the start to produce only relative results, and religion rested on a reality wholly different from that with which one dealt in metaphysics. Individuals therefore had to choose for themselves the path they would follow.

In this process Herrmann pleaded for a proper understanding of the relativistic role of facts. Christians, he concluded, should not be so foolish as to think that their beliefs were forced on them by such facts as the resurrection. Even a child could see that unless one had already come to belief from different quarters, such facts were but fairy tales, or at best doubtful stories. Christians could not demand that the social democrat should be constrained by what to the religious person were the facts of their belief. Social democrats might hear our words, wrote Herrmann, but they simply could not see the facts. Christians could not even expect that social democrats should want to recognize these facts, for they would be lying if they said that was their wish.

But, continued Herrmann, none of this meant that Christians should not appeal for themselves "to facts which exert on us the compulsion of the real, even when our faith appears to have left us in the distress of life." When social democrats and others mocked Christians for doing this, he urged, they did not realize what they were doing. They were correct, as all of us are, to acknowledge what to them possessed the power of facts. What they did not discern was that religion was *not* a realm of mere fantasy far removed from a repository of its own facts.[99]

In the end Herrmann's message was as simple or complex as the listener's capacity to absorb it. His urgent and passionate confession of Christ as personal savior could be understood by the most uneducated member of German society, while his defense of his intellectual right to make the confession carried only scholars with him into the intricacies of Kantian philosophy. His writing and teaching proved attractive to more and more budding theologians, who, like the young Karl Barth and Rudolf Bultmann, came to Marburg to study with him.

And yet at the heart of the theological stance that was so enticing to those who shared a presentiment for the "defiant self-assertion"[100] of the existential thought just beginning to flower in Germany was the complete loss of nature from the theological enterprise. Herrmann could not have known that the haunting question raised by this loss was whether or not his theology could survive the inevitable "failure of the existentialist revolt."[101]

Epilogue: The Future Challenge of Religion and Science

WRITING ABOUT German academic theology in the twentieth century, Ueli Hasler claims that "Herrmann's recommendation that one [can] exclude all possibility of conflict between theology and natural science by clarifying the epistemological jurisdiction of science has acquired almost canonical significance."[1] Convinced of the provisionality of scientific knowledge, some German theologians of the new century proceeded as if they had been freed from the burdens of the warfare of days just past. Gone from the works of Bultmann, Barth, Tillich, and others was the defensive posture so often forced upon theologians during the waning decades of the nineteenth century. A young pastor from Kiel who, as a student, had found in Herrmann's works the anchor he so desperately needed to hold him firm amid the tumultuous waves of modernism, wrote about Herrmann's significance for the church on the occasion of his mentor's seventieth birthday. It was his teacher, noted Johannes Iversen, who had shown that Christianity was a living historical religion which could not simply be cast aside by adopting a rationalistic viewpoint.[2]

With newly found confidence in the intellectual autonomy of a personalistic interpretation of the world, theologians in German universities tended to devote their energies more and more to the investigation of meaning as determined by personal value, paying less and less attention to the cognitive dimensions of human experience. Although they were unaware of the categories future philosophers of science would employ, they were nevertheless anti-realists by instinct; that is, their outlook was no longer determined by the One Truth toward which humans allegedly strove. They had abandoned the perspective in which rational conclusions, when correct, corresponded to external reality. The crucial studies of twentieth century German theology were defined with reference to the individual and social self, especially in light of the shock inflicted on the Western human psyche by the unprecedented horror of world war.

If Germany's leading professional theologians were successful in proclaiming the positive message of a theology of personal value in the decades following the Great War, they were less successful in communicating its stance regarding the provisionality of scientific knowledge. They proceeded as if Herrmann's critique of our knowledge of the world *(Welterkennen)* had established a permanent and radical separation of religion from natural science. They saw confirmation of their perception in the work of positivistic historians of science like George Sarton and in the writings of the philosophers associated with the Vienna Circle, all of which suggested that the sharp separation of science and religion was likely to endure. They agreed with these thinkers about what constituted scientific explanation, about the elimination of metaphysics as a repository of truth, and about the need to demarcate cognitive issues from questions of ethics and value.[3] Where the theologians did disagree with the philosophers—namely over which of the two spheres, cognition or value, held the greater importance—the disparity expressed itself by each side actively ignoring what the other chose to emphasize.

Yet these academics did not speak in a language that was understandable to most people. Laypeople, including practicing scientists, no more accepted the notion of the radical separation of science and religion than they embraced a coherence understanding of truth. Nature may have been lost to the German existential theologians of the twentieth century, but it certainly was not lost to the majority of those attempting to determine what it was they still believed. For individuals not caught up in the trappings of academic existential theology, truth was still a matter of establishing a correspondence between what one thought and what was real.

The theological traditions represented by Strauss, Zöckler, and Schmid continued throughout the twentieth century as the most meaningful articulations of the relationship between religion and natural science. Most people of faith living at the end of the twentieth century have not lost nature; on the contrary, they insist on keeping nature in the heart of their religious understanding. The descendants of Otto Zöckler exist everywhere as modern creationists, who demand that truths regarding nature are a vital part of their theological systems. The progeny of David Friedrich Strauss are readily visible in contemporary naturalists, who regularly make a religion out of their celebration of the incredible improbability of the development of life on earth. And the lineage of Rudolf Schmid is recognizable in religious liberals, who doggedly maintain that there is no actual problem if one really understands natural science or if one really understands religion.

If German existential theology has had difficulty speaking to laypeople in terms they can understand, it has also suffered opposition from its academic peers. The breakdown of the older positivistic understanding of science and the history of science characteristic of scholarship in the second half of the twentieth century has made a facile and complete separation of science and religion far more difficult to maintain. Historians of science for some time have urged that a more comprehensive account of science in history must recognize the nonpermanent nature of scientific theory and its susceptibility to factors external to its rational structure.[4] Without attempting to take sides in the debates currently engaging historians and philosophers of science, one can say that the surrender of nature by theologians in its own way involves assumptions no longer shared by numerous historians and philosophers of science. While some still find it satisfactory to assign natural science the question How, leaving to religion the Why, others have begun to argue that religion cannot avoid the How nor science the Why.[5]

Nor is the loss of nature to theology of merely academic or scholarly interest. Ole Jensen has asserted that a theology like Herrmann's or Bultmann's, in which technical mastery of the world is cut off from an understanding of issues of value, has no ground on which to judge the potentially fatal consequences that might result from pursuing technological control.[6] Bultmann's theology in particular proves useless as an alternative to a view of nature tied to the capacity for technical mastery, since for Bultmann, as for his teacher Herrmann, nature always remains an object at the disposal of humans. Nature is a means which can never be its own purpose.[7] Such a theology allegedly contains no grounds on which the destruction of nature through exploitation or neglect of the environment can be denounced.

What Herrmann and the ensuing existential theological school wished at all costs to avoid, writes Jensen, was the misunderstanding of faith as knowledge. The gain of Herrmann's solution was clear and must not be lost: belief was reclaimed as a free and responsible action of the self. But, he continues, the gain came at too high a price. For if the mistaking of faith for knowledge was illusion, Herrmann dealt with the illusion by means of a vicious restriction of religion to the realm of personal value that rendered humanity theologically impotent where nature's interests are concerned.[8]

Jensen calls for existential theologians to find a new mean between illusion and restriction, in effect to regain nature for themselves. If successful, nature would represent both something pre-culturally given to human experience and something which affects human experience. In this harmony between humanity and nature the Kantian distinctions between God and the world, history and nature must somehow be overcome.[9]

In the face of the ecological crisis the radical separation of science from religion appears to be an intellectual luxury the human race can no longer afford to retain. For all of the urgency behind the call for a new mean between illusion and restriction, however, the discovery of such a mean will surely not be easy. It may involve nothing less than an understanding of science akin to what some scholars have identified as the ecological or humane science that characterized the ideal toward which Goethe's science strove.[10] Or it may involve a variety of the pluralism Isaiah Berlin embraced in the moral and cultural sphere to escape the relativism his loss of the Platonic ideal seemed to require.[11] However the new mean appears, it will have to blend together the realms of person and mechanism in a manner which possesses normative power and is simultaneously consonant with human value, utility, and diversity. To accomplish such a Herculean task may well require what Albert Einstein, in the face of the nuclear threat to humanity, declared to be necessary if the human race was to survive. Perhaps we shall require a substantially new manner of thinking.

Bibliography

Notes

Index

Bibliography

Altholtz, Joseph L. *The Churches in the Nineteenth Century.* New York: Bobbs-Merrill Co., Inc., 1967.
Amrine, F., F. J. Zucker, and H. Wheeler, eds. *Goethe and the Sciences: A Reappraisal.* Dordrecht: Reidel, 1987.
Aner, Karl. *Die Theologie der Lessingzeit.* Halle: Niemeyer, 1929.
Bachmann, Johann. "Ernst Wilhelm Hengstenberg." Pp. 670–674 in *Realencyklopädie für protestantische Theologie und Kirche,* vol. 7. Leipzig: Hinrichs, 1899.
Barth, Karl. *Protestant Theology in the Nineteenth Century: Its Background and History.* London: SCM Press, 1972.
Berlin, Isaiah. *The Crooked Timber of Humanity: Chapters in the History of Ideas,* ed. Henry Hardy. New York: Knopf, 1991.
Bigler, Robert. *The Politics of German Protestantism: The Rise of the Protestant Church Elite in Prussia, 1815–1848.* Berkeley: University of California Press, 1972.
Boller, Paul F., Jr. *American Thought in Transition: The Impact of Evolutionary Naturalism, 1865–1900.* Chicago: Rand McNally and Co., 1969.
Bowler, Peter. *The Eclipse of Darwinism: Anti-Darwinian Evolution Theories in the Decades around 1900.* Baltimore: Johns Hopkins University Press, 1983.
——— *Evolution: The History of an Idea.* Berkeley: University of California Press, 1984.
Brandt, Richard. *The Philosophy of Schleiermacher: The Development of His Theory of Scientific and Religious Knowledge.* New York: Harper and Brothers, 1941.
Brazill, William. *The Young Hegelians.* New Haven: Yale University Press, 1970.

Brent, Peter. *Charles Darwin: A Man of Enlarged Curiosity*. New York: W. W. Norton, 1981.
Busch, Eberhard. *Karl Barth: His Life from Letters and Autobiographical Texts*, trans. John Bowden. Philadelphia: Fortress Press, 1975.
Carnap, Rudolf. "The Elimination of Metaphysics through Logical Analysis of Language." Pp. 60–81 in A. J. Ayer, ed., *Logical Positivism*. New York: Free Press, 1959.
Carstens. "Claus Harms." Pp. 607–611 in *Allgemeine Deutsche Biographie*, vol. 10. Leipzig: Duncker and Humblot, 1879.
Cube, Felix. *Die Auffassungen Jakob Friedrich Fries und seiner Schule über die philosophischen Grundlagen der Mathematik und ihr Verhältnis zur Grundlagentheorie*. Stuttgart: Dissertation an der Technischen Hochschule Stuttgart, 1957.
Cunningham, Andrew, and Nicholas Jardine, eds. *Romanticism and the Sciences*. Cambridge: Cambridge University Press, 1990.
Dawkins, Richard. *The Blind Watchmaker*. New York: W. W. Norton and Co., 1986.
Dietsch, Steffen, ed. *Natur-Kunst-Mythos: Beiträge zur Philosophie F. W. J. Schellings*. Berlin: Akademie-Verlag, 1978.
Dillenberger, John. *Protestant Thought and Natural Science: A Historical Interpretation*. Nashville: Abingdon Press, 1960.
Dillenberger, John, and Claude Welch. *Protestant Christianity*. New York: Charles Scribner's Sons, 1954.
Dilthey, Wilhelm. *Leben Schleiermachers*, vol. 2, pt. 1, *Schleiermachers System als Philosophie*, ed. Martin Redeker. Berlin: de Gruyter, 1966.
Dorner, Isaak. *Briefwechsel zwischen H. L. Martensen und I. A. Dorner, 1839–1881*, 2 vols. Berlin: Reuther, 1888.
DuBois-Reymond, Emil. "Über die Grenzen des Naturerkennens." *Ein Vortrag in der zweiten öffentlichen Sitzung der 45. Versammlung deutscher Naturforscher und Ärzte zu Leipzig am 14. August 1872*. Leipzig: Veit, 1872.
Duke of Argyll. *The Reign of Law*, 11th ed. London: Strahan, 1884.
Ehrenfeuchter, Friedrich. *Christentum und moderne Weltanschauung*. Göttingen: Vandenhoeck und Ruprecht, 1876.
Ende, H. *Der Konstruktionsbegriff im Urkreis des deutschen Idealismus*. Meisenheim am Glan: Hain, 1973.
Falcke, Heino. *Theologie und Philosophie der Evolution: Grundaspekte der Gesellschaftslehre F. Schleiermachers*. Zürich: Theologischer Verlag, 1977.
Feuerbach, Ludwig. *Das Wesen des Christentums*, vol. 7 in *Sämtliche Werke*, 13 vols. in 12, ed. Wilhelm Bolin and Friedrich Jodl. Stuttgart: Frommann, 1959–1964.

Fischer-Appelt, Peter. *Metaphysik im Horizont der Theologie Wilhelm Herrmanns*. Munich: Kaiser, 1965.
Frei, Hans W. *The Eclipse of Biblical Narrative: A Study in Eighteenth and Nineteenth Century Hermeneutics*. New Haven: Yale University Press, 1974.
Fries, Jakob Friedrich. *Knowledge, Belief, and Aesthetic Sense*, ed. Frederick Gregory, trans. Kent Richter. Cologne: Dinter Verlag, 1989.
——— *Neue oder anthropologische Kritik der Vernunft*, vols. 4–6 in *Sämtliche Schriften*, 24 vols., ed. Lutz Geldsetzer and Gert König. Aalen: Scientia Verlag, 1967.
——— *Wissen, Glaube, und Ahndung*, vol. 3 in *Sämtliche Schriften*, 24 vols., ed. Lutz Geldsetzer and Gert König. Aalen: Scientia Verlag, 1968.
Goodman, Nelson. *Ways of Worldmaking*. Indianapolis: Hackett Publishing Co., 1978.
Graue, Georg. "Zur Abwehr," *Protestantische Kirchenzeitung für das evangelische Deutschland* 24 (1877): 492–501.
Greg, W. R. "Life at High Pressure," *Contemporary Review* 25 (1875): 623–635.
Gregory, Frederick. "The Historical Investigation of Science in North America," *Zeitschrift für allgemeine Wissenschaftstheorie* 16 (1985): 151–166.
——— "The Impact of Darwinian Evolution on Theology in the Nineteenth Century." Pp. 369–390 in David C. Lindberg and Ronald L. Numbers, eds., *God and Nature: Historical Essays in the Encounter between Christianity and Science*. Berkeley: University of California Press, 1986.
——— "Kant's Influence on Natural Science in the German Romantic Period." Pp. 53–66 in R. P. W. Visser et al., eds., *New Trends in the History of Science*. Amsterdam: Rodopi, 1989.
——— "Kant, Schelling, and the Administration of Science in the Romantic Era," *Osiris*, 2d ser., 5 (1989): 17–35.
——— *Scientific Materialism in Nineteenth Century Germany*. Dordrecht: Reidel, 1977.
——— "Theology and the Sciences in the German Romantic Period." Pp. 69–81 in Andrew Cunningham and Nicholas Jardine, eds., *Romanticism and the Sciences*. Cambridge: Cambridge University Press, 1990.
Groh, John E. *Nineteenth Century German Protestantism: The Church as Social Model*. Washington: University Press of America, 1982.
Gussmann, Karl. "Zwei Schwäbische Freischärler," *Der Schwabenspiegel* 4 (1911): 233–234, 245–246.

Hacking, Ian. *Representing and Intervening*. Cambridge: Cambridge University Press, 1988.

Haeckel, Ernst. *Die Welträtsel: Gemeinverständliche Studien über monistische Philosophie*. Bonn: Emil Strauss, 1899.

Hanne, J. W. "Ideen über den Ursprung des Menschen," *Zeitschrift für wissenschaftliche Theologie* 8 (1868): 1–21, 117–132.

Harris, Horton. *David Friedrich Strauss and His Theology*. Cambridge: Cambridge University Press, 1973.

——— *The Tübingen School*. Oxford: Clarendon Press, 1975.

Hasler, Ueli. *Beherrschte Natur: Die Anpassung der Theologie an die bürgerliche Naturauffassung im 19. Jahrhundert*. Bern: Peter Lang, 1982.

Hayward, John, ed. *Complete Poetry and Selected Prose of John Donne*. Bloomsbury: Nonesuch Press, 1929.

Heilbron, John. *Electricity in the Seventeenth and Eighteenth Centuries*. Berkeley: University of California Press, 1979.

Helmreich, Ernest. *Religious Education in German Schools: An Historical Approach*. Cambridge, Mass.: Harvard University Press, 1959.

Hendry, George S. *Theology of Nature*. Philadelphia: The Westminster Press, 1980.

Hermelink, Heinrich. *Das Christentum in der Menschheitsgeschichte*, 3 vols. Tübingen: Metzler, 1951–1955.

Herrmann, Wilhelm. "Die Auffassung der Religion in Cohens und Natorps Ethik." Pp. 377–405 in F. W. Schmidt, ed., *Gesammelte Aufsätze*. Tübingen: Mohr, 1923.

——— *Die Dogmatik*. Gotha: Perthes, 1925.

——— "Die Erlösung durch Jesus Christus und die Wissenschaft." Pp. 336–344 in F. W. Schmidt, ed., *Gesammelte Aufsätze*. Tübingen: Mohr, 1923.

——— "Der evangelische Glaube und die Theologie Albrecht Ritschls." Pp. 1–25 in F. W. Schmidt, ed., *Gesammelte Aufsätze*. Tübingen: Mohr, 1923.

——— "Faith as Ritschl Defined It." Pp. 7–62 in *Faith and Morals*, trans. D. Matheson and R. W. Stewart. New York: Putnam's Sons, 1904.

——— *Die Gewissheit des Glaubens und die Freiheit der Theologie*, 2d ed. Freiburg: Mohr, 1889.

——— "Der Glaube an Gott und die Wissenschaft unserer Zeit." Pp. 189–213 in F. W. Schmidt, ed., *Gesammelte Aufsätze*. Tübingen: Mohr, 1923.

——— "Glaube und Erkennen," *Zeitschrift für Theologie und Kirche* 17 (1907): 152–154.

——— "Die Lage und Aufgabe der evangelischen Dogmatik in der Gegenwart." Pp. 95–188 in F. W. Schmidt, ed., *Gesammelte Aufsätze*. Tübingen: Mohr, 1923.

——— "Die Lage der evangelischen Dogmatik in der Gegenwart," *Zeitschrift für Theologie und Kirche* 16 (1907): 1–33, 172–201, 315–351.

——— *Die Metaphysik in der Theologie*. Halle: Niemeyer, 1876.

——— "The Moral Law as Understood in Romanism and in Protestantism." Pp. 71–193 in *Faith and Morals*, trans. D. Matheson and R. W. Stewart. New York: Putnam's Sons, 1904.

——— "Religion und Sozialdemokratie." Pp. 463–489 in F. W. Schmidt, ed., *Gesammelte Aufsätze*. Tübingen: Mohr, 1923.

——— *Die Religion im Verhältnis zum Welterkennen und zur Sittlichkeit*. Halle: Niemeyer, 1879.

——— Review of O. Pfleiderer, *Die Ritschl'sche Theologie*, in *Theologische Literaturzeitung* 17 (1892): 383–387.

——— Review of R. A. Lipsius, *Lehrbuch der evangelish-protestantischen Dogmatik*, in *Theologische Studien und Kritiken* 50 (1877): 521–554.

——— *Warum bedarf unser Glaube geschichtliche Thatsachen?* Halle: Niemeyer, 1884.

"Der Herzog von Argyll und sein schwäbischer Erzieher," *Schwäbischer Merkur*, no. 209 (1914): 5.

Hilgenfeld, Adolf. "Der alte und der neue Glaube, nach den neuesten Schriften von D. F. Strauss," *Zeitschrift für wissenschaftliche Theologie* 16 (1873): 304–340.

Himmelfarb, Gertrude. *Darwin and the Darwinian Revolution*. New York: W. W. Norton, 1962.

Hodge, Charles. *What Is Darwinism?* New York: Scribner, Armstrong Co., 1874.

Hoffmann, A. Review of R. Schmid, *Das Naturwissenschaftliche Glaubensbekenntnis eines Theologen*, in *Protestantische Monatshefte* 10 (1906): 74–76.

Iversen, Johannes. "Zu Wilhelm Herrmanns 70. Geburtstag," *Evangelische Freiheit* 16 (1916): 414–418.

Jäger, Gustav. *Die darwinische Theorie und ihre Stellung zu Moral und Religion*. Stuttgart: Hoffmann, 1869.

Jagnow, Albert A. "Karl Barth and Wilhelm Herrmann: Pupil and Teacher," *Journal of Religion* 16 (1936): 300–316.

Jensen, Ole. *Theologie zwischen Illusion und Restriktion: Analyse und Kritik der existenz-kritizistischen Theologie bei dem jungen Wilhelm Herrmann und bei Rudolf Bultmann*. Munich: Kaiser, 1975.

Jordan, Herrmann. "Otto Zöckler." Pp. 148–151 in *Biographisches Jahrbuch und deutscher Nekrolog*, vol. 11. Berlin: Reimer, 1908.

——— "Verzeichnis der literarischen Veröffentlichungen Otto Zöcklers." Pp. 707–47 in Otto Zöckler, *Geschichte der Apologie Christentums*. Gütersloh: Bertelsmann, 1907.

Kaftan, Julius. Review of W. Herrmann, *Die Metaphysik in der Theologie*, in *Theologische Literaturzeitung* 2 (1877): 63–65.

Kant, Immanuel. *Die Kritik der reinen Vernunft*, 2d ed., vol. 3 in *Gesammelte Schriften*. Berlin: Reimer, 1911. Trans. Norman Kemp Smith as *The Critique of Pure Reason*. New York: St. Martin's Press, 1965.

——— *Metaphysiche Anfangsgründe der Naturwissenschaft*. Pp. 465–565 in *Gesammelte Schriften*, vol. 4. Berlin: Reimer, 1911.

——— *Prolegomena zu jeder künftigen Metaphysik*, Pp. 253–383 in *Gesammelte Schriften*, vol. 4. Berlin: Reimer, 1911.

——— *Religion innerhalb der Grenzen der blossen Vernunft*. Pp. 1–202 in *Gesammelte Schriften*, vol. 6. Berlin: Reimer, 1914.

Kaufmann, Walter. *Hegel: A Reinterpretation*. Garden City, N.Y.: Anchor Books, 1966.

Keeser, K. "Rudolf Schmid," *Schwäbischer Merkur*, no. 401 (1907): 5.

Keuth, Herbert. *Realität und Wahrheit: Zur Kritik des kritischen Rationalismus*. Tübingen: Mohr, 1978.

Kohak, Erazim. *The Embers and the Stars: A Philosophical Inquiry into the Moral Sense of Nature*. Chicago: University of Chicago Press, 1984.

Köhnke, Klaus Christian. *Die Entstehung und Aufstieg des Neukantianismus*. Frankfurt: Suhrkamp, 1986.

Kohut, Adolph. *David Friedrich Strauss als Denker und Erzieher*. Leipzig: Kröner, 1908.

Krauss, Alfred. "Sendschrieben an Herrn Professor W. Herrmann in Marburg," *Jahrbücher für protestantische Theologie* 9 (1883): 193–240.

Lakatos, Imre. "Falsification and the Methodology of Scientific Research Programmes." Pp. 91–195 in Imre Lakatos and Alan Musgrave, eds., *Criticism and the Growth of Knowledge*. Cambridge: Cambridge University Press, 1974.

Lang, Heinrich. *Die Religion im Zeitalter Darwins*. Berlin: Luderlitz, 1873.

Lenoir, Timothy. *The Strategy of Life: Teleology and Mechanics in Nineteenth Century Biology*. Dordrecht: Reidel Publishing Co., 1982.

Leplin, Jarrett, ed. *Scientific Realism*. Berkeley: University of California Press, 1984.

Lindberg, David C., and Ronald L. Numbers, eds. *God and Nature: His-*

torical Essays on the Encounter between Christianity and Science. Berkeley: University of California Press, 1986.

Lipsius, R. A. "Neue Beiträge zur wissenschftlichen Grundlegung der Dogmatik," *Jahrbücher für protestantische Theologie* 11 (1885): 177–288, 369–453, 550–671.

—— Review of W. Herrmann, *Die Metaphysik in der Theologie,* in *Protestantische Kirchenzeitung für das evangelische Deutschland* 23 (1876): 650.

—— "Vorwort zu einem Vorwort," *Protestantische Kirchenzeitung für das evangelische Deutschland* 23 (1876): 641–651.

Löwith, Karl. *From Hegel to Nietzsche: The Revolution in Nineteenth Century Thought.* Garden City, N.Y.: Anchor Books, 1967.

Lovejoy. *The Great Chain of Being: A Study of the History of an Idea.* New York: Harper and Row, 1960.

Mackintosh, Hugh Ross. *Types of Modern Theology.* New York: Charles Scribner's Sons, 1937.

Marsden, George. *Fundamentalism and American Culture: The Shaping of Twentieth-Century Evangelicalism, 1870–1925.* Oxford: Oxford University Press, 1980.

Massey, Marilyn Chapin. *Christ Unmasked: The Meaning of the Life of Jesus in German Politics.* Chapel Hill: University of North Carolina Press, 1983.

Mildenberger, Friedrich. *Geschichte der deutschen evangelischen Theologie im 19. und 20. Jahrhundert.* Stuttgart: Kohlhammer, 1981.

Moleschott, Jakob. *Ursache und Wirkung in der Lehre vom Leben.* Giessen, 1867.

Morris, Henry. *Scientific Creationism.* San Diego: Creation-Life Publishers, 1974.

Natorp, Paul. "Über das Verhältnis des theoretischen und praktischen Erkennens zur Begründung einer nichtempirischen Realität," *Zeitschrift für Philosophie und philosophische Kritik* 79 (1881): 242-259.

Nenon, Thomas. *Objectivität und endliche Erkenntnis.* Munich: Karl Alber, 1986.

Nietzsche, Friedrich. "David Friedrich Strauss as Confessor and Writer." Pp. 3–55 in *Untimely Meditations,* trans. R. J. Hollingdale. New York: Cambridge University Press, 1983.

Ollig, Hans-Ludwig. *Der Neukantianismus.* Stuttgart: Metzler, 1979.

Ott, Günther. "Johann Georg Wilhelm Herrmann." Pp. 691–692 in *Neue Deutsche Biographie,* vol. 8. Berlin: Duncker and Humblot, 1968.

Otto, Rudolf. *The Philosophy of Religion Based on Kant and Fries.* London: Williams and Norgate, Ltd., 1931.
Overbye, Dennis. *Lonely Hearts of the Cosmos: The Scientific Quest for the Secret of the Universe.* New York: Harper Collins, 1991.
Peake, A. S. "The History of Theology." Pp. 131-184 in *Germany in the Nineteenth Century.* Freeport, N.Y.: Books for Libraries Press, 1967.
Perriraz, Louis. *Histoire de la théologie protestante au xixme siècle, surtout en Allemagne,* vol. 1, *Les doctrines: De Kant à Karl Barth.* Neuchâtel: Messeiller, 1949.
Pfleiderer. Otto. "Evolution and Theology." Pp. 1-26 in *Evolution and Theology and Other Essays.* London: A. and C. Black, 1900.
——— "Introduction." Pp. v-xxvi in D. F. Strauss, *The Life of Jesus,* trans. George Eliot. London: Swan Sonnenschein, 1898.
——— *Die Ritschl'sche Theologie kritisch beleuchtet.* Braunschweig: Schwetschke, 1891.
——— "Silhouetten aus der Religionswissenschaft der Gegenwart," *Protestantische Kirchenzeitung für das evangelische Deutschland* 24 (1877): 461-468, 477-492.
Popper, Karl. *Objective Knowledge: An Evolutionary Approach.* Oxford: Clarendon Press, 1983.
——— *The Logic of Scientific Discovery.* New York: Harper, 1968.
Powell, Baden. *The Unity of Worlds and of Nature.* London: Longman, 1856.
Prior, A. N. "The Correspondence Theory of Truth." Pp. 223-232 in *The Encyclopedia of Philosophy,* vol. 2. New York: Macmillan Publishing Co. and Free Press, 1967.
Putnam, Hilary. "What is Realism?" Pp. 140-153 in Jarrett Leplin, ed., *Scientific Realism.* Berkeley: University of California Press, 1984.
Rade, Martin. "Akademische Gedächtnisrede auf Wilhelm Herrmann." Pp. viii-xxi in Wilhelm Herrmann, *Dogmatik.* Gotha: Perthes, 1925.
Rapp, Adolf. *Briefwechsel zwischen Strauss und Vischer,* 2 vols. Stuttgart: Klett, 1953.
Redeker, Martin. *Schleiermacher: Life and Thought,* trans. John Wallhausser. Philadelphia: Fortress Press, 1973.
Rescher, Nicholas. *The Coherence Theory of Truth.* Oxford: Clarendon Press, 1973.
Reuschle, Carl Gustav. *Philosophie und Naturwissenschaft: Zur Erinnerung an D. F. Strauss.* Bern, 1874.
Ritschl, Albrecht. *Die christliche Lehre der Rechtfertigung und Versöhnung,* 3 vols. Bonn: Marcus, 1870, 1874.

―――― *Theologie und Metaphysik: Zur Verständigung und Abwehr*, 2d ed. Bonn: Marcus, 1887.
―――― *Three Essays*, ed. and trans. Philip Hefner. Philadelphia: Fortress Press, 1972.
Ritschl, Otto. *Albrecht Ritschls Leben*, 2 vols. Freiburg: Mohr, 1892, 1896.
Rosenkranz, K. "Der deutsche Materialismus und die Theologie," *Zeitschrift für wissenschaftliche Theologie* 7 (1864): 225–287.
Rückriem, Georg, et al., eds. *Historischer Materialismus und menschliche Natur*. Cologne: Pahl-Rugenstein, 1978.
Schellong, Dieter. *Bürgertum und Christliche Religion: Anpassungsprobleme der Theologie seit Schleiermacher*. Munich: Kaiser, 1975.
Schleiermacher, Friedrich. *The Christian Faith*. Edinburgh: Clark, 1928.
―――― *Das Leben Jesu: Vorlesungen an der Universität zu Berlin im Jahr 1832*. Berlin: Reimer, 1864. Trans. as *The Life of Jesus*, and ed. with an Introduction by Jack C. Verheyden. Philadelphia: Fortess Press, 1975.
―――― *Über die Religion: Reden an die Gebildeten unter ihren Verächtern*. Berlin: Unger, 1799.
Schmid, Rudolf. *Der alttestamentliche Religionsunterricht im Seminar und Obergymnasium: Seine Schwierigkeiten und der Weg zu ihrer Überwindung*. Tübingen: Fues, 1889.
―――― *Die Darwin'schen Theorien und ihre Stellung zur Philosophie, Religion und Moral*. Stuttgart: Moser, 1876. Trans. as *The Theories of Darwin and Their Relation to Philosophy, Religion, and Morality*, 2d ed., trans. G. A. Zimmermann. Chicago: Jansen, McChurg and Co., 1885.
―――― "Die durch Darwin angeregte Entwickelungsfrage, Ihr Gegenwärtiger Stand und ihre Stellung zur Theologie," *Theologische Studien und Kritiken* 48 (1875): 7–60.
―――― *Lebenserinnerungen aus meinem Leben: Aufzeichnungen für die Seinigen*. Konstanz: Komissionsverlag von Karl Gess, 1909.
―――― *Das naturwissenschaftliche Glaubensbekenntnis eines Theologen: Ein Wort zur Verständigung zwischen Naturforschung und Christentum*, 2d ed. Stuttgart: Kilemann, 1906. Trans. as *The Scientific Creed of a Theologian*. London: Hodder and Stoughton, 1906.
―――― Review of F. A. Lange, *Geschichte des Materialismus und Kritik seiner Bedeutung in der Gegenwart*, in *Theologische Studien und Kritiken* 50 (1877): 194–207.
―――― "Robert Mayer, der grosse Förderer unserer heutigen wissenschaftlichen Welterkenntnis, Seine wissenschaftliche Entdeckung

und sein religiöser Standpunkt," *Theologische Studien und Kritiken* 51 (1878): 677–692.

——— "Die Tage in Genesis 1–2, 4a," *Jahrbücher für protestantische Theologie* 13 (1889): 688–714.

Schmidt, Friedrich Wilhelm. "Johann Wilhelm Herrmann." Pp. 96–104 in *Deutsches Biographisches Jahrbuch*, vol. 4. Berlin: Deutsche Verlags-Anstalt Stuttgart, 1929.

——— *Wilhelm Herrmann: Ein Bekenntnis zu seiner Theologie.* Tübingen: Mohr, 1922.

Schmidt, Friedrich Wilhelm, ed. *Gesammelte Aufsätze.* Tübingen: Mohr, 1923.

Schmidt, Wilhelm. Review of *Die Darwin'schen Theorien*, in *Theologische Studien und Kritiken* 50 (1877): 554–573.

Schnabel, Franz. *Deutsche Geschichte im neunzehnten Jahrhundert.* (*Die protestantischen Kirchen in Deutschland*, vol. 8). Freiburg: Herder, 1965.

Schönborn, P. Christoph. "Schöpfungskatechese und Evolutionstheorie: Vom Burgfrieden zum konstruktiven Konflikt." Pp. 91–116 in R. Spaemann, R. Löw, and P. Koslowski, eds., *Evolutionismus und Christentum*. Weinheim: VCH, 1986.

Schultze, Victor. "Ansprache am Sarge." Pp. 124–126 in T. Zöckler, *Erinnerungsblätter*. Gütersloh: Bertelsmann, 1906.

——— "Otto Zöckler." Pp. 704–708 in *Realencyklopädie für protestantische Theologie und Kirche*, vol. 21. Leipzig: Hinrichs, 1908.

Schütz, W. *Das Grundgefüge der Herrmannschen Theologie, ihre Entwicklung und ihre geschichtlichen Wurzeln.* Berlin: Ebering, 1926.

Schweitzer, Albert. *The Quest of the Historical Jesus.* New York: Macmillan, 1961.

Seeberg, R. "Otto Pfleiderer." Pp. 316–323 in *Realencyklopädie für protestantische Theologie und Kirche*, vol. 24. Leipzig: Hinrichs, 1913.

Senft, Christoph. *Wahrhaftigkeit und Wahrheit: Die Theologie des 19. Jahrhunderts zwischen Orthodoxie und Aufklärung.* Tübingen: Mohr, 1956.

Smith, Norman Kemp. *A Commentary on Kant's Critique of Pure Reason.* New York: Humanities Press, 1962.

Solms, Ludwig Fürst zu. "Recht und Unrecht der Metaphysik," *Jahrbücher für protestantische Theologie* 6 (1880): 581–593.

Stephan, Horst. "Am Sarge Wilhelm Herrmanns: Im Namen der Fakultät," *Die Christliche Welt* 36 (1922): 76–79.

Steude, E. G. "Der Apologet." Pp. 95–118 in T. Zöckler, *Erinnerungsblätter*. Gütersloh: Bertelsmann, 1906.

——— "Der Darwinismus die Erfüllung des Christentums?" *Beweis des Glaubens* 31 (1895): 121–133.
Strauss, David Friedrich. *Der alte und der neue Glaube: Ein Bekenntniss.* Leipzig: Hirzel, 1872.
——— "Die Asteroiden und die Philosophen." Pp. 402–407 in D. F. Strauss, *Kleine Schriften.* Leipzig, Brockhaus, 1862.
——— *Die Christliche Glaubenslehre in ihrer geschichtlichen Entwicklung und im Kampfe mit der modernen Wissenschaft dargestellt,* 2 vols. Tübingen: Osiander, 1840–1841.
——— *Krieg und Friede: Zwei Briefe von David Friedrich Strauss an Ernst Renan und dessen Antwort.* Leipzig: Insel, 1870.
——— *Das Leben Jesu für das deutsche Volk bearbeitet,* 4th ed., vol. 3 in *Gesammelte Schriften,* ed. Eduard Zeller. Bonn: Verlag von Emil Strauss, 1877.
——— *Das Leben Jesu kritisch bearbeitet,* 4th ed., 2 vols. Tübingen: Osiander, 1840.
——— "Ein Nachwort als Vorwort." Pp. 255–278 in *Der alte und der neue Glaube,* 4th ed., vol. 6 in Eduard Zeller, ed., *Gesammelte Schriften.* Bonn: Verlag von Emil Strauss, 1877.
——— *Streitschriften zur Verteidigung meiner Schrift über das Leben Jesu und zur Charakteristik der gegenwärtigen Theologie..* Tübingen: Osiander, 1837.
——— *Ulrich von Hütten: His Life and Times,* trans. Jane Sturge. Reprint of the 1874 edition. New York: AMS Press, 1970.
Swing, Albert. *The Theology of Albrecht Ritschl.* New York: Longmans, Green, and Co., 1901.
Tillich, Paul. *The Shaking of the Foundations.* New York: Scribner's Sons, 1948.
Toews, John Edward. *Hegelianism: The Path toward Dialectical Humanism, 1805–1841.* New York: Cambridge University Press, 1980.
Traub, Friedrich. "Ritschls Erkenntnistheorie," *Zeitschrift für Theologie und Kirche* 4 (1894): 91–129.
Turner, James. *Without God, without Creed: Origins of Unbelief in America.* Baltimore: Johns Hopkins University Press, 1985.
Turner, R. Stephen. "The Growth of Professional Research in Prussia, 1818–1848: Causes and Context," *Historical Studies in the Physical Sciences* 3 (1971): 137–182.
——— "The Prussian Professoriate and the Research Imperative." Pp. 109–121 in Hans N. Jahnke and Michael Otte, eds., *Epistemological and Social Problems of the Sciences in the Early Nineteenth Century.* Dordrecht: Reidel, 1981.

Ulrici, Hermann. *Gott und die Natur*, 3d ed. Leipzig: Weigel, 1875.
Voelkel, Robert T. *The Shape of the Theological Task*. Philadelphia: Westminster Press, 1968.
Walker, Ralph C. S. *The Coherence Theory of Truth: Realism, Anti-Realism, Idealism*. New York: Routledge, 1989.
Walsh, W. H. *An Introduction to the Philosophy of History*. London: Hutchinson and Co., 1961.
Weizsäcker, C. Review of W. Herrmann, *Die Religion im Verhältnis zum Welterkennen und zur Sittlichkeit*, in *Theologische Literaturzeitung* 4 (1879): 590–598.
Welch, Claude, ed. and trans. *God and Incarnation in Mid-Nineteenth Century German Theology*. New York: Oxford University Press, 1965.
———. *Protestant Thought in the Nineteenth Century*, 2 vols. New Haven: Yale University Press, 1972, 1985.
Willey, Thomas E. *Back to Kant: The Revival of Kantianism in German Social and Historical Thought*. Detroit: Wayne State University Press, 1978.
Wrzecionko, Paul. *Die philosophischen Wurzeln der Theologie Albrecht Ritschls: Ein Beitrag zum Problem des Verhältnis von Theologie und Philosophie im 19. Jahrhundert*. Berlin: Töpelmann, 1964.
Yandell, Keith E. "Protestant Theology and Natural Science in the Twentieth Century." Pp. 448-471 in David C. Lindberg and Ronald L. Numbers, eds., *God and Nature: Historical Essays on the Encounter between Christianity and Science*. Berkeley: University of California Press, 1986.
Zeller, Eduard. *David Friedrich Strauss in His Life and Writings*. London: Smith, Elder and Co., 1874.
Zeller, Eduard, ed. *Ausgewählte Briefe*. Bonn: Emil Strauss, 1895.
Ziegler, Theobald. *David Friedrich Strauss*, 2 vols. Tübingen: Trübner, 1908.
Zöckler, Otto. "Die Civilizationsfähigkeit der Wilden," *Beweis des Glaubens* 22 (1886): 355–358.
———. "Christentum und Darwinismus," *Theologisches Literaturblatt* 15 (1894): 529–531.
———. "Der Darwinismus und seine Gegner," *Zeitschrift für die gesamte lutherische Theologie und Kirche* 32 (1871): 247–271.
———. "Darwins Grossvater als Arzt, Dichter und Naturphilosoph." Pp. 127–158 in W. Frommel and F. Pfaff, eds., *Sammlung von Vorträgen für das deutsche Volk*, vol. 3. Heidelberg: Winter, 1880.

——— "Die einheitliche Abstammung des Menschengeschlechts," *Jahrbücher für deutsche Theologie* 8 (1863): 51–90.

——— "Gegen den Evolutionismus auf dem Gebiete der Religionsgeschichte, insbesondere der alttestamentlichen," *Beweis des Glaubens* 40 (1904): 56–63.

——— *Geschichte der Apologie Christentums.* Gütersloh: Bertelsmann, 1907.

——— *Geschichte der Beziehungen zwischen Theologie und Naturwissenschaft mit besonderer Rücksicht auf Schöpfungsgeschichte*, 2 vols. Gütersloh: Bertelsmann, 1877, 1879.

——— "Der Himmel des Naturforschers und der Himmel des Christen." Pp. 181–205 in W. Frommel and F. Pfaff, eds., *Sammlung von Vorträgen für das deutsche Volk*, vol. 7. Heidelberg: Winter, 1882.

——— *Kritische Geschichte der Askese: Ein Beitrag zur Geschichte christlicher Sitte und Kultur.* Frankfurt: Heyder und Zimmer, 1863.

——— "Die Lehre Darwins," *Theologisches Literaturblatt* 13 (1892): 265–268.

——— *Die Lehre vom Urstand des Menschen, geschichtlich und dogmatisch-apologetisch untersucht.* Gütersloh: Bertelsmann, 1879.

——— "Die Moral des Darwinismus," *Zeitschrift für die gesamte lutherische Theologie und Kirche* 34 (1873): 76–93.

——— "Der Mosaische Schöpfungsbericht und die neue Naturwissenschaft," *Evangelische Kirchenzeitung* 106/107 (1880): 473–486, 501–506.

——— "Neue Flut-Phantasien," *Beweis des Glaubens* 31 (1895): 202–208.

——— "Neue Flut-Theorien," *Beweis des Glaubens* 30 (1894): 432–437.

——— "Seit wann leben Menschen auf der Erde?" *Beweis des Glaubens* 32 (1896): 208–212.

——— *Theologia Naturalis: Entwurf einer systematischen Naturtheologie vom offenbarungsgläubigen Standpunkte aus.* Frankfurt: Heyder und Zimmer, 1860.

——— "Über die Anfänge der menschlichen Geschichte," *Zeitschrift für die gesamte lutherische Theologie und Kirche* 30 (1869): 209–231.

——— "Über die neueste Physikotheologie der Engländer, verglichen mit verwandten Bestrebungen und Leistungen der Deutschen," *Jahrbücher für deutsche Theologie* 5 (1860): 760–798.

——— "Über die notwendige Einigung von kirchlichem Bekenntnis und christlichem Leben," *Pastoral-theologische Blätter* 7 (1864): 84–103.

——— *Über Schöpfungsgeschichte und Naturwissenschaft*. Gotha: Perthes, 1869.

——— "Über die Speziesfrage nach ihrer theologischen Bedeutung mit besonderer Rücksicht auf die Ansichten von Agassiz und Darwin," *Jahrbücher für deutsche Theologie* 6 (1861): 659–713.

——— *Die Urgeschichte der Erde und des Menschen*. Gütersloh: Bertelsmann, 1868.

——— "Zur Lehre der Schöpfung: Der theistische Schöpfungsbegriff im Kampfe mit den Theorien des Materialismus, Pantheismus und Deismus," *Jahrbücher für deutsche Theologie* 9 (1864): 688–759.

Zöckler, Theodor, et al. *Erinnerungsblätter*. Gütersloh: Bertelsmann, 1906.

Notes

1. Historiographical Approaches to German Religion and Science

1. James Turner, *Without God, without Creed: Origins of Unbelief in America* (Baltimore: Johns Hopkins University Press, 1985).
2. Ueli Hasler, *Beherrschte Natur: Die Anpassung der Theologie an die bürgerliche Naturauffassung im 19. Jahrhundert* (Bern: Peter Lang, 1982), p. 295. All translations from the German are my own unless otherwise indicated.
3. Keith E. Yandell, "Protestant Theology and Natural Science in the Twentieth Century," pp. 448–471 in David C. Lindberg and Ronald L. Numbers, eds., *God and Nature: Historical Essays on the Encounter between Christianity and Science* (Berkeley: University of California Press, 1986), p. 450.
4. George S. Hendry, *Theology of Nature* (Philadelphia: Westminster Press, 1980), p. 16.
5. Ibid., p. 11.
6. Put this way one sees why the revision of classical mechanical explanations that occurs in quantum theory has had relatively little impact on existential theology. The profound epistemological implications of quantum theory are not seen by theologians to have transformed the task of natural science from exploring the How to addressing the Why.
7. Erazim Kohak, *The Embers and the Stars: A Philosophical Inquiry into the Moral Sense of Nature* (Chicago: University of Chicago Press, 1984), p. 126. Kohak adds: "There really is no third."
8. John Dillenberger, *Protestant Thought and Natural Science: A Historical Interpretation* (Nashville: Abingdon Press, 1960), p. 253.

9. R. Stephen Turner, "The Prussian Professoriate and the Research Imperative," pp. 109–121 in Hans N. Jahnke and Michael Otte, eds., *Epistemological and Social Problems of the Sciences in the Early Nineteenth Century*. (Dordrecht: Reidel, 1981), p. 110; and Turner, "The Growth of Professional Research in Prussia, 1818–1848—Causes and Context," *Historical Studies in the Physical Sciences* 3 (1971): 142, 147, 153, 156, 172.
10. Arthur Lovejoy, *The Great Chain of Being: A Study of the History of an Idea* (New York: Harper and Row, 1960), chap. 3.
11. Frederick Gregory, "Kant, Schelling, and the Administration of Science in the Romantic Era," *Osiris*, 2d ser., 5 (1989): 19.
12. The phrase is borrowed from the title of Paul Tillich's book of sermons, *The Shaking of the Foundations* (New York: Scribner's Sons, 1948).
13. Hasler, *Beherrschte Natur*, pp. 8–10.
14. Ibid., pp. 10–13.
15. Ibid., p. 13.
16. Ibid., p. 18.
17. Ibid. Schellong's study is *Bürgertum und Christliche Religion: Anpassungsprobleme der Theologie seit Schleiermacher* (Munich: Kaiser, 1975).
18. Hasler, *Beherrschte Natur*, pp. 30–31.
19. Cited in ibid., p. 26, from Franz Unger, "Natur als Legitimationskategorie im Gesellschaftsdenken der Neuzeit," in Georg Rückriem et al., eds., *Historischer Materialismus und menschliche Natur* (Cologne: Pahl-Rugenstein, 1978), p. 22.
20. Hasler, *Beherrschte Natur*, p. 26.
21. Ibid., p. 19.
22. Ibid., p. 21.
23. Quoted in ibid., p. 27, from Peter Ruben, "Schelling und die romantische deutsche Naturphilosophie," in Steffen Dietsch, ed., *Natur-Kunst-Mythos: Beiträge zur Philosophie F. W. J. Schellings* (Berlin: Akademie-Verlag, 1978), p. 26.
24. Fichte surely does not represent the most important figure in the historical development of German thought whose work served as the foundation of the de-naturing of theology. For one thing, Fichte's influence on German theology was not noteworthy. Second, Fichte's hostility to nature was hardly an attitude that would one day generate the "amicable juxtaposition" between theology and natural science that eventually emerged.

The intellectual stimulus to remove nature from theology found its immediate source in the Kantian critical philosophy. The impact of Kant's efforts was to make explicit what the limits of rational inquiry were. Genuine science, he argued, was present only to the extent that mathematical descriptions of natural phenomena were possible (*Metaphysiche Anfangsgründe der Naturwissenschaft*, in *Gesammelte Schriften*, vol. 4, Berlin: Reimer, 1911, p. 470). By restricting knowledge to our encounter with nature, Kant was denying that what we call knowledge was possible for the traditional objects of theology. The soul, a free will, and God Himself could not be objects of knowledge. Although Kant denied that we could have knowledge of God, that did not mean we were not justified in presupposing God's existence. Indeed, the presence of the moral law within not only presupposed the existence of a supreme being but also justified us in postulating it (*Kritik der reinen Vernunft*, 2d ed., 1787, in *Gesammelte Schriften*, vol. 3, Berlin: Reimer, 1911, pp. 421–422). In the section of his famous *Critique of Pure Reason* entitled "Critique of All Theology Based upon Speculative Principles of Reason," Kant wrote: "I maintain that all attempts to employ reason in theology in any merely speculative manner are altogether fruitless and by their nature null and void, and that the principles of its employment in the study of nature do not lead to any theology whatsoever. Consequently, the only theology of reason which is possible is that which is based upon moral laws or seeks guidance from them" (*Gesammelte Schriften*, III, 423; translated in Norman Kemp Smith, *Immanuel Kant's Critique of Pure Reason*, New York: St. Martin's Press, 1965, p. 528). The intent of Kant's critique, which was confirmed in his later book on religion, *Religion within the Limits of Reason Alone* (1793), was that theology was restricted solely to the moral realm. Kant was hardly hostile to nature and natural science; his investigation vindicated the central importance of our encounter with nature for knowledge. Nor was he hostile to religion, in spite of the perception of the Prussian King Frederick William II, who censured Kant for debasing some of the fundamentals of Holy Writ and of Christianity. (Cf. Gregory, "Kant, Schelling, and the Administration of Science," p. 21.) His whole motivation was to win a place for religion. In the Preface to the second edition of the *Critique* he wrote: "I have therefore found it necessary to deny knowledge, in order to make room for faith" (*Gesammelte Schriften*, III, 19; Smith, *Kant's Critique of Pure Reason*, p. 29). Theology, in effect, had been de-natured. In one important sense,

then, the appearance of de-natured theology is the story of the eventual triumph of Kant's view.
25. Hasler, *Beherrschte Natur*, pp. 45–46. See below, Chap. 2, n. 32.
26. The German theologian Ernst Hengstenberg complained in 1856 of the lack of a tradition of natural theology in Germany compared to that of the English. See below, Chap. 4, n. 12.
27. Hasler, *Beherrschte Natur*, p. 18.
28. I borrow this phrase from the work of Eldon Turner, whose forthcoming book, *Echoes of Dead Voices: Societies of Knowledge and Aesthetics in Early New England*, investigates epistemological assumptions among early American settlers.
29. Karl Popper, *Objective Knowledge: An Evolutionary Approach* (Oxford: Clarendon Press, 1983), p. 309; Nelson Goodman, *Ways of Worldmaking* (Indianapolis: Hackett Publishing Co., 1978), pp. 17–19; Ian Hacking, *Representing and Intervening* (Cambridge: Cambridge University Press, 1988), pp. 2, 112–128. And yet philosophers cannot quite let go of the notion of truth either. In spite of his doubt Hacking has a great deal to say about the issue of truth in contemporary philosophy of science, as has Ralph Walker in his thorough and captivating recent analysis of the question in the history of philosophy. Cf. Ralph C. S. Walker, *The Coherence Theory of Truth: Realism, Anti-Realism, Idealism* (New York: Routledge, 1989).
30. Let me be clear about what the difference in interest means in this study. No critique of the "three main theories of truth," as Popper calls them (*Objective Knowledge*, p. 308), will be given here, no criticism of the inconsistent manner in which the leading figures in this study may have represented one or another of the theories will be attempted, and only a minimal acknowledgment of the numerous varieties of the theories will be advanced. I will draw only on the main force of the theories, for no more than that was usually present in the sentiments of the theologians. These men were not philosophers, and the assumptions they made about the nature of truth reflected little concern with the complex implications that rightly command the attention of the philosopher. Even the great German philosophers of the nineteenth century did not set out to formulate a systematic theory of truth. For the most part their positions on truth must be recreated by contemporaries.
31. In what follows I do not have in mind the meaning of moral truth, although it may well be possible to define the correspondence theory in such a way as to include it.

32. A. N. Prior, "The Correspondence Theory of Truth," *The Encyclopedia of Philosophy*, 8 vols. (New York: Macmillan Publishing Co. and Free Press, 1967), II, 223.
33. Karl Popper, *The Logic of Scientific Discovery* (New York: Harper, 1968), p. 274, n. Prior suggests that Aristotle took it over from Plato's *Sophist*, and that the use of the word "correspondence" to denote the relation between thought and reality in which the truth of a thought consists appears to come from the Middle Ages. See Prior, "Correspondence Theory of Truth," p. 224.
34. Isaiah Berlin, *The Crooked Timber of Humanity: Chapters in the History of Ideas*, ed. Henry Hardy (New York: Knopf, 1991), pp. 5–6.
35. Jacob Moleschott, *Ursache und Wirkung in der Lehre vom Leben* (Giessen, 1867), p. 8.
36. Popper, *Objective Knowledge*, p. 317. Putnam observes: "Whatever else realists say, they typically say that they believe in a 'correspondence theory of truth.' " See Hilary Putnam, "What is Realism?" pp. 140–153 in Jarrett Leplin, ed., *Scientific Realism* (Berkeley: University of California Press, 1984), p. 140.
37. The correspondence theory is sometimes called "the copy theory of truth," following William James's characterization of it. Cf. Hacking, *Representing and Intervening*, pp. 104, 113.
38. The fact that opinions vary among philosophers concerning the origin of the so-called coherence theory of truth, which is discussed below, does not detract in any substantial manner from the role assigned to Kant's achievement. For one thing, many appear to agree with Norman Kemp Smith's succinct assertion that "Kant is the real founder of the coherence theory of truth." See *A Commentary on Kant's Critique of Pure Reason* (New York: Humanities Press, 1962), p. 36; G. E. Moore, as approvingly cited by A. N. Prior in "The Correspondence Theory of Truth," p. 224; Nicholas Rescher, *The Coherence Theory of Truth* (Oxford: Clarendon Press, 1973), p. 9; Hacking, *Representing and Intervening*, pp. 59, 98. Moreover, although Walker argues that Spinoza embraced a coherence theory prior to Kant, he does so with an acknowledgment that in the traditional understanding of Spinoza one does not associate a new conception of truth with him. Cf. Walker, *The Coherence Theory of Truth*, pp. 41, 48–49. Finally, even when Kant's view is characterized as containing a correspondence theory of truth, as in the work of Thomas Nenon, there is no attempt to deny that Kant may be properly counted among those who antici-

pated the coherence theory. Cf. Thomas Nenon, *Objectivität und endliche Erkenntnis* (Munich: Karl Alber, 1986), pp. 212–218.

39. Kant did not mean, by "laws involved in a nature in general," laws such as Newton's inverse square law. That was determined empirically in conjunction with the general laws. But because our understanding *(Verstand)* exercises a prescriptive function for all possible empirically discovered laws, we dictate to nature the general conditions under which it is to be known. Cf. Kant's *Prolegomena zu jeder künftigen Metaphysik*, in *Gesammelte Schriften*, IV, 306.

40. Jakob Friedrich Fries, *Wissen, Glaube, und Ahndung* (1805), in Lutz Geldsetzer and Gert König, eds., *Sämtliche Schriften*, 24 vols. (Aalen: Scientia Verlag, 1967), III, 457. Cf. Fries, *Knowledge, Belief, and Aesthetic Sense*, ed. Frederick Gregory, trans. Kent Richter (Cologne: Dinter Verlag, 1989), p. 31. These sentiments of Fries support G. E. Moore's suspicion that the coherence theory of truth was a comparatively late invention, most probably owing its origin to the stimulus of Kant's work. Cf. Prior, "The Correspondence Theory of Truth," p. 224. It was Fries, according to the philosopher of science Imre Lakatos, who first convincingly asserted that the truth of a proposition could not be proven by an appeal to facts, thereby destroying classical empiricism. See Imre Lakatos, "Falsification and the Methodology of Scientific Research Programmes," pp. 91–195 in Imre Lakatos and Alan Musgrave, *Criticism and the Growth of Knowledge* (Cambridge: Cambridge University Press, 1974), p. 99, n.

41. Cited from the third edition of Lotze's *Microcosmus*, bk. 8, ch. 1, "Truth and Science," in William Woodward, *From Mechanism to Value: Hermann Lotze and Nineteenth Century German Thought* (forthcoming), chap. 11.

42. Walker, *The Coherence Theory of Truth*, p. 2.

43. Ibid., p. 64. The truth of these beliefs, therefore, would have to consist in their *correspondence* with the independent reality of the noumenal world if such a correspondence could be established. Since it could not, Kant could not establish the truth or falsity of his beliefs about things-in-themselves. A pure coherence theory, such as that of Hegel, did not require a correspondence relation at all. Hegel dispensed with the thing-in-itself, positing that thought was the thought of Absolute Spirit itself. Cf. Walker, *The Coherence Theory of Truth*, pp. 94–96. On Hegel and the coherence theory, see also Herbert Keuth, *Realität und Wahrheit: Zur Kritik des kritischen Rationalismus* (Tübingen: Mohr, 1978), pp. 7, 38.

44. The philosopher W. H. Walsh overstates the case when he asserts that according to the coherence view all truth is essentially relative. For Kant and Herrmann the presuppositions are hardly arbitrary, since without the most important of them experience itself would not be possible. This requirement allows Kantians to retain a sense of objectivity in knowledge. See W. H. Walsh, *An Introduction to the Philosophy of History* (London: Hutchinson and Co., 1961), p. 86.
45. Anti-realism is the view that there is nothing to the truth of a statement over and above its being recognizable as true by us. According to Walker it is a version of the coherence theory of truth. Cf. Walker, *The Coherence Theory of Truth*, pp. 19, 35, 38.
46. There is another theory of truth which we shall not consider here. The pragmatic theory identifies truth by the usefulness or utility of a proposition. Unlike the first two theories, the pragmatic theory of truth makes no assertion, metaphysical or hypothetical, regarding the rationality of nature. Even though a systematic account of phenomena which we bring to nature might contain logical inconsistencies, if the system is useful, which is usually determined by its ability to make successful predictions, then we must conclude that we have stumbled up against truths of nature. Nature's truth simply will not be forced into the logically consistent categories of human reason.

    The pragmatic theory of truth shares with the classical correspondence theory its belief in the determining role of external reality but differs from it in not insisting that what is real can also be shown to be rational. It shares the rejection of metaphysics with the neo-Kantian coherence theory but differs in that it feels no requirement to make our explanations of natural phenomena coherent, even though that requirement is made as a hypothesis. This theory of truth was articulated by Charles Sanders Peirce in late nineteenth century America and had no substantial effect on nineteenth century German theology. This is by no means to suggest that American pragmatism bears no resemblance to the stance taken by Herrmann, especially where the rejection of metaphysics was concerned. It means merely that an investigation of the resemblance lies outside the German theological scene and therefore beyond the scope of this study.

2. The Shape of German Protestant Theology in the Nineteenth Century

1. Karl Barth, *Protestant Theology in the Nineteenth Century: Its Background and History* (London: SCM Press, 1972), p. 163.
2. Quoted by Barth, *Protestant Theology*, p. 164, from Karl Aner, *Die Theologie der Lessingzeit* (Halle: Niemeyer, 1929), p. 245.
3. John Dillenberger and Claude Welch, *Protestant Christianity* (New York: Charles Scribner's Sons, 1954), pp. 155–156.
4. Albert Schweitzer, *The Quest of the Historical Jesus* (New York: Macmillan, 1961), p. 28; Dillenberger and Welch, *Protestant Christianity*, pp. 153–154; Dillenberger, *Protestant Thought and Natural Science* (New York: Abingdon Press, 1960), pp. 173–178. Schweitzer names Johann Hess, Franz Reinhard, and Johann Herder among others, while Dillenberger identifies Johann Buddeus, Lorenz von Mosheim, Sigmund Baumgarten, and Christoph Pfaff as members of this group.
5. Hugh Ross Mackintosh, *Types of Modern Theology* (New York: Charles Scribner's Sons, 1937), p. 14; Claude Welch, *Protestant Thought in the Nineteenth Century*, vol. 1, *1799–1870* (New Haven: Yale University Press, 1972), pp. 37–38.
6. See Dillenberger's analysis of Jerusalem's *Betrachtungen über die vornehmsten Wahrheiten* of 1775 in his *Protestant Thought and Natural Science*, pp. 180–181.
7. The *Apology* had been composed prior to 1768, when Reimarus died, and was not intended for public dissemination. Cf. Barth, *Protestant Theology*, p. 238.
8. Welch, *Protestant Thought*, I, 39–41.
9. Barth, *Protestant Theology*, p. 474.
10. Ibid., p. 477.
11. Ibid.
12. Franz Schnabel, *Deutsche Geschichte im neunzehnten Jahrhundert*, vol. 8, *Die protestantischen Kirchen in Deutschland* (Freiburg: Herder, 1965), p. 121.
13. See Schweitzer, *Quest of the Historical Jesus*, pp. 51, 53.
14. Ibid., p. 103.
15. Schnabel, *Protestantischen Kirchen*, p. 70.
16. Ibid., p. 71. The king's personal motives also played a key role. As a member of the Reformed church, Frederick William III wished to take communion with his Lutheran wife. See Martin Redeker, *Schleiermacher: Life and Thought*, trans. John Wallhausser (Philadelphia: Fortress Press, 1973), p. 189.

17. See Carstens, "Claus Harms," *Allgemeine Deutsche Biographie,* vol. 10 (Leipzig: Duncker and Humblot, 1879), p. 607.
18. Quoted in Carstens, "Claus Harms," p. 608. For Schleiermacher's involvement in the union question, see Redeker, *Schleiermacher,* pp. 189–199.
19. Carstens, "Claus Harms," p. 609.
20. Church union was not denounced because the editor, Ernst Hengstenberg, had endorsed union. Cf. Friedrich Mildenberger, *Geschichte der deutschen evangelischen Theologie im 19. und 20. Jahrhundert* (Stuttgart: Kohlhammer, 1981), p. 243.
21. John E. Groh, *Nineteenth Century German Protestantism: The Church as Social Model* (Washington: University Press of America, 1982), pp. 175–177. See also Joseph L. Altholtz, *The Churches in the Nineteenth Century* (New York: Bobbs-Merrill Co., 1967), pp. 109–111.
22. Mildenberger, *Geschichte der deutschen Theologie,* p. 244.
23. Friedrich Schleiermacher, *Über die Religion: Reden an die Gebildeten unter ihren Verächtern* (Berlin: Unger, 1799).
24. Hasler, *Beherrschte Natur,* p. 138. See also Martin Redeker, *Schleiermacher,* pp. 123–124.
25. Wilhelm Dilthey, *Leben Schleiermachers,* vol. 2, part 1, *Schleiermachers System als Philosophie,* ed. Martin Redeker (Berlin: de Gruyter, 1966), p. 451.
26. Quoted by Horton Harris in *The Tübingen School* (Oxford: Clarendon Press, 1975), p. 21, from Ernst Schneider, *Ferdinand Christian Baur in seiner Bedeutung für die Theologie* (Munich: Lehmann, 1909), pp. 9–10.
27. Hasler, *Beherrschte Natur,* p. 121.
28. Ibid., pp. 41, 70.
29. Ibid., pp. 72–73; see esp. n. 16.
30. Ibid., pp. 72, 75–78. For an analysis of the role evolution played in Schleiermacher's social theory, see Heino Falcke, *Theologie und Philosophie der Evolution: Grundaspekte der Gesellschaftslehre F. Schleiermachers* (Zürich: Theologischer Verlag, 1977). Falcke's effort is intended to complement the work of those few who have turned to Schleiermacher as opposed to the many who have dealt with Hegel in this connection.
31. Hasler, *Beherrschte Natur,* pp. 80–81.
32. Hasler's class analysis of Schleiermacher's position points to his membership in that part of the educated middle class in Prussia which wished to oppose the anarchism of the French Revolution and the counterrevolution of the Restoration. The middle class sought to legitimize social and political reform by calling on the

structural correspondence between revolution in nature and social progress. According to Hasler, the middle class, with its bourgeois principle of self-purposefulness, thought nature was on its side against the church as the cornerstone of feudalism and divine purposefulness. A middle class theology like Schleiermacher's had to incorporate nature into itself in a way that was consistent with the social and political needs of the time. In the precapitalistic Romantic Period these needs required above all the consolidation of the gains of the French Revolution without producing anarchy; society was not yet ready for a theology based on a ruthlessly bourgeois self-purposefulness, with God as the "corpse of absolute understanding" (E. Bloch). Society could, however, accept Schleiermacher's bourgeois idea—the naturizing *(Naturwendung)* of the divine spirit. *(Beherrschte Natur,* pp. 41–44.)

33. Quoted in Richard Brandt, *The Philosophy of Schleiermacher: The Development of His Theory of Scientific and Religious Knowledge* (New York: Harper and Brothers, 1941), p. 325.
34. Ibid. and pp. 118–123. See also John Edward Toews, *Hegelianism: The Path toward Dialectical Humanism, 1805–1841* (New York: Cambridge University Press, 1980), pp. 62–63. For a similar but more complete account of the issue of religious knowledge in Schleiermacher, see Frederick Gregory, "Theology and the Sciences in the German Romantic Period," pp. 69–81 in Andrew Cunningham and Nicholas Jardine, eds., *Romanticism and the Sciences* (Cambridge: Cambridge University Press, 1990).
35. Schnabel reports that intermarriage between the two Protestant denominations had become common as doctrinal disputes gave way more and more to a "reasonable Christianity" based on selected portions of various Protestant confessions. See Schnabel, *Die protestantischen Kirchen in Deutschland,* p. 72.
36. Quoted from volume 1 by Friedrich Mildenberger, *Geschichte der deutschen Theologie,* p. 240.
37. Fries, *Wissen, Glaube, und Ahndung* in *Sämtliche Schriften,* III, 413–755. *Ahndung* is an old form of the modern *Ahnung.* For bibliographical information concerning the recent translation of this work, see n. 40 to Chap. 1 above.
38. Cf. Kent Richter, Translator's Introduction to J. F. Fries, *Knowledge, Belief, and Aesthetic Sense,* p. 11.
39. Fries means by "intuitive knowledge" only the knowledge of mathematics and sense perception. He was vehemently opposed to Schelling's intellectual intuition, which escaped from the frame-

work of space and time. In this he followed Kant, who also opposed all intuition that was not tied to sensibility. Cf. Kant, *Kritik der reinen Vernunft*, in *Gesammelte Schriften*, III, 9–10, 209–210.
40. Fries makes clear what he means in his later *Neue Kritik der Vernunft* when he observes: "When I say, for example, every substance persists, every change has a cause . . . I acknowledge laws of nature . . . But these very laws, of which I again become conscious in the judgment, must lie in my reason as immediate knowledge, only I need the judgment in order to become conscious of them." *Sämtliche Schriften*, IV, 341–342.
41. Fries, *Wissen, Glaube, und Ahndung*, p. 119. Regarding Kant, see Chap. 1 above, n. 24.
42. Fries, *Wissen, Glaube, und Ahndung*, p. 118.
43. Ibid., pp. 122, 129.
44. Ibid., p. 251.
45. Cf. Gregory, "Kant, Schelling, and the Administration of Science," pp. 31–32.
46. Claude Welch, ed. and trans., *God and Incarnation in Mid-Nineteenth Century German Theology* (New York: Oxford University Press, 1965), p. 9.
47. Cf. Karl Löwith, *From Hegel to Nietzsche: The Revolution in Nineteenth Century Thought* (Garden City, N.Y.: Anchor Books, 1967), pp. 13–28, 43–49.
48. Lovejoy, *The Great Chain of Being*, p. 317.
49. Quoted in ibid., p. 318. Goethe's attempt to contrast his understanding of reason with that of Kant is evident from his statement, "Reason depends on what becomes, the understanding on what has become." Quoted in Walter Kaufmann, *Hegel: A Reinterpretation* (Garden City, N.Y.: Anchor Books, 1966), p. 25, n. 20.
50. Barth, *Protestant Theology*, p. 401. For a reinterpretation of the background and meaning of Hegel's dialectic, including the claim that there is no dialectical method in Hegel, see Kaufmann, *Hegel: A Reinterpretation*, pp. 153–162.
51. Welch, *Protestant Thought*, I, 89; Barth, *Protestant Theology*, pp. 398–399. Welch argues that the reconciling role of love in religion was the source of Hegel's later idea of the reconciliation of opposites (p. 93).
52. Löwith, *From Hegel to Nietzsche*, pp. 59–60. Löwith cites approvingly the historian K. Fischer, who was an acquaintance of several of the Young Hegelians.
53. Cf. Schweizer, *Quest of the Historical Jesus*, chap. 8, for a discussion

of how Strauss characterized the deficiencies of the Gospel of John and how he stood on the so-called synoptic question.
54. Welch, *Protestant Thought*, I, 150–151.
55. Baur also owed a debt to Schleiermacher, whose influence did not simply cease once Baur had begun to make use of Hegelian categories.
56. Cf. the Preface to the second edition of *Das Wesen des Christentums* in Feuerbach's *Sämtliche Werke*, ed. Wilhelm Bolin and Friedrich Jodl, 13 vols., 1903–1911; reprint, 13 vols. in 12 (Stuttgart: Frommann, 1959–1964), VII, 9–10.
57. Ibid., p. 12.
58. Löwith, *From Hegel to Nietzsche*, p. 333.
59. Cf. Frederick Gregory, *Scientific Materialism in Nineteenth Century Germany* (Dordrecht: Reidel, 1977), chap. 1.
60. Feuerbach, *Sämtliche Werke*, VII, 11.
61. Groh, *Nineteenth Century German Protestantism*, p. 179.
62. Cited from volume 1, p. 61, in Mildenberger, *Geschichte der deutschen Theologie*, p. 243.
63. Welch, *God and Incarnation*, p. 6. For a summary of the Erlangen theology see his *Protestant Thought*, I, 218–227.
64. Altholtz, *Churches in the Nineteenth Century*, p. 111.
65. Cited from Kähler's *Geschichte der protestantischen Dogmatik im neunzehnten Jahrhundert* by Welch in *Protestant Thought*, I, 190.
66. On Hengstenberg and the *Evangelische Kirchenzeitung*, see Johann Bachmann, "Ernst Wilhelm Hengstenberg," pp. 670–674 in *Realencyklopädie für protestantische Theologie und Kirche*, vol. 7 (Leipzig: Hinrichs, 1899), pp. 671–672.
67. Welch, *Protestant Thought*, I, 190.
68. Groh, *Nineteenth Century German Protestantism*, pp. 364–366.
69. Mildenberger, *Geschichte der deutschen Theologie*, p. 245.
70. George Marsden, *Fundamentalism and American Culture: The Shaping of Twentieth-Century Evangelicalism, 1870–1925* (Oxford: Oxford University Press, 1980), p. 18.
71. Heinrich Hermelink, *Das Christentum in der Menschheitsgeschichte*, vol. 3, *Nationalismus und Sozialismus, 1870–1914* (Tübingen: Metzler, 1955), p. 243.
72. Quoted in Harris, *The Tübingen School*, p. 248.
73. Hermelink, *Das Christentum in der Menschheitsgeschichte*, vol. 1, *Revolution und Restauration, 1789–1835* (Tübingen: Metzler, 1951), p. 423.

74. Barth, *Protestant Theology*, p. 425; Welch, *God and Incarnation*, p. 13.
75. Quoted in Mildenberger, *Geschichte der deutschen Theologie*, p. 246. Yet it was from this issue that the name derives. Cf. Hermelink, *Das Christentum in der Menschheitsgeschichte*, I, 423. Another journal identified as an organ of the mediating approach was Schleiermacher's *Theologische Zeitschrift*, which ran only three years from 1819 to 1822. (Mildenberger, *Geschichte der deutschen Theologie*, p. 246.)
76. Welch, *Protestant Thought*, I, 271, n. 3. See also Welch, *God and Incarnation*, p. 13.
77. Cf. Gregory, *Scientific Materialism in Nineteenth Century Germany*.
78. Cf. Frederick Gregory, "The Impact of Darwinian Evolution on Theology in the Nineteenth Century," chap. 16 in David C. Lindberg and Ronald L. Numbers, eds., *God and Nature: Historical Essays in the Encounter between Christianity and Science* (Berkeley: University of California Press, 1986).
79. Hermelink, *Das Christentum in der Menschheitsgeschichte*, vol. 2, *Liberalismus und Konservatismus, 1835–1870* (Tübingen: Metzler, 1953), p. 475.
80. Ibid., I, 426.
81. Welch, *Protestant Thought*, I, 270.
82. Ibid., pp. 276–277.
83. Quoted from Dorner's *System of Christian Doctrine* in Welch, *Protestant Thought*, I, 281. For a more thorough treatment of this subject, see Welch's introduction and translation of the relevant texts in *God and Incarnation*, part 2.
84. Friedrich Ehrenfeuchter, *Christentum und moderne Weltanschauung* (Göttingen: Vandenhoeck und Ruprecht, 1876), p. 7.
85. Dorner's remark occurs in a letter to Martensen in 1862. See I. Dorner, *Briefwechsel zwischen H. L. Martensen und I. A. Dorner, 1839–1881*, 2 vols. (Berlin: Reuther, 1888), I, 372. See also p. 359.
86. Groh, *Nineteenth Century German Protestantism*, p. 85.
87. Cited in Welch, *Protestant Thought*, I, 273, n. 7.
88. Philip Hefner, Introduction to *Albrecht Ritschl: Three Essays* (Philadelphia: Fortress Press, 1972), p. 4.
89. Ibid., pp. 4–5.
90. Quoted in Otto Ritschl, *Albrecht Ritschls Leben*, 2 vols. (Freiburg: Mohr, 1892, 1896), I, 31. For the letter to his father regarding Strauss as the endpoint to which one would come if one studied Hegel, see p. 57. Ritschl was drawn to Rothe precisely because of his speculative yet anti-Strauss position; see p. 100. Strauss's influ-

ence on Ritschl is clear in Ritschl's 1846 work on the gospels; see pp. 114–115.
91. In 1844 Ritschl put together a brief vita, which is given in O. Ritschl, *Ritschls Leben*, I, 426–431. Information about the summer of 1843 is found on p. 431. See also I, 79. Otto Ritschl notes that Kant was not influential on any of his father's early works. Not until the lectures of the winter semester of 1858–59 did Kant figure prominently, and then only in his course on ethics (I, 245, 346).
92. The courses Ritschl taught at both universities are listed by year in O. Ritschl, *Ritschls Leben*, I, 448–50; II, 533–535. When Ritschl first introduced the study of theological ethics it bore the title "Theological Morality."
93. The historical treatment of justification and reconciliation constituted volume 1 of Ritschl's well-known study of those doctrines. It appeared in 1870 and was followed by an analysis of their biblical basis (vol. 2, 1874) and an investigation of their positive development (vol. 3, 1874). Ritschl's three-volume work on the history of pietism began with a Prolegomena in *Zeitschrift für Kirchengeschichte* 2 (1877): 1–55, and was succeeded by volumes 1 through 3 in 1880, 1884, and 1886.
94. Cf. O. Ritschl, *Ritschls Leben*, I, 178.
95. Hefner, *Three Essays*, p. 11.
96. Ibid., p. 13.
97. Ritschl's break with Baur in 1856 was largely due to his insensitivity to Baur's feelings. His inability to respond to the friendly overtures of the theologian Richard Lipsius later made any likely reconciliation between them impossible. For an example of this side of his personality, see O. Ritschl, *Ritschls Leben*, II, 305–307.
98. W. R. Greg, "Life at High Pressure," *Contemporary Review* 25 (1875): 623–635.
99. Cf. John Hayward, ed., *Complete Poetry and Selected Prose of John Donne* (Bloomsbury: Nonesuch Press, 1929), p. 202.
100. Groh, *Nineteenth Century German Protestantism*, p. 422. Hermelink observes that Ritschl always remained loyal to his Bismarck. See Hermelink, *Das Christentum in der Menschheitsgeschichte*, III, 220.
101. Cf. above, Chapter 1.
102. Hefner, *Three Essays*, p. 21.
103. Welch, *Protestant Thought in the Nineteenth Century*, vol. 2, *1870–1914* (New Haven: Yale University Press, 1985), p. 19; Hefner, *Three Essays*, p. 22. Hefner draws here on the work of Norman Metzler. Hefner notes that it was Ritschl who made the centrality

of the kingdom of God unavoidable for those who came after him.
104. Albrecht Ritschl, *Theologie und Metaphysik: Zur Verständigung und Abwehr*, 2d ed. (Bonn: Marcus, 1887), pp. 40–41. English translation, "Theology and Metaphysics," in Hefner, *Three Essays*, p. 187. See also pp. 60–62; English translation, pp. 204–205.
105. Ibid., p. 65; English trans., p. 209. James Turner's analysis of the historical development of English religious thought in the post-Reformation period finds confirmation in Ritschl's views. See Turner, *Without God*, pp. 22–25, 29–31, 35–43.
106. Ibid., p. 46; English trans., p. 192.
107. Hefner, *Three Essays*, p. 26; Hermelink, *Das Christentum in der Menschheitsgeschichte*, III, 225.
108. Ritschl, *Die christliche Lehre der Rechtfertigung und Versöhnung*, 3 vols. (Bonn: Marcus, 1870–1874), I, 413.
109. *Rechtfertigung und Versöhnung*, III, 537.
110. Ibid., p. 538.
111. Ibid., p. 539.
112. Ibid., pp. 539–540.
113. Ibid., pp. 542–543. Hasler's claim that Ritschl believed the mastering of nature could be a means of attaining freedom from nature would seem to be at odds with the sentiments quoted here. See Hasler, *Beherrschte Natur*, p. 55.
114. *Rechtfertigung und Versöhnung*, III, 543.
115. Ibid., p. 544. The third edition was identical except for the following: "So likewise is a perspective employing an active cause not abstracted from our experience, but is a presupposition of our thinking that makes experience possible" (Bonn: Marcus, 1883, III, 572).
116. Hefner, *Three Essays*, p. 26. Hermelink expresses similar sentiments. See Hermelink, *Das Christentum in der Menschheitsgeschichte*, III, 217.
117. Hans-Ludwig Ollig, in his *Der Neukantianismus* (Stuttgart: Metzler, 1979), does not even mention Ritschl, though he does include a section on the philosophy of religion of neo-Kantianism that deals with philosophers rather than theologians. Thomas E. Willey, *Back to Kant: The Revival of Kantianism in German Social and Historical Thought* (Detroit: Wayne State University Press, 1978), mentions Ritschl mistakenly as a philosopher, noting merely that he embraced Lotze's defense of theism (p. 41). Finally, Klaus Christian Köhnke's definitive *Entstehung und Aufstieg des Neukan-*

tianismus (Frankfurt: Suhrkamp, 1986) devotes considerable space to the philosopher Hermann Cohen, who frequently is included among the members of the Ritschl school, but his examination of Cohen in no way categorizes him in this manner.

118. *Abhandlungen der Fries'schen Schule* 1 (1847). The *Abhandlungen* ceased publication during the political upheaval of 1848, but a second Friesian school began with its revival by Leonard Nelson in Göttingen in 1904 and ran until 1937. In 1958 Julius Kraft began the contemporary journal *Ratio* to continue the phiolosophical aims of the *Abhandlungen*.

119. D. F. Strauss, *Streitschriften zur Verteidigung meiner Schrift über das Leben Jesu und zur Charakteristik der gegenwärtigen Theologie*, vol. 3 (Tübingen: Osiander, 1837), p. 3.

3. The New Hegelian Faith of David Friedrich Strauss

1. Adolph Kohut, *David Friedrich Strauss als Denker und Erzieher* (Leipzig: Kröner, 1908), p. 95. Strauss's recent biographer agrees. Cf. Horton Harris, *David Friedrich Strauss and His Theology* (Cambridge: Cambridge University Press, 1973), p. 248.
2. Eduard Zeller, *David Friedrich Strauss in His Life and Writings* (London: Smith, Elder and Co., 1874), p. 6.
3. Ibid., pp. 10–11; Harris, *Strauss and His Theology*, p. 3. An account of Strauss's formative years may also be found in Toews, *Hegelianism*, pp. 165–170.
4. Harris, *The Tübingen School*, pp. 25–26.
5. Quoted by Harris in *Strauss and His Theology*, p. 125.
6. Cf. Frederick Gregory, "Kant's Influence on Natural Science in the German Romantic Period," pp. 53–66 in R. P. W. Visser et al., eds., *New Trends in the History of Science* (Amsterdam: Rodopi, 1989), pp. 54–56.
7. Quoted by Harris in *Strauss and His Theology*, p. 13.
8. Quoted by Harris in ibid., p. 18.
9. William Brazill, *The Young Hegelians* (New Haven: Yale University Press, 1970), p. 46.
10. Strauss had access to notes on Schleiermacher's course on the life of Jesus. Here Schleiermacher said clearly what he had penned more obliquely in the *Christmas Eve Celebration: A Dialogue of 1806:* that the external history of Jesus' life was of secondary importance to the internal and could well have involved different circumstances. Cf. Friedrich Schleiermacher, *The Life of Jesus,* ed. with an

Introduction by Jack C. Verheyden (Philadelphia: Fortress Press, 1975), pp. xxi-xxii, 8–9. This work is a translation of a publication of 1864, a reconstruction of Schleiermacher's teaching on the life of Jesus compiled from student notes by K. A. Rutenick, *Das Leben Jesu: Vorlesungen an der Universität zu Berlin im Jahr 1832* (Berlin: Reimer, 1864). Schleiermacher relegated Jesus' miracles to external history. They were of no value at all if one was interested in distinguishing Christ inwardly from all other men, "for they would have no relation to what makes Christ the object of our faith" (p. 163; cf. pp. 156–163, 190–229). After a thorough analysis of Christ's miracles Schleiermacher concluded that it was not necessary to try to explain them naturally, as rationalist theologians had felt obligated to do, but, nevertheless, he did "not need to assume anything supernatural, anything that is at the same time contrary to nature" (p. 209). When it came to the resurrection, however, Schleiermacher could not bring himself to deny that nothing supernatural had occurred, even though in his famous study, *The Christian Faith*, he maintained that the historicity of the resurrection was theologically inconsequential! The disciples knew Jesus was the Son of God before the resurrection took place, hence "the facts of the resurrection and the ascension of Christ, and the prediction of his return to judgment, cannot be laid down as properly constituent parts of the doctrine of his person." See *The Christian Faith* (Edinburgh: Clark, 1928), sections 99, 119–120; *Life of Jesus*, pp. 415–416, 432–434, esp. pp. 455–456, 479–480.

11. Toews, *Hegelianism*, pp. 170–172, 268; Harris, *Strauss and His Theology*, p. 22.
12. Quoted by Harris in ibid., p. 31.
13. Harris, *The Tübingen School*, p. 26.
14. Quoted by Harris in *Strauss and His Theology*, p. 87.
15. Schweitzer, *The Quest of the Historical Jesus*, p. 78.
16. D. F. Strauss, *Das Leben Jesu kritisch bearbeitet*, 4th ed., 2 vols. (Tübingen: Osiander, 1840), II, 663.
17. Quoted in Harris, *Strauss and His Theology*, p. 19. On Schleiermacher's influence on Strauss as a demystifier, see Carl Gustav Reuschle, *Philosophie und Naturwissenschaft: Zur Erinnerung an D. F. Strauss* (Bern, 1874), p. 112.
18. Strauss, *Das Leben Jesu*, I, 83–84. Cf. Harris's critique of this, *Strauss and His Theology*, pp. 89, 283.
19. See Chapter 1, above, n. 43.
20. Strauss, *Das Leben Jesu*, II, 663. In fact, Strauss's motivation was to

take contradictions within the historical record seriously. See Toews, *Hegelianism,* pp. 257–258, 260.
21. In his defense of the book to officials at Tübingen University, Strauss complained that the rationalists, who also destroyed the supernatural character of history, were allowed to remain in the seminaries because, in arguing that there was some historical basis behind the myths, they falsely encouraged people to continue to believe in their historicity. He, by contrast, was more honest. See the quotation from Strauss in Harris, *Strauss and His Theology,* p. 62.
22. Barth, *Protestant Theology,* p. 546.
23. Schweitzer, *Quest of the Historical Jesus,* p. 85.
24. Ibid., p. 111.
25. Quoted by Strauss, from Hengstenberg's *Evangelische Kirchenzeitung,* in Strauss's reply to Hengstenberg in *Streitschriften zur Vertheidigung meiner Schrift über das Leben Jesu,* III, 7.
26. Ibid., p. 8. See also pp. 12–13.
27. Quoted by Schweizer from Hengstenberg's critique of Strauss's reply to Hengstenberg in *Quest of the Historical Jesus,* p. 106.
28. A. S. Peake, "The History of Theology," pp. 131–184 in *Germany in the Nineteenth Century* (Freeport, N.Y.: Books for Libraries Press, 1967), p. 145.
29. Zeller, *David Friedrich Strauss,* p. 6.
30. Strauss, *Streitschriften,* III, 57. His comment on Hegel's lack of appreciation of historical criticism comes four pages later. For arguments against any essential and lasting Hegelian motivation in Strauss's theology, a position set down in 1952 in C. Hartlich and W. Sachs, *Der Ursprung des Mythosbegriffes in der modernen Bibelwissenschaft,* see Hans W. Frei, *The Eclipse of Biblical Narrative: A Study in Eighteenth and Nineteenth Century Hermeneutics* (New Haven: Yale University Press, 1974), pp. 242, 336, n. 5; Harris, *Strauss and His Theology,* pp. 135–137, 271; and Peake, "History of Theology," p. 145.
31. Quoted by Harris in *Strauss and His Theology,* p. 60.
32. Brazill, *The Young Hegelians,* p. 102.
33. Strauss, *Das Leben Jesu,* II, 709.
34. Toews, *Hegelianism,* p. 266.
35. Cf. Harris, *Strauss and His Theology,* p. 80.
36. Marilyn Chapin Massey, *Christ Unmasked: The Meaning of the Life of Jesus in German Politics* (Chapel Hill: University of North Carolina Press, 1983), pp. 92–93.

37. In the second and especially the third edition Strauss made some concessions. He softened his claims about the lack of authenticity of John's gospel and even allowed some rationalistic explanations of portions of the gospel narratives. When the concessions were taken as signs of inconsistency and contradiction, Strauss dropped them in the fourth edition of 1840. See Toews, *Hegelianism*, pp. 276–278, 281–282.
38. Gregory, *Scientific Materialism*, pp. 22–33.
39. Ludwig Feuerbach, *Das Wesen des Christentums*, in *Sämtliche Werke*, VIII, 9–11.
40. Cf. Harris, *Strauss and His Theology*, p. 137.
41. Zeller, *David Friedrich Strauss*, p. 72. On the role Feuerbach played in Strauss's abandoning of Hegel's claim of the equality of religion and philosophy, see Reuschle, *Philosophie und Naturwissenschaft*, p. 22; Brazill, *The Young Hegelians*, p. 114.
42. D. F. Strauss, *Die Christliche Glaubenslehre in ihrer geschichtlichen Entwicklung und im Kampfe mit der modernen Wissenschaft dargestellt*, 2 vols. (Tübingen: Osiander, 1840–1841), I, 356.
43. Cf. Gregory, *Scientific Materialism*, p. 41.
44. Strauss, *Glaubenslehre*, I, 623–25.
45. Ibid., p. 679.
46. Ibid., pp. 681–682.
47. Ibid., p. 682. Replies to objections to this theory are discussed on pp. 682–684.
48. Ibid., pp. 685–686.
49. Harris, *Strauss and His Theology*, p. 141.
50. Ibid., p. 194. The quotation is taken from the *Gespräche* of 1864.
51. There was considerable activity on this topic in the 1860s alone. For a brief summary of it, see Jack C. Verheyden's Introduction to Schleiermacher's *Life of Jesus*, pp. xv–xvi.
52. Strauss agreed with the views of the Tübingen school that Matthew was the oldest synoptic gospel, Mark the most recent, with Luke between the two. All emerged in the second century A.D. For Baur's view on the dates of the synoptic gospels, see Harris, *The Tübingen School*, pp. 209–213.
53. Strauss conceded that the mythologizing could not have been as unconscious as he had once thought. Baur's work had convinced him that deliberate mythologizing played a greater role than he had realized. *Das Leben Jesu für das deutsche Volk bearbeitet*, 4th ed., in Eduard Zeller, ed., *Gesammelte Schriften*, 12 vols. (Bonn: Verlag von Emil Strauss, 1876–1878), III, 200–201.

54. Gregory, *Scientific Materialism*, chap. 1.
55. Ibid., pp. 156, 213.
56. Friedrich Nietzsche, "David Friedrich Strauss as Confessor and Writer," pp. 3–55 in *Untimely Meditations*, trans. R. J. Hollingdale (New York: Cambridge University Press, 1983), p. 27.
57. Strauss, "Ein Nachwort als Vorwort," pp. 255–278 in *Der alte und der neue Glaube*, 4th ed. (1872), in *Gesammelte Schriften*, VI, 271.
58. Nietzsche, "David Friedrich Strauss," pp. 8–9.
59. Quoted from a letter to Adolf Rapp in J. P. Stern, Introduction to F. Nietzsche, *Untimely Meditations*, p. xiv.
60. D. F. Strauss, *Der alte und der neue Glaube: Ein Bekenntniss* (Leipzig: Hirzel, 1872), p. 9. Strauss's brother some years earlier had asked him to write a catechism for liberally minded people. At the time Strauss said it was not a task he could do. Cf. Theobald Ziegler, *David Friedrich Strauss*, 2 vols. (Tübingen: Trübner, 1908), II, 676.
61. Adolf Rapp, *Briefwechsel zwischen Strauss und Vischer*, 2 vols. (Stuttgart: Klett, 1953), II, 294; Zeller, *David Friedrich Strauss*, pp. 133–134. In the poems Strauss left behind there is the following revealing statement: "Carelessly it seems expressed, / Though not carelessly achieved; / Within a few weeks is compressed, / That which was through years conceived" (Zeller, p. 135).
62. D. F. Strauss, *Krieg und Friede: Zwei Briefe von David Friedrich Strauss an Ernst Renan und dessen Antwort* (Leipzig: Insel, 1870), pp. 12–13. See also Strauss's biography of Ulrich von Hütten, in which France is characterized as Germany's "oppressor for centuries." Strauss, *Ulrich von Hütten: His Life and Times*, trans. Jane Sturge, reprint of the 1874 edition (New York: AMS Press, 1970), p. vi.
63. Strauss, *Der alte und der neue Glaube*, pp. 5–6.
64. Ibid., pp. 17–18.
65. Ibid., pp. 16, 20–21.
66. Ibid., pp. 27–28, 30–31.
67. Ibid., pp. 36–37.
68. Ibid., pp. 42–46. Strauss was particularly impatient with Schleiermacher's defense of the authenticity of the Gospel of John while conceding the inauthenticity of the first three gospels (pp. 49–51).
69. Ibid., p. 79.
70. Ibid., p. 85.
71. Ibid., p. 90.
72. Ibid., p. 92.
73. Ibid., pp. 94–96.

74. Ibid., p. 103.
75. Ibid., p. 106.
76. Ibid., p. 111.
77. Ibid., pp. 114–116.
78. Ibid., p. 119.
79. Ibid., pp. 119–120. Strauss's favorable citation of Schleiermacher here is balanced by his earlier strong criticism of Schleiermacher's defense of prayer as important for its subjective influence on the one praying rather than for any possible objective result. Strauss viewed this as an inexcusable justification of deliberate illusion (pp. 108–110).
80. Ibid., pp. 132–133.
81. Nietzsche, *Untimely Meditations*, p. 17.
82. Strauss, *Der alte und der neue Glaube*, pp. 140–141.
83. Ibid., p. 145.
84. Ibid., p. 146.
85. Ibid., pp. 146–147.
86. Ibid., p. 149.
87. Ibid., pp. 156–165. In his account of the original motion of the nebulous matter of the solar system Strauss appealed to the very same reason Copernicus used to justify the rotation of the earth: rotation was natural to a sphere (p. 156).
88. Ibid., p. 168.
89. Ibid.
90. Ibid., pp. 170–171.
91. Cf. Emil DuBois-Reymond, "Über die Grenzen des Naturerkennens," in *Ein Vortrag in der zweiten öffentlichen Sitzung der 45. Versammlung deutscher Naturforscher und Ärzte zu Leipzig am 14. August 1872* (Leipzig: Veit, 1872).
92. Strauss, *Der alte und der neue Glaube*, p. 205.
93. Ibid., p. 207.
94. Strauss, "Ein Nachwort als Vorwort," p. 268.
95. Ibid., pp. 269–270.
96. Ziegler, *David Friedrich Strauss*, II, 677.
97. Ibid., p. 693.
98. Eduard Zeller, ed., *Ausgewählte Briefe* (Bonn: Emil Strauss, 1895), p. 506.
99. Strauss, *Der alte und der neue Glaube*, pp. 175–176. See also pp. 181–182.
100. Ibid., p. 177.
101. Ibid., p. 175. See also pp. 174–175.

102. Ibid., p. 184.
103. Ibid.
104. Ibid., p. 186.
105. Ibid., p. 187.
106. See Peter Bowler's summary of it in *Evolution: The History of an Idea* (Berkeley: University of California Press, 1984), pp. 52 and esp. 170–171.
107. Strauss, *Der alte und der neue Glaube*, pp. 189–191.
108. Ibid. p. 193.
109. Ibid., p. 200.
110. Ibid., p. 194.
111. Ibid., pp. 210–211.
112. Ibid., pp. 220–221.
113. Ibid., p. 222. See also Ziegler, *David Friedrich Strauss*, II, 702–703; Reuschle, *Philosophie und Naturwissenschaft*, p. 96.
114. Strauss, *Der alte und der neue Glaube*, pp. 208–209.
115. Ibid., p. 210.
116. Ibid., pp. 226–231.
117. Ibid., p. 239.
118. Ibid., p. 241.
119. Ibid., p. 252. On Strauss's inability to obtain a divorce from Agnese von Schebst, see Harris, *Strauss and His Theology*, p. 231.
120. Strauss, *Der alte und der neue Glaube*, pp. 268–269, 278.
121. Ibid., pp. 280–293.
122. Letter quoted in Harris, *Strauss and His Theology*, pp. 99–100.
123. Rapp, *Briefwechsel zwischen Strauss und Vischer*, II, 262. His depression in 1868 was no doubt exacerbated by his final breakup with Marie Locher in the fall of 1867. Locher was a woman with whom he had developed an intimate relationship after 1866, but whom he could never marry because Agnese von Schebst refused to grant him a divorce. See also n. 119 above; Ziegler, *David Friedrich Strauss*, II, 672–675; Harris, *Strauss and His Theology*, pp. 230–232.
124. Strauss, "Ein Nachwort als Vorwort," pp. 259–260, 263–264.
125. Zeller, *David Friedrich Strauss*, pp. 141–142.
126. Strauss, "Ein Nachwort als Vorwort," p. 278.
127. Ziegler, *David Friedrich Strauss*, II, 724.
128. Rapp, *Briefwechsel zwischen Strauss und Vischer*, II, 295. In spite of Vischer's personal visit to Strauss to explain his position, Vischer's silence led to the final break between them.
129. See n. 56 above.

130. Zeller, *David Friedrich Strauss*, p. 544.
131. Zeller, *Ausgewählte Briefe*, p. 468. Even Theobald Ziegler, for whom Strauss was a hero, acknowledged that in *The Old Faith and the New* Strauss had overlooked a critical examination of the limits of reason at a time when his friends were sharing a widespread sympathy for the neo-Kantian revival in Germany. Ziegler, *David Friedrich Strauss*, II, 700–702.
132. D. F. Strauss, "Die Asteroiden und die Philosophen," pp. 402–407 in Strauss, *Kleine Schriften* (Leipzig: Brockhaus, 1862).
133. Rapp, *Briefwechsel zwischen Strauss und Vischer*, II, 295.
134. In early 1869 Biedermann sent Strauss a copy of his *Dogmatics*, which, Strauss confessed, he could not understand. He speculated to Biedermann that his problem might be that he was "not ensured against the sirens of materialism through the firm support of a philosophical system like you are," adding that he often viewed materialism as the equally justified brother "of our Hegelian idealism" (Ziegler, *David Friedrich Strauss*, II, 697).
135. Ibid., p. 722. Nevertheless Biedermann did not oppose Strauss in writing until after the latter's death.
136. Ibid., pp. 716–717. Ziegler conceded that Strauss was being overly materialistic at a time when scientific materialism à la Karl Vogt, Ludwig Büchner, and Jakob Moleschott was no longer the mode (pp. 69, 695). The series of articles Ziegler composed in Strauss's defense gave Strauss much joy (pp. 724–727), but, Ziegler claimed, later in his career he did not receive a post at the gymnasium in Ulm because of his backing of Strauss (ibid., I, 232, n. 2).
137. Reuschle, *Philosophie und Naturwissenschaft*, pp. vi, 97–98.
138. Ibid., p. 108.
139. E. G. Steude, "Der Darwinismus die Erfüllung des Christentums?" *Beweis des Glaubens* 31 (1895): 122.
140. Ernst Haeckel, *Die Welträtsel: Gemeinverständliche Studien über monistische Philosophie* (Bonn: Emil Strauss, 1899), pp. 357–358.
141. Ziegler, *David Friedrich Strauss*, II, 724.

4. Otto Zöckler, the Orthodox School, and the Problem of Creation

1. On this delay see Gregory, "Impact of Evolution on Protestant Theology," pp. 372–374.
2. Victor Schultze, "Otto Zöckler," pp. 704–708 in *Realencyklopädie*

für protestantische Theologie und Kirche, vol. 21 (Leipzig: Hinrichs, 1908), p. 706.
3. Theodor Zöckler et al., *Erinnerungsblätter* (Gütersloh: Bertelsmann, 1906), pp. 5, 7.
4. Ibid., p. 8. The electrophore, a device which Volta claimed supplied electricity inexhaustibly, was invented by the Italian eletrician in 1775. See John Heilbron, *Electricity in the Seventeenth and Eighteenth Centuries* (Berkeley: University of California Press, 1979), pp. 416–417.
5. Schultze, "Zöckler," p. 704.
6. Barth, *Protestant Theology,* p. 625.
7. Hermann Jordan, "Otto Zöckler," pp. 148–151 in *Biographisches Jahrbuch und deutscher Nekrolog,* vol. 11 (Berlin: Reimer, 1908), p. 149; See also Schultze, "Zöckler," p. 705.
8. Jordan confirmed from conversation with the scholar who was dean of the faculty of philosophy at Giessen in 1854 that Zöckler graduated without writing a dissertation. See H. Jordan, "Verzeichnis der literarischen Veröffentlichungen Otto Zöcklers," pp. 707–747 in Otto Zöckler, *Geschichte der Apologie Christentums* (Gütersloh: Bertelsmann, 1907), p. 709, n. 1. On the cool reception at Giessen, see Schultze, "Zöckler," p. 705, and T. Zöckler, *Erinnerungsblätter,* p. 23. Theodor Zöckler maintains it was his father's orthodox views plus the little time Zöckler spent with the Protestant theologians in Giessen that soured them on him.
9. T. Zöckler, *Erinnerungsblätter,* pp. 20–21.
10. Ibid., p. 35.
11. Ibid., pp. 32–34.
12. Quoted from Hengstenberg's *Evangelische Kirchenzeitung* by Zöckler in "Über die neueste Physikotheologie der Engländer, verglichen mit verwandten Bestrebungen und Leistungen der Deutschen," *Jahrbücher für deutsche Theologie* 5 (1860): 762.
13. Zöckler, *Theologia Naturalis: Entwurf einer systematischen Naturtheologie vom offenbarungsgläubigen Standpunkte aus* (Frankfurt: Heyder und Zimmer, 1860), p. iv.
14. Ibid., pp. 159, 174.
15. Ibid., p. 159.
16. Ibid., pp. 167, 174.
17. Ibid., p. vi.
18. Zöckler, "Über die neueste Physikotheologie der Engländer," p. 764, n. 1.
19. Ibid., p. 798.

20. Ibid., pp. 763–765. Curiously Zöckler did see in the plurality-of-worlds controversy a hint that English natural theology was breaking out from its confinement within the old argument from design to a higher standpoint which addressed the broader question of the relation between Christian revelation and natural facts (p. 798).
21. T. Zöckler, *Erinnerungsblätter,* pp. 41–42. He also received encouraging letters from Dorner and his old gymnasium teacher Vilmar. On Zöckler's assumption of the editorship of the *Evangelische Kirchenzeitung,* see p. 71.
22. Ibid., pp. 39–40.
23. Schultze, "Zöckler," p. 705.
24. Zöckler, "Über die neueste Physikotheologie der Engländer," p. 769. Zöckler recommended the review of the *Origin* "by an opponent of the transmutation theory" in the *Edinburgh Review.* The review, though not signed, is known to have been written by Richard Owen. Cf. Gertrude Himmelfarb, *Darwin and the Darwinian Revolution* (New York: W. W. Norton, 1962), p. 277; Peter Brent, *Charles Darwin: A Man of Enlarged Curiosity* (New York: W. W. Norton, 1981), p. 433.
25. Baden Powell, *The Unity of Worlds and of Nature* (London: Longman, 1856), p. 424.
26. Ibid., pp. 431–443. For Zöckler's view of Powell, see "Über die Speziesfrage nach ihrer theologischen Bedeutung mit besonderer Rücksicht auf die Ansichten von Agassiz und Darwin," *Jahrbücher für deutsche Theologie* 6 (1861): 667.
27. Zöckler, "Über die Speziesfrage," p. 668.
28. Ibid.
29. Ibid., p. 675.
30. Ibid.
31. Ibid., p. 677.
32. Ibid., p. 684.
33. Ibid., pp. 685, 687–688. Zöckler, following Darwin's German translator H. G. Bronn, preferred the German word *Züchtung* to *Auswahl;* hence he referred here to nature's "unbewusste züchtende Tätigkeit" (p. 685).
34. Ibid., p. 688.
35. Ibid., p. 689, n. 3.
36. Ibid., pp. 692–695.
37. Ibid., p. 698.
38. Ibid., p. 702.
39. Ibid., pp. 707–708.

40. Ibid., p. 707.
41. Ibid., p. 708.
42. Ibid., p. 710. See also p. 709. Creationists in the twentieth century also argue that geological processes proceeded at a faster rate in the past than at present. They declare that "the present is not the key to the past" to establish that the earth's geological features cannot be accounted for by uniformitarianism. Cf. Henry Morris, *Scientific Creationism* (San Diego: Creation-Life Publishers, 1974), pp. 101–102.
43. Zöckler, "Über die Speziesfrage," p. 704.
44. Ibid., p. 713.
45. For example, the Göttingen anatomist Rudolph Wagner was criticized by colleagues for permitting himself to be drawn into debate with Karl Vogt in a common newspaper. See Gregory, *Scientific Materialism,* p. 230, n. 92.
46. Schultze, "Zöckler," p. 705.
47. Zöckler, *Kritische Geschichte der Askese: Ein Beitrag zur Geschichte christlicher Sitte und Kultur* (Frankfurt: Heyder und Zimmer, 1863). Because of its standing, it was reissued in 1897 under the title *Askese und Mönchtum.*
48. Quoted from "Über die notwendige Einigung von kirchlichem Bekenntnis und christlichem Leben" by Schultze, "Zöckler," p. 706.
49. T. Zöckler, *Erinnerungsblätter,* pp. 80–81. On Zöckler's irenic position regarding union and the function of confession, see also pp. 20, 39–40, 57–58, 68–69, 71, 79, 92.
50. Quoted from the *Katholische Zeitschrift* endorsement of *Beweis des Glaubens* in an advertisement of Bertelsmann, the publisher of *Beweis des Glaubens.*
51. This according to Schultze, "Zöckler," p. 706. Theologians in Greifswald included Ludwig Diestel (Old Testament history), Hermann Reuter (church history), Johann Wilhelm Hanne (Christology), and Karl Wieseler (New Testament history). See the index of German theologians in Mildenberger, *Geschichte der deutschen evangelischen Theologie,* supplement.
52. Zöckler, *Über Schöpfungsgeschichte und Naturwissenschaft* (Gotha: Perthes, 1869), p. 29.
53. See Gregory, "Impact of Evolution on Protestant Theology," pp. 375–378.
54. Joseph Altholtz remarks that "biblical criticism, largely a German product, was brought to England relatively late in the century,"

Notes to Pages 130–133 | 307

noting that the roots of the movement in Germany went back to the eighteenth century. Altholtz, *Churches in the Nineteenth Century,* pp. 134–135.

55. Cf. Gregory, *Scientific Materialism,* pp. 70, 170, 176.
56. Cf. above, n. 12.
57. Zöckler, "Zur Lehre der Schöpfung: Der theistische Schöpfungsbegriff im Kampfe mit den Theorien des Materialismus, Pantheismus und Deismus," *Jahrbücher für deutsche Theologie* 9 (1864): 688–759.
58. Ibid., p. 688.
59. Ibid., p. 690. Zöckler cited H. G. Volger's *Naturalische Geschichte der Erde* (1857) as another work that defended similar claims. See also Zöckler's treatment of Czolbe in "Über die Speziesfrage," pp. 683ff. On Czolbe's so-called stability theory, see Gregory, *Scientific Materialism,* pp. 130–131.
60. Zöckler, "Zur Lehre der Schöpfung," p. 693.
61. Ibid., pp. 693–694.
62. Zöckler explained the retreat of Hegelian idealism as the result of "the realistic tendency of present-day natural science, the basic orientation of our time, evident everywhere, to the concrete in nature and to sense empiricism" (ibid., p. 694).
63. Ibid., p. 703. German's work, *Schöpfergeist und Weltstoff, oder die Welt im Werden,* appeared in 1862. Other pantheistic works reviewed included Theodor Rohmer's *Gott und seine Schöpfung* of 1857 (pp. 695-697), Adolph Cornhill's *Materialismus und Idealismus in ihren gegenwärtigen Entwicklungskrisen* of 1858 (pp. 698–699), Carl Gustav Carus's *Natur und Idee oder das Werdende und sein Gesetz* of 1861 (pp. 704–706).
64. The review is found on pages 704–722 of "Zur Lehre der Schöpfung." See esp. pp. 714, 720.
65. Zöckler, *Die Urgeschichte der Erde und des Menschen* (Gütersloh: Bertelsmann, 1868), p. 73.
66. "Zur Lehre der Schöpfung," pp. 725, 730.
67. Ibid., p. 733.
68. Two such were C. H. Weisse in volume 2 of his *Philosophische Dogmatik* of 1860 and Hermann Ulrici's *Gott und die Natur* of 1862. See "Zur Lehre der Schöpfung," pp. 733–735 and 744–746. Zöckler's concern about Ulrici was well placed, for the dismissal of Darwinism here would be replaced by a qualified willingness to consider it in the third edition of 1875. Cf. Hermann Ulrici, *Gott und die Natur,* 3d ed. (Leipzig: Weigel, 1875), pp. 387, 410.
69. Zöckler, "Zur Lehre der Schöpfung," p. 758.

70. Ibid., p. 759.
71. Zöckler, *Über Schöpfungsgeschichte und Naturwissenschaft*, p. 50. See also *Die Urgeschichte der Erde und des Menschen*, p. 71.
72. Zöckler, *Über Schöpfungsgeschichte und Naturwissenschaft*, pp. 36–37.
73. Zöckler, *Die Urgeschichte der Erde und des Menschen*, pp. 4, 21, 31. Some of Zöckler's contemporaries did feel obligated to assert a geocentric view, an attitude Zöckler called curious (p. 4).
74. Ibid., pp. 33–34.
75. Ibid., p. 46.
76. Ibid., p. 47.
77. Ibid., pp. 43, 47–48, 52, n. 16. Zöckler also provided exegetical difficulties for the restitutionists (p. 48).
78. Zöckler, "Der Mosaische Schöpfungsbericht und die neue Naturwissenschaft," *Evangelische Kirchenzeitung* 106/107 (1880): 475–476. In his *Geschichte der Beziehungen zwischen Theologie und Naturwissenschaft mit besonderer Rücksicht auf Schöpfungsgeschichte* the old and new concordance hypotheses are treated in II, 497–510 and 538–553, respectively. On the *Geschichte der Beziehungen zwischen Theologie und Naturwissenschaft*, see below, n. 106.
79. Zöckler, *Die Urgeschichte der Erde und des Menschen*, p. 42.
80. Ibid., pp. 64–65.
81. Ibid., pp. 57–58. All other life, however, was invertebrate.
82. Ibid., p. 60.
83. Ibid., p. 61.
84. Ibid., p. 63.
85. Ibid.
86. Zöckler, "Über die Anfänge der menschlichen Geschichte," *Zeitschrift für lutherische Theologie und Kirche* 30 (1869): 218.
87. Zöckler, *Die Lehre vom Urstand des Menschen, geschichtlich und dogmatisch-apologetisch Untersucht* (Gütersloh: Bertelsmann, 1879), p. 304. In *Schöpfungsgeschichte und Naturwissenschaft*, Zöckler referred to "the 6,000 years that Holy Writ permits since the creation of the first humans" (p. 352). This reference is ambiguous, since it is not clear if Zöckler meant that the Genesis account permitted 6,000 years prior to the composition of the Mosaic record or 6,000 years from his own day. That he intended the latter is clear when he questions extending "the 4,000 year span from Adam to Christ" to 5,000 years based on alternative texts of the Old Testament. See "Über die Anfänge der menschlichen Geschichte," p. 218, and below, n. 101. For Zöckler's later reviews of treatments of primitive flood accounts both by those sympathetic and those unfavorable to

descendance theory, see "Neue Flut-Theorien," *Beweis des Glaubens* 30 (1894):432–437, and "Neue Flut-Phantasien," *Beweis des Glaubens* 31 (1895): 208–212.
88. Zöckler, *Die Urgeschichte der Erde und des Menschen*, pp. 140–141. Earlier in 1863 Zöckler indicated that the Florida skull was reputed to be 185,000 years old. Cf. "Die einheitliche Abstammung des Menschengeschlechts," *Jahrbücher für deutsche Theologie* 8 (1863): 55.
89. Zöckler, *Die Urgeschichte der Erde und des Menschen*, pp. 143–145; *Schöpfungsgeschichte und Naturwissenschaft*, p. 40.
90. Zöckler, *Schöpfungsgeschichte und Naturwissenschaft*, pp. 38–39; emphasis his.
91. Zöckler, *Die Urgeschichte der Erde und des Menschen*, pp. 146–147.
92. Zöckler, "Über die Anfänge der menschlichen Geschichte," p. 213.
93. Ibid., p. 214.
94. Ibid., pp. 215–216.
95. Ibid., p. 216.
96. Ibid., p. 219.
97. Zöckler, *Die Urgeschichte der Erde und des Menschen*, p. 149.
98. See Gregory, *Scientific Materialism*, pp. 72–75.
99. Zöckler, "Die einheitliche Abstammung des Menschengeschlechts," pp. 87–88.
100. Zöckler, "Über die Anfänge der menschlichen Geschichte," p. 227. And yet Zöckler did not hold that all blacks were necessarily inferior or without a high religious capacity. See his review of Wilhelm Schneider's *Die Naturvölker* entitled "Die Civilizationsfahigkeit der Wilden," *Beweis des Glaubens* 22 (1886): 355–358.
101. Zöckler, *Die Urgeschichte der Erde und des Menschen*, p. 137. See also p. 109. Zöckler conceded that even then these investigations generally demanded anywhere from 10,000 to 100,000 years for humanity's existence, while the whole scope of Darwinian transmutation required millions or billions of years. Later Zöckler approached the whole question of the age of the earth from an exegetical point of view. Depending on the Old Testament text used and the theological authority consulted, different figures emerged for the number of years since the creation. Less well known than Archbishop Usher's calculation of the date of creation (4004 B.C.) was that of Rabbi Hillel ha Nassi, who, using the Greek Septuagint text of the Old Testament, determined around A.D. 350 that there had been 5,640 years since creation.

Zöckler examined numerous different versions of the text and demonstrated that, based on an exegesis of Holy Writ, the outside limit for the amount of time since the creation was 7,500 years. Even if this estimate were selected as the one most compatible with Scripture, argued Zöckler, there was no real gain for those scientists calling for hundreds of thousands of years. (*Die Lehre vom Urstand des Menschen,* pp. 290–292.)

102. Zöckler, *Schöpfungsgeschichte und Naturwissenschaft,* p. 6.
103. Ibid., p. 7.
104. Ibid.
105. Ibid., pp. 9, 13.
106. Zöckler, *Geschichte der Beziehungen zwischen Theologie und Naturwissenschaft mit besonderer Rücksicht auf Schöpfungsgeschichte,* 2 vols. (Gütersloh: Bertelsmann, 1877–1879), I, 16, n. 4. The popular piece, written by W. Hoffmann and entitled "Die biblische Schöpfungskunde in ihrer Auslegung," appeared in the periodical *Deutschland* (1872), pp. 191–287.
107. Ibid., I, 11. See also I, vi–viii, 6–10.
108. Ibid., 1–2. Later Zöckler absolved the Englishman David Brewster, author of *The Martyrs of Science,* from sharing the views of "that shallow and biased 'conflict-literature' " (p. 14, n. 2).
109. Ibid., I, 12, n. 1. Other authors and works put into the same category included the second edition of J. W. Draper, *History of the Conflict between Religion and Science* (1875), G. H. Lewes, *History of Philosophy from Thales to Comte* (1873, 1876), the Frenchman L. Figuier's studies of the Renaissance and the seventeenth century, and the followers of Haeckel who wrote on history in the monist journal *Kosmos.* Cf. ibid., I, 13–14, n. 1.
110. Cf. Gregory, "Impact of Evolution on Protestant Theology," pp. 384–385.
111. Charles Hodge, *What Is Darwinism?* (New York: Scribner, Armstrong Co., 1874), p. 142.
112. Zöckler, *Die Lehre vom Urstand des Menschen,* p. 293.
113. See White's copy of the *Geschichte der Beziehungen zwischen Theologie und Naturwissenschaft,* II, 358, located in the Cornell University Library.
114. Zöckler, "Der Mosaische Schöpfungsbericht und die neue Naturwissenschaft," pp. 475–476.
115. Zöckler, *Die Lehre vom Urstand des Menschen,* p. 293.
116. Zöckler, "Die Lehre Darwins," *Theologisches Literaturblatt* 13 (1892): 265–266.

117. Gregory, "Impact of Evolution on Protestant Theology," pp. 376, 384.
118. Zöckler, *Geschichte der Beziehungen zwischen Theologie und Naturwissenschaft*, II, 599. One year after volume 2 of the *History* appeared Zöckler published a short study of Erasmus Darwin and his relationship to Charles. There he again referred to the growth of "a remarkable branch of literature devoted to monographical treatment of individual 'precursors of Darwinism.' " Cf. "Darwins Grossvater als Arzt, Dichter und Naturphilosoph," pp. 127–158 in W. Frommel and F. Pfaff, eds., *Sammlung von Vorträgen für das deutsche Volk*, vol. 3 (Heidelberg: Winter, 1880), p. 129. Zöckler's portrait of Erasmus showed him as one who anticipated many specific concerns of Charles's later study, including such phenomena as the inheritance of instinct, insect mimicry, rudimentary organs, and the common origin of species. Further, Erasmus's deism was described in terms not unlike those Zöckler had used for Charles. The latter did not need Lamarck, wrote Zöckler, for his relation to Erasmus was one of the most significant cases of the inheritance of mental characteristics and talents (pp. 148–154).
119. Zöckler, "Die Lehre Darwins," p. 268.
120. Zöckler, *Geschichte der Beziehungen zwischen Theologie und Naturwissenschaft*, II, 618.
121. Ibid., p. 615. In a piece from 1872 Zöckler drew an even sharper distinction between final scientific truth and temporary results. Not only had experiments chased away the 1,500-year-old illusion of scholastic philosophy, but the achievements of Newton had forever removed heliocentrism from the ranks of the hypothetical. Cf. "Der Himmel des Naturforschers und der Himmel des Christen," pp. 181–205 in W. Frommel and F. Pfaff, eds., *Sammlung von Vorträgen für das deutsche Volk*, vol. 7 (Heidelberg: Winter, 1882), pp. 183–184.
122. Zöckler, "Der Darwinismus und seine Gegner," *Zeitschrift für die gesamte lutherische Theologie und Kirche* 32 (1871): 256. See also p. 249.
123. Zöckler, "Die Moral des Darwinismus," *Zeitschrift für die gesamte lutherische Theologie und Kirche* 34 (1873): 76.
124. Ibid., p. 78. The phrase "literally brutal theory of human origin" was borrowed from a critic of Darwin named Andree (see ibid., n. 1).
125. Zöckler, *Geschichte der Beziehungen zwischen Theologie und Naturwissenschaft*, II, 622–624.

126. Ibid., p. 632. See also pp. 620–621.
127. Ibid., p. 623.
128. Ibid., p. 618.
129. The four groups included those who held dogmatically to the fixity of species, those who permitted some form of development from inner causes, those who held that Darwin's hypothesis had great scientific value even if it were not yet finally established, and radical dogmatists of descendance. See ibid., II, 658–667.
130. Ibid., pp. 670–671.
131. Ibid., p. 668. Eight years earlier Zöckler quoted with approval Friedrich Michelis's similar views. Cf. "Der Darwinismus und seine Gegner," p. 247.
132. See Zöckler, *Geschichte der Beziehungen zwischen Theologie und Naturwissenschaft*, II, 697–709, 711–717. On the English propensity toward reconciliation, see "Christentum und Darwinismus," *Theologisches Literaturblatt* 15 (1894): 529.
133. Zöckler, *Geschichte der Beziehungen zwischen Theologie und Naturwissenschaft*, II, 717. Much later, when reviewing a critique of Julius Wellhausen's dating of Deuteronomy and the Priestly Codex of the Old Testament, Zöckler made the following observation in passing: "For our century the philosophical (Hegel-Vatke) and natural scientific (Darwin) idea of development from lower to higher is perhaps unconsciously so deeply embedded in our blood that one sacrifices everything else for this theory, eliminating the thought of revelation in preference to it, or fashioning revelation in such a way that only a caricature is left." See "Gegen den Evolutionismus auf dem Gebiete der Religionsgeschichte, insbesondere der alttestamentlichen," *Beweis des Glaubens* 40 (1904): 58–59.
134. Zöckler, *Geschichte der Beziehungen zwischen Theologie und Naturwissenschaft*, II, 719.
135. Ibid., pp. 719–730.
136. Ibid., p. 730.
137. Ibid., p. 731.
138. Ibid., pp. 731–791.
139. Ibid., p. 792.
140. Ibid., p. 794. Zöckler claimed that this "psychophysics of morality" and "anthropology of crime," where prisons became insane asylums, began in Austria, where he hoped it would stay.
141. Ibid., p. 795. See also "Die Moral des Darwinismus," p. 90.
142. Zöckler, "Die Moral des Darwinismus," pp. 82–83, 89. See also

Geschichte der Beziehungen zwischen Theologie und Naturwissenschaft, II, 793.
143. Zöckler, *Geschichte der Beziehungen zwischen Theologie und Naturwissenschaft,* II, 797. See also "Die Moral des Darwinismus," p. 81.
144. Zöckler, "Christentum und Darwinismus," p. 531. A more detailed presentation of the same theme was carried out by Zöckler's collaborator E. G. Steude in the *Beweis des Glaubens* in 1895 under the title "Der Darwinismus die Erfullung des Christentums?" *Beweis des Glaubens* 31 (1895): 121–133.
145. Zöckler, *Geschichte der Beziehungen zwischen Theologie und Naturwissenschaft,* II, 798–799.
146. E. G. Steude, "Der Apologet," pp. 95–118 in T. Zöckler, *Erinnerungsblätter,* p. 117.
147. Ibid., p. 96.
148. Victor Schultze, "Ansprache am Sarge," pp. 124–126 in T. Zöckler, *Erinnerungsblätter,* p. 124.

5. Rudolf Schmid and the Reconciliation of Science and Religion

1. Quoted by Paul F. Boller, Jr., from *The American Church Quarterly Review* in Boller's *American Thought in Transition: The Impact of Evolutionary Naturalism, 1865–1900* Chicago: Rand McNally and Co., 1969), p. 24.
2. Rudolf Schmid, *Lebenserinnerungen aus meinem Leben: Aufzeichnungen für die Seinigen* (Konstanz: Komissionsverlag von Karl Gess, 1909), p. 110. See also K. Keeser, "Rudolf Schmid," *Schwäbischer Merkur,* no. 401 (1907): 5a.
3. Schmid, *Lebenserinnerungen,* pp. 9, 24.
4. Latin schools in Germany originated during the Middle Ages to educate the lay citizenry. At that time they stood in contrast to church schools. During the Reformation, under the influence of Melanchthon, the study of Latin was emphasized as the only means of coming to an understanding of Scripture and of true religion. See Ernest Helmreich, *Religious Education in German Schools: An Historical Approach* (Cambridge, Mass.: Harvard University Press, 1959), pp. 5, 18–19.
5. Schmid, *Lebenserinnerungen,* p. 11.
6. Ibid., p. 13.
7. Ibid., pp. 33–34.

8. Ibid., pp. 36–37.
9. Ibid., p. 37.
10. Ibid., p. 38.
11. Ibid., p. 39.
12. Ibid., p. 40.
13. The account of Schmid's participation in the events of 1848–49 is given in chapter 4 of his *Lebenserinnerungen* and is also related by Karl Gussmann in "Zwei Schwäbische Freischärler," *Der Schwabenspiegel* 4 (1911): 233–234, 245–246.
14. Schmid, *Lebenserinnerungen*, p. 48.
15. Ibid., pp. 50–51.
16. Ibid., pp. 51–52. Schmid added that he realized then that there simply was not going to be a social relief program from above. At the end of his life he noted that only through the advent of social democracy and Kaiser Wilhelm I's open declaration of his support for the idea of social assistance was anything accomplished (p. 52).
17. Julius Köstlin established himself as a Luther scholar and should not be mistaken for Karl Köstlin, the higher critic and member of the Tübungen school. On Karl Köstlin, see Horton Harris, *The Tübingen School*, chap. 6. For Schmid's recollection of Karl Köstlin, see *Lebenserinnerungen*, p. 27.
18. Schmid, "Der Herzog von Argyll und sein schwäbischer Erzieher," *Schwäbischer Merkur*, no. 209 (1914): 5.
19. Keeser, "Rudolf Schmid," p. 5a.
20. See *Lebenserinnerungen*, pp. 66–79, for extended descriptions of the duke, the duchess, the children, and the famous individuals listed. The duke of Sutherland was the father of the duchess of Argyll.
21. Schmid, "Die durch Darwin Angeregte Entwickelungsfrage, Ihr Gegenwärtiger Stand und ihre Stellung zur Theologie," *Theologische Studien und Kritiken* 48 (1875): 7–60; *Die Darwin'schen Theorien und ihre Stellung zur Philosophie, Religion und Moral* (Stuttgart: Moser, 1876). Schmid's friend the duke of Argyll provided an Introduction to the English translation of the *Die Darwin'schen Theorien*, which appeared in 1883 and in a second edition in 1885.
22. Schmid, "Robert Mayer, der grosse Förderer unserer heutigen wissenschaftlichen Welterkenntnis, Seine wissenschaftliche Endeckung und sein religiöser Standpunkt," *Theologische Studien und Kritiken* 51 (1878): 677–692.
23. Schmid, *Der alttestamentliche Religionsunterricht im Seminar und Obergymnasium: Seine Schwierigkeiten und der Weg zu ihrer Überwindung* (Tübingen: Fues, 1889); "Die Tage in Genesis 1–2, 4a," *Jahrbücher für protestantische Theologie* 13 (1889): 689–691; *Das naturwissenschaft-*

liche Glaubensbekenntnis eines Theologen: Ein Wort zur Verständigung zwischen Naturforschung und Christentum, 2d ed.(Stuttgart: Kilemann, 1906).
24. Keeser, "Rudolf Schmid," p. 5a.
25. Ibid., p. 5b.
26. Schmid, *Lebenserinnerungen,* p. 85.
27. Schmid, *Religionsunterricht,* pp. 74–75.
28. Schmid, *Glaubensbekenntnis,* p. vi.
29. Ibid., p. 1.
30. Schmid, *Die Darwin'schen Theorien,* p. 311; *Glaubensbekenntnis,* p. 33.
31. Schmid, "Die durch Darwin Angeregte Entwicklungsfrage," pp. 58–59.
32. Ibid., pp. 191–194.
33. Quoted by Keeser, "Rudolf Schmid," p. 5b. Looking back at the attacks he had had to endure, Schmid later noted that they had never really pained him. See Schmid's obituary notice in the *Kirchliche Anzeiger* (1907), p. 285.
34. Schmid, *Religionsunterricht,* pp. 69–71; *Glaubensbekenntnis,* pp. 41–43.
35. Schmid, *Glaubensbekenntnis,* p. 26. Even the Kant-Laplace nebular hypothesis was not necessarily incompatible with his religion; cf. ibid., pp. 49–50.
36. See also *Die Darwin'schen Theorien,* p. 7.
37. Schmid, *Religionsunterricht,* p. 71.
38. Schmid, "Die durch Darwin Angeregte Entwickelungsfrage," pp. 7–8. The move into what Schmid saw as metaphysics was made so naturally and automatically that he simply could not permit it to go unchallenged. It was also done with such fire and certainty of victory that Schmid could not decide if he should be more astounded at the naive scientific self-confidence of those behind it or at the good-natured believability of a great portion of the public (pp. 46–47).
39. In his *What Is Darwinism?* of 1874, Hodge distinguished between evolution, selection, and natural selection without design, noting that both William Wells and Patrick Matthew had described ideas similar to Darwin's well before Darwin himself. See Charles Hodge, *What Is Darwinism?* (New York: Scribner, Armstrong, and Co., 1874), pp. 48–51.
40. Wilhelm Schmidt, review of *Die Darwin'schen Theorien,* pp. 554–573 in *Theologische Studien und Kritiken* 50 (1877): 566–567.
41. One difference between the long article of 1875 and the book of

1876 was the way he categorized "the Darwinian theories." In the article the theory of decendance comprised what in the book was separated into two theories, descendance and development.
42. Schmid, *Die Darwin'schen Theorien*, pp. 92, 244.
43. Ibid., pp. 91–92, 243.
44. Ibid., pp. 46f, 51–56.
45. Ibid., pp. 44–45.
46. Ibid., p. 45.
47. Ibid., pp. 60–65.
48. Ibid., pp. 65–68.
49. Cf. Schmid, *Glaubensbekenntnis*, pp. 82–91.
50. Schmid, *Die Darwin'schen Theorien*, pp. 71–83. Schmid made clear that it was for natural scientists alone to discover these forces. While philosophers did properly look for metaphysical principles underlying all physical appearances, they could not begin their investigation until after the physical phenomena had been identified. See *Glaubensbekenntnis*, p. 94.
51. Schmid, *Die Darwin'schen Theorien.*, pp. 248–249.
52. Ibid., pp. 254–255.
53. Ibid., p. 23; see also *Glaubensbekenntnis*, pp. 60–62.
54. Schmid, *Die Darwin'schen Theorien*, pp. 84–85.
55. Schmid, *Glaubensbekenntnis*, p. 75.
56. Cf. ibid., pp. 63–64; and *Die Darwin'schen Theorien*, p. 34.
57. Schmid, *Glaubensbekenntnis*, p. 76. On the existence of neo-Lamarckian theories around the turn of the century, see Peter Bowler, *The Eclipse of Darwinism: Anti-Darwinian Evolution Theories in the Decades around 1900* (Baltimore: Johns Hopkins University Press, 1983).
58. Schmid, *Die Darwin'schen Theorien*, pp. 86–87.
59. Ibid., p. 88.
60. Ibid., p. 160.
61. See Richard Dawkins, *The Blind Watchmaker* (New York: W. W. Norton and Co., 1986), chap. 3.
62. Schmid, *Die Darwin'schen Theorien*, p. 193.
63. Ibid., p. 256. This categorical statement was compromised somewhat by Schmid's concession in the *Glaubensbekenntnis* that those who wanted to read Darwin as eliminating purpose naturally saw variation as aimless, while those who admitted purposive causality saw variation to be determined. *Glaubensbekenntnis*, p. 77.
64. Schmid, *Die Darwin'schen Theorien*, pp. 88–89.
65. Ibid., pp. 256–257.

66. Ibid., p. 256.
67. Duke of Argyll, *The Reign of Law*, 11th ed. (London: Strahan, 1884), p. 219. Schmid referred to the duke's *Reign of Law* in *Glaubensbekenntnis*, p. 103.
68. Schmid, *Die Darwin'schen Theorien*, pp. 257–258.
69. See Gregory, *Scientific Materialism*, p. 186.
70. Schmid, *Die Darwin'schen Theorien*, p. 91.
71. Ibid., p. 121. Cf. pp. 117–121. The same was true of the suggestion that life arrived on earth in organic germs from elsewhere (p. 124).
72. Ibid., p. 124.
73. Ibid., p. 143. Cf. pp. 138–140. Schmid reflected the traditionally German view of the nature of matter as the product of an active elementary force. Hence his dynamical conception was incompatible with Ludwig Büchner's popular equation of force and matter as two equal expressions of the one underlying reality. Cf. pp. 128–129.
74. Ibid., pp. 137, 144–145.
75. Schmid, *Glaubensbekenntnis*, p. 5; see also *Die Darwin'schen Theorien*, p. 153.
76. Schmid, *Die Darwin'schen Theorien*, pp. 214–215.
77. References to Helmholtz and Haeckel come in 1876, in *Die Darwin'schen Theorien*, pp. 146–147; DeVries is cited in the *Glaubensbekenntnis*, p. 71. On Haeckel's *Riddle*, see *Glaubensbekenntnis*, p. 15.
78. Schmid, *Glaubensbekenntnis*, p. 71.
79. Ibid., p. 159.
80. Ibid., p. 158.
81. Ibid., pp. 163–164.
82. Ibid., p. 205. Schmid explained Darwin's inconsistency in these works by saying that "the strength of this naturalist does not seem to lie in logical philosophical thought" (p. 205).
83. Ibid., p. 159.
84. Duke of Argyll, *The Reign of Law*, p. 80.
85. Schmid, *The Theories of Darwin and Their Relation to Philosophy, Religion, and Morality*, 2d ed., trans. G. A. Zimmermann (Chicago: Jansen, McChurg and Co., 1885), p. 5.
86. See Timothy Lenoir, *The Strategy of Life: Teleology and Mechanics in Nineteenth Century Biology* (Dordrecht: Reidel Publishing Co., 1982), *passim*.
87. Schmid, *Glaubensbekenntnis*, p. 9. See also pp. 12–13.
88. Ibid., p. 2. For Zöckler's similar sentiments, see above, Chap. 4,

n. 71. Other ways the complementarity represented by the causal and teleological viewpoints was commonly expressed, according to Schmid, were the contrasts between science and theology, knowledge and faith, and modern culture and the traditional Christian conviction. The two sides could not be in opposition, however, "because Truth can only be one" (p. 1). Contradictions emerged, he explained, when one side or the other overstepped its limits out of lack of respect for the other (pp. 4–5).

89. A. Hoffmann, review of *Das Naturwissenschaftliche Glaubensbekenntnis eines Theologen*, pp. 74–76 in *Protestantische Monatshefte* 10 (1906): 74.
90. Duke of Argyll, *Reign of Law*, pp. 56–57.
91. Schmid, *Die Darwin'schen Theorien*, pp. 210–211.
92. Ibid., pp. 199–201. Jäger's book was *Die darwinische Theorie und ihre Stellung zu Moral und Religion* (Stuttgart: Hoffmann, 1869).
93. Schmid, *Die Darwin'schen Theorien*, pp. 199–201. Schmid was responding to Lang's book, *Die Religion im Zeitalter Darwins* (Berlin: Luderlitz, 1873).
94. Schmid, *Die Darwin'schen Theorien*, pp. 235–236.
95. Schmid, *Glaubensbekenntnis*, p. 13. Schmid's son applied the definition, which involved more than the portion cited here, to his father as an apt tribute to his attitude. Schmid, *Lebenserinnerungen*, p. 109.
96. Schmid, *Lebenserinnerungen*, p. 25.
97. Ibid., pp. 90–91. Quoted by Schmid's son Rudolf.
98. Schmid, *Religionsunterricht*, p. 44.
99. Quoted in Schmid's obituary in the *Kirchliche Anzeiger* (1907), p. 284.
100. Schmid, *Glaubensbekenntnis*, p. 93; See also *Religionsunterricht*, p. 37.
101. Schmid reported the few places in Mayer's work in which he discussed science and religion. He was, of course, delighted to cite Mayer's claim that any view, including Darwin's, which attacked purposefulness and freedom had no scientific value. See Schmid, "Robert Mayer," pp. 689–690.
102. Ibid., p. 682.
103. Ibid., pp. 680–681.
104. Schmid, *Glaubensbekenntnis*, p. 37.
105. Ibid., p. 38.
106. Ibid., p. 40. An exception Schmid was willing to make concerned the origin of sensation and self-consciousness from a material source that was neither sentient nor conscious. At present, he de-

clared, we have to assert DuBois-Reymond's ignoramus ("we do not know"), and it was better in these questions to employ DuBois-Reymond's ignorabimus ("we shall never know") than to "plunge into the bottomless ocean of hypotheses." *Die Darwin-'schen Theorien*, pp. 115–116.
107. Schmid, "Robert Mayer," p. 687.
108. Schmid, *Glaubensbekenntnis*, pp. 65–66, 69–73, 81.
109. Schmid, *Religionsunterricht*, p. 48.
110. Schmid, *Glaubensbekenntnis*, pp. 3–4.
111. Ibid., pp. 139–143.
112. Ibid., pp. 149–163.
113. Ibid., pp. 50–51. See esp. pp. 61–68.
114. Schmid, "Die Tage in Genesis," pp. 689–691.
115. Ibid., pp. 697–702.
116. Ibid., pp. 337–338. See also *Glaubensbekenntnis*, pp. 125–133.
117. Schmid, *Glaubensbekenntnis*, p. 122.
118. Ibid., pp. 118, 122–124.
119. Schmid, *Die Darwin'schen Theorien*, pp. 269–273.
120. Ibid., pp. 237–240.
121. Ibid., p. 237.
122. Ibid., p. 93.
123. Ibid., p. 258.
124. Quoted from Schmid's obituary notice in the *Kirchliche Anzeiger* (1907), p. 284.
125. Cf. the announcement of the train schedule in the *Staats Anzeiger* (1907), p. 1265.
126. Schmid, *Lebenserinnerungen*, p. 100.
127. A list of the numerous honors can be found in the entry for Schmid in the catalog of the Landeskirchliches Archiv in Stuttgart.
128. Schmid, *Lebenserinnerungen*, p. 109.
129. Ibid., p. 111.
130. *Staats Anzeiger* (1907), p. 1279.
131. *Kirchliche Anzeiger* (1907), pp. 262, 285.

6. Wilhelm Herrmann's Encounter with the Theology of Albrecht Ritschl

1. The *Primal History* was published in the same year as the lectures; the *Doctrine of the Origin of Humankind* appeared in 1879. See T. Zöckler, *Erinnerungsblätter*, p. 56.
2. Welch, *Protestant Thought*, II, 12.

3. See below, n. 36.
4. Cf. above, Chap. 2, n. 116.
5. Numerous authors confirm that the status of natural science at the end of the century was a major motivation for the program of the Ritschl school. See, for example, Friedrich Wilhelm Schmidt, *Wilhelm Herrmann: Ein Bekenntnis zu seiner Theologie* (Tübingen: Mohr, 1922), pp. 65–66; Hermelink, *Das Christentum in der Menschheitsgeschichte*, III, 220; Paul Wrzecionko, *Die philosophischen Wurzeln der Theologie Albrecht Ritschls: Ein Beitrag zum Problem des Verhältnis von Theologie und Philosophie im 19. Jahrhundert* (Berlin: Töpelmann, 1964), p. 261; Peter Fischer-Appelt, *Metaphysik im Horizont der Theologie Wilhelm Herrmanns* (Munich: Kaiser, 1965), p. 103, n. 3. Fischer-Appelt emphasizes the impact on Herrmann of Emil DuBois-Reymond's famous lecture on the limits of natural scientific knowledge of 1872.
6. Herrmann, *Die Gewissheit des Glaubens und die Freiheit der Theologie*, 2d ed. (Freiburg: Mohr, 1889), p. 3.
7. Ibid., p. 2.
8. Herrmann, *Die Dogmatik* (Gotha: Perthes, 1925), p. 1.
9. Ritschl, *Rechtfertigung und Versöhnung*, III, 542. See Chap. 2, above.
10. Martin Rade, "Akademische Gedächtnisrede auf Wilhelm Herrmann," pp. viii-xxi in Herrmann, *Dogmatik*, p. viii.
11. Ibid., p. ix.
12. Ibid.
13. Ibid., p. ix, n.
14. Friedrich Schmidt reports that Herrmann was in Halle from 1864 to 1870, but Rade maintains he entered in the fall of 1866. See Schmidt, *Wilhelm Herrmann*, p. 3; Rade, "Gedächtnisrede," p. x.
15. Werner Schütz asserts that Herrmann himself included the two and one half years spent in Tholuck's house among the influences that shaped him, and that the reference to the "other theological presuppositions" which Herrmann makes at the beginning of his *Religion im Verhältnis zum Welterkennen und zur Sittlichkeit* is to Tholuck. See W. Schütz, *Das Grundgefüge der Herrmannschen Theologie, ihre Entwicklung und ihre geschichtlichen Wurzeln* (Berlin: Ebering, 1926), p. 93. Schütz concedes that Tholuck's occasional embrace of supernatural apologetics was foreign to Herrmann, but he insists that Tholuck's linking of belief and feeling, in which faith rested on "the facts of our inner life," affected Herrmann and was consonant with his eventual position (pp. 95–99). Rade dismisses Tholuck's influence on Herrmann, as he does the

possible influence of all the Halle theologians ("Gedächtnisrede," p. x).
16. Rade, "Gedächtnisrede," p. x.
17. O. Ritschl, *Ritschls Leben*, II, 267. The phrase was "Ultra posse nemo obligatur."
18. Ibid. See also Rade, "Gedächtnisrede," p. xii, where the letter is quoted.
19. O. Ritschl, *Ritschls Leben*, II, 269.
20. Rade, "Gedächtnisrede," p. xii.
21. Quoted from the 1875 issue of *Beweis des Glaubens* by O. Ritschl in *Ritschls Leben*, II, 300.
22. Ritschl, *Rechtfertigung und Versöhnung*, III, 11.
23. Ibid.
24. Ibid., p. 12.
25. Ibid., p. 13.
26. Ibid., p. 14. Because later editions of the Introduction incorporate and revolve around sections of Ritschl's *Theology and Metaphysics* of 1881, one must be aware that the initial effect of Ritschl's work on Herrmann cannot be ascertained from them.
27. Ibid., pp. 170–171, 182–183.
28. Ibid., p. 174. See also p. 171.
29. Ibid., p. 179.
30. Ibid., p. 183.
31. Ibid., p. 184.
32. Ibid., p. 186.
33. Ibid., pp. 544–545.
34. Ibid., p. 545.
35. Fischer-Appelt, *Metaphysik im Horizont der Theologie*, p. 103, n. 3.
36. Rade, "Gedächtnisrede," pp. xii-xiii. Günther Ott unhesitatingly identifies Herrmann as "founder of the Ritschl school." "Johann Georg Wilhelm Herrmann," *Neue Deutsche Biographie*, vol. 8 (Berlin: Duncker and Humblot, 1968), p. 691. Regarding the "seductive" capacity of Ritschl's assumption that the moral will was just as real as natural processes and that therefore the knowledge coming from it was in no way of a second order, see Christoph Senft, *Wahrhaftigkeit und Wahrheit: Die Theologie des 19. Jahrhunderts zwischen Orthodoxie und Aufklärung* (Tübingen: Mohr, 1956), p. 155, and Wrzecionko, *Die philosophischen Wurzeln der Theologie Albrecht Ritschls*, p. 257.
37. Rade, "Gedächtnisrede," p. xi.
38. As a student Karl Barth heard Herrmann say that the first four of

Schleiermacher's speeches were the most important writings that had appeared since the New Testament canon itself. See Eberhard Busch, *Karl Barth: His Life from Letters and Autobiographical Texts,* trans. John Bowden (Philadelphia: Fortress Press, 1975), p. 44. For a comparison of the thought of Herrmann and Barth, see Albert A. Jagnow, "Karl Barth and Wilhelm Herrmann: Pupil and Teacher," *Journal of Religion* 16 (1936): 300–316.

39. Herrmann, *Die Metaphysik in der Theologie* (Halle: Niemeyer, 1876), pp. 1–2.
40. In his review the theologian Julius Kaftan identified Herrmann's tract as "obviously the fruit of the stimulation which theology owes Ritschl's work on justification and reconciliation." He noted that even Ritschl's characteristic manner of expression had been taken over, though in general he complained in earnest about Herrmann's horrendous use of the German language. See Kaftan's review in *Theologische Literaturzeitung* 2 (1877): 63, 65.
41. Herrmann, *Metaphysik in der Theologie,* p. 2. Dennis Overbye reinforces Herrmann's observation in his *Lonely Hearts of the Cosmos* when he writes: "It is probably part of the human condition that cosmologists (or the shamans of any age) always think they are knocking on eternity's door, that the final secret of the universe is in reach." Overbye adds that "it may also be part of the human condition that they are always wrong." Dennis Overbye, *Lonely Hearts of the Cosmos: The Scientific Quest for the Secret of the Universe* (New York: Harper Collins, 1991), p. 3.
42. Ibid., pp. 4–5.
43. Ibid., pp. 5–6.
44. Ibid.
45. Ibid., p. 7.
46. Ibid., p. 8.
47. On Fries's *Wissen, Glaube, und Ahndung* see Chap. 1 above. That Herrmann was at least aware of Fries's thought can be inferred from his citation of the work of De Wette and from a letter in which he links the work of his mentor, Tholuck, to Fries. See Rudolf Otto, *The Philosophy of Religion Based on Kant and Fries* (London: Williams and Norgate, Ltd., 1931), p. 221, n. 1. Otto concurs that Fries influenced Tholuck (pp. 219–221).
48. Herrmann, *Metaphysik in der Theologie,* p. 9.
49. Ibid., p. 11.
50. Ibid., p. 12. Herrmann added that even if metaphysics were to succeed, the moral question would still remain unsolved.

51. Ibid., p. 14.
52. Ibid.
53. Ibid., p. 17.
54. Ibid., p. 18.
55. Ibid., pp. 16–17. Herrmann noted that the theologian could relate as easily to a deistic, pantheistic, or theistic philosophy (pp. 21–22).
56. Ibid., p. 20.
57. Ibid., p. 23.
58. Ibid., pp. 23–24.
59. Ibid., pp. 80–81. The last third of Herrmann's tract was given over to an examination of Christology, in which, Herrmann remarked, mixing up the theological and metaphysical tasks was even more condemnable. Cf. p. 51.
60. Ibid., p. 48.
61. Ibid., p. 50.
62. O. Ritschl, *Ritschls Leben*, II, 107.
63. Ibid., pp. 305–306.
64. Ibid., p. 306. The letter is dated July 16.
65. Cf. Georg Graue, "Zur Abwehr," *Protestantische Kirchenzeitung für das evangelische Deutschland* (1877), p. 501. Lipsius's remarks about the danger of Ritschl's position being mishandled by others, namely by Herrmann in the *Metaphysik in der Theologie,* were published in this same journal in the volume for 1876: "Vorwort zu einem Vorwort," *Protestantische Kirchenzeitung für das evangelische Deutschland* (1876), pp. 649–650.
66. O. Ritschl, *Ritschls Leben*, II, 308.
67. Rade, "Gedächtnisrede," p. xiv. For Herrmann's deep regret "to have fallen into an animated tone," see the Preface to his 1879 *Religion im Verhältnis zum Welterkennen und zur Sittlichkeit* (Halle: Niemeyer, 1879), p. vii.
68. Review of R. A. Lipsius, *Lehrbuch der evangelish-protestantischen Dogmatik,* in *Theologische Studien und Kritiken* 50 (1877): 522–523.
69. Ibid., p. 526.
70. Ibid., p. 545. See also pp. 523–524.
71. Herrmann, "Die Auffassung der Religion in Cohens und Natorps Ethik," pp. 377–405 in F. W. Schmidt, ed., *Gesammelte Aufsätze* (Tübingen: Mohr, 1923), p. 379.
72. Review of Lipsius's *Lehrbuch,* pp. 528–529. Later Herrmann wrote regarding William James's psychological analysis of religion that one could not learn about religion from a statistical study of it;

rather, one had to see it originate in one's self. See "Die Lage und Aufgabe der evangelischen Dogmatik in der Gegenwart," pp. 95–188 in *Gesammelte Aufsätze*, p. 137. This essay originally appeared in *Zeitschrift für Theologie und Kirche* in 1907.

73. Graue, "Zur Abwehr," p. 493. Lipsius's treatment of the compatibility of science and religion was in fact closer to Herrmann's than it was to that of the mediating school. It is therefore easy to see why Herrmann later regretted making an enemy of him. On Lipsius's rejection of objectively adequate knowledge of religion, see Hermelink, *Das Christentum in der Menschheitsgeschichte*, III, 237.

74. Ibid., p. 494.

75. Quoted in Rade, "Gedächtnisrede," p. xv.

76. Graue, "Zur Abwehr," p. 499.

77. Ibid., p. 501. After Herrmann's review of Lipsius the rumor circulated that it had been done on Ritschl's order. A friend common to both Ritschl and Lipsius (probably Diestel) assured Lipsius that such was not the case, prompting a letter from Lipsius to Ritschl in which Lipsius indicated that he had disapproved of Graue's entry into the fray and that he wanted an end to the matter in light of the substantial agreement he shared with Ritschl. Lipsius's defense of his textbook in the *Jahrbücher für protestantische Theologie* two years later (1878, nos. 1–4) was far more dignified than Graue's had been, although he did make reference to the critique of an "unsolicited spokesman of the Ritschl school." Ritschl took fresh offense at this remark in a letter to Holtzmann. Ritschl assumed that Lipsius expected him not only to have disapproved of Herrmann's attack on him but also to have control over those in his school. That, however, Ritschl said he would not do. See O. Ritschl, *Ritschls Leben*, II, 308–309.

78. Herrmann drew on Pfleiderer's 1869 work, *Die Religion, ihr Wesen und ihre Geschichte*, which Herrmann cited merely as *Wesen der Religion*.

79. Rade, "Gedächtnisrede," pp. xi, xiv.

80. R. Seeberg, "Otto Pfleiderer," pp. 316–323 in *Realencyklopädie für protestantische Theologie und Kirche*, vol. 24 (Leipzig: Hinrichs, 1913), p. 318.

81. Pfleiderer, "Silhouetten aus der Religionswissenschaft der Gegenwart," *Protestantische Kirchenzeitung für das evangelische Deutschland*, no. 22 (1877): 462. Part 1 is found on pp. 461–468; part 2 appeared in the following number (23) on pp. 477–492.

82. Ibid., no. 22, pp. 463–464.

83. Ibid., p. 467.
84. Ibid.
85. Ibid., p. 485, n., p. 492. Ritschl's dislike of the Swabian historian of philosophy Eduard Zeller is mentioned on p. 488. Seeberg specifically denies the charge of Otto Ritschl that Pfleiderer's caustic remarks were due to his father's first-place standing over Pfleiderer's third place in the competition for a post in Tübingen. See Seeberg, "Pfleiderer," p. 323, and O. Ritschl, *Ritschls Leben*, II, 301–304.
86. Pfleiderer, "Silhouetten," p. 487.
87. Ibid.
88. Ibid., pp. 487–488.
89. Herrmann, *Metaphysik in der Theologie*, p. 51. For Herrmann's characterization of Pfleiderer as a rationalist, see p. 76.
90. Pfleiderer, "Silhouetten," p. 491. The Ebionite heresy urged that Jesus, who was the human son of Joseph and Mary, received his divine authority at his baptism when the Holy Spirit descended on him. By contrast the adherents of the teaching of the British monk Pelagius denied that God raised human nature to a supernatural level in Christ and that the life and death of Jesus possessed exemplary as opposed to intrinsic value.
91. The university associations listed represent the last post occupied by each individual.

7. The Existential Critique of Science and Theology

1. Two exceptions to this are Fischer-Appelt, *Metaphysik im Horizont der Theologie*, and Hasler, *Beherrschte Natur*. Herrmann himself later criticized his own work and refused to premit a new edition after it had been long out of print. Cf. Rade, "Gedächtnisrede," p. xiv.
2. Herrmann, *Religion im Verhältnis zum Welterkennen*, p. iii.
3. Ibid., p. 2.
4. O. Ritschl, *Ritschls Leben*, II, 107.
5. Cf. above, Chapter 6, n. 75.
6. Herrmann, *Religion im Verhältnis zum Welterkennen*, p. 4.
7. Ibid., p. 5.
8. Ibid., p. 12.
9. Ibid., p. 16.
10. Not everyone agreed that Herrmann's "pure knowing" as a state uninfluenced by emotion or will was possible. Cf. the review of Herrmann's book by the philosopher Paul Natorp, who believed

that such a state was unthinkable. "Über das Verhältnis des theoretischen und praktischen Erkennens zur Begründung einer nichtempirischen Realität," *Zeitschrift für Philosophie und philosophische Kritik* 79 (1881): 244, n.

11. Herrmann, *Religion im Verhältnis zum Welterkennen*, pp. 16–17. Herrmann promised that he would show how the use of this freedom was the necessary condition for procuring a suitable theological grounding for Protestant Christianity. He noted that if he was successful, he would have established a link between Kant and the Reformation, a connection that should have won more serious attention for Kant's ethics and religion than they had usually received from the Romantic Era on.

12. Ibid., pp. 22–23.

13. Walker, *Coherence Theory of Truth*, p. 33. Walker continues: "It would be natural to think that the totality of truths was indeterminate if one thought there was nothing more to truth than what we can recognize as true, and if one regarded as somehow open-ended what we can recognize as true: as something that cannot be firmly delimited in advance of investigation. This is the line taken by those who call themselves anti-realists, and as we shall see it was also taken by Kant."

14. Herrmann, *Religion im Verhältnis zum Welterkennen*, p. 24. In an address about Ritschl delivered in the fall of 1890, just after his death in March of that year, Herrmann characterized the conceptions of science as "boundlessly changeable." See "Faith as Ritschl Defined It," pp. 7–62 in *Faith and Morals*, trans. D. Matheson and R. W. Stewart (New York: Putnam's Sons, 1904), p. 11.

15. Herrmann, *Religion im Verhältnis zum Welterkennen*, p. 25.

16. Ibid., p. 29.

17. Herrmann quoted this from Kant. Fries's primary concern in this connection is less with natural science than with mathematics, where Kant's insight influenced Fries's understanding of construction. See the *Neue Kritik der Vernunft*, II, 134. See also Felix Cube, *Die Auffassungen Jakob Friedrich Fries und seiner Schule über die philosophischen Grundlagen der Mathematik und ihr Verhältnis zur Grundlagentheorie* (Stuttgart: Dissertation an der Technischen Hochschule Stuttgart, 1957), pp. 39–40; and H. Ende, *Der Konstruktionsbegriff im Urkreis des deutschen Idealismus* (Meisenheim am Glan: Hain, 1973), pp. 34, 37.

18. Herrmann, *Religion im Verhältnis zum Welterkennen*, p. 27.

19. Ibid., p. 32.

20. Ibid., p. 33.
21. Ibid., p. 35. Although the phrase "feeling and willing" came ultimately from Lotze, Ritschl had taken it over with approval as well. Cf. Hasler, *Beherrschte Natur*, p. 45. Herrmann's critic Lipsius summarized Herrmann's claim here by saying that according to Herrmann the comprehensibility of nature "could neither be derived epistemologically nor proven empirically, hence it owed its origin to the practical intersts of feeling and willing beings." Cf. R. A. Lipsius, "Neue Beiträge zur wissenschaftlichen Grundlegung der Dogmatik," *Jahrbücher für protestantische Theologie* 11 (1885): 219.
22. Herrmann, "Glaube und Erkennen," pp. 152–154 in *Zeitschrift für Theologie und Kirche* 17 (1907): 152.
23. Herrmann, *Religion im Verhältnis zum Welterkennen*, pp. 35–37.
24. Ibid., p. 40. See also p. 47. Herrmann quoted Lotze here to say that we commonly fail to see the contradiction present when we use the understanding to connect individual things without realizing that the assumption of a world order that makes this possible is made at a different level (pp. 40–41).
25. Ibid., p. 65.
26. Ibid., p. 66.
27. Ibid., p. 69. Hilary Putnam has suggested that to an anti-realist the successes of science in explaining the world can be justified only as miracles. See "What Is Realism?" pp. 140–153 in Leplin, *Scientific Realism*, p. 41.
28. Herrmann, *Religion im Verhältnis zum Welterkennen*, pp. 70–71.
29. Ibid., p. 71.
30. "Dem unsere Erkenntniss sich gleichsam asymptotisch annähere." Natorp, "Über das Verhältnis des theoretischen und praktischen Erkennens," p. 252. For Natorp's endorsement of a Platonic correspondence theory of truth, see p. 257.
31. Herrmann, *Religion im Verhältnis zum Welterkennen*, p. 80.
32. Hasler goes so far as to suggest that for Herrmann the more nature is mastered, the more humanity is enslaved. Such an extreme interpretation is more characteristic of the later Herrmann. See Hasler, *Beherrschte Nature*, pp. 55, 299–301.
33. Herrmann, *Religion im Verhältnis zum Welterkennen*, p. 82; emphasis mine.
34. Ibid.
35. Herrmann, *Dogmatik*, p. 48, n.
36. Ibid., p. 83.
37. Ibid., pp. 83, 100, 114–118. The two false paths of religion are

dealt with again in Herrmann's address on Ritschl. Cf. "Faith as Ritschl Defined It," pp. 18, 34–35.
38. Herrmann, *Dogmatik,* pp. 86–89.
39. Herrmann, "Faith as Ritschl Defined It," p. 11.
40. Herrmann, *Religion im Verhältnis zum Welterkennen,* pp. 97, 99.
41. Ibid., p. 93. Herrmann quoted Pfleiderer's 1877 review.
42. Ibid., pp. 348–349.
43. Ibid., pp. 93–94. In a note Herrmann labeled Pfleiderer's viewpoint a variety of modern orthodoxy because of its blind commitment to one way of seeing.
44. Ibid., pp. 121–131. See esp. p. 130. Herrmann's use of the verb bracket *(klammern)* reminds one of the phenomenological tradition where it is also employed. See, for example, Erazim Kohak, *The Embers and the Stars.* Paul Wrzecionko refers to Ritschl's "bracketing" of the dimension of faith; see Wrzecuionko, *Die philosophischen Wurzeln der Theologie Albrecht Ritschls,* pp. 258, 264.
45. Herrmann, *Religion im Verhältnis zum Welterkennen,* pp. 281–310.
46. Ibid., p. 311.
47. Herrmann eventually brought considerable attention to Marburg. In addition to students from all over the world who came to study with him, including, according to Welch, a generation of Americans who pursued graduate study in pre–World War I Germany, a Norwegian and an American university bestowed honorary doctorates on him, as did the philosophical faculty at Marburg itself. See Welch, *Protestant Thought,* II, 45, and Horst Stephan, "Im Namen der Fakultät," *Die Christiliche Welt* 36 (1922): 77. Herrmann remained at Marburg for the rest of his life in spite of numerous offers from other universities. What held him there, according to F. W. Schmidt, was his role as the soul of the theology faculty and the possibility of interchange with the neo-Kantians of the philosophy faculty. See F. W. Schmidt, "Johann Wilhelm Herrmann," pp. 96–104 in *Deutsches Biographisches Jahrbuch,* vol. 4 (Berlin: Deutsche Verlags-Anstalt Stuttgart, 1929), p. 99.
48. Ludwig Fürst zu Solms, "Recht und Unrecht der Metaphysik," *Jahrbücher für protestantische Theologie* 6 (1880): 590–591.
49. Ibid., p. 581. "It would seem that every future dogmatics must assume that there can be but *one* unconditionally true explanation of the world which a dogmatician will enhance to the same extent that he succeeds in demonstrating and confirming the manifest and fixed limits of the knowledge of the world possible to humans, the limiting concepts of all metaphysics" (p. 593).

50. C. Weizsäcker in *Theologische Literaturzeitung*, no. 25 (1879): 592–596. Weizsäcker identified himself as one who was not a professional theologian (p. 593).
51. Alfred Krauss, "Sendschrieben an Herrn Professor W. Herrmann in Marburg," *Jahrbücher für protestantische Theologie* 9 (1883): 207, 209.
52. Ibid., pp. 206, 227.
53. Ibid., pp. 238, 240.
54. Lipsius, "Neue Beiträge zur wissenschaftlichen Grundlegung der Dogmatik," *Jahrbücher für protestantische Theologie* 11 (1885): 178.
55. "Scientifically, that is for purely theoretical knowing, the one is as provable as the other. Materialistic 'monism' has no cause at all to raise itself above theological faith; it is faith just like the other, and one asks oneself indeed whether teleological faith is not the 'more reasonable.'" Ibid., p. 287.
56. Lipsius, "Neue Beiträge," p. 181. For his discussion of the determining role of the given, see pp. 182–185.
57. Berlin, *Crooked Timber of Humanity*, p. 6.
58. Lipsius, "Neue Beiträge," pp. 374–375.
59. O. Ritschl, *Ritschls Leben*, II, 304.
60. Ibid., p. 388.
61. Ibid., pp. 388–389.
62. Ibid., p. 388. Philipp Jakob Spener was one of the pioneers of German pietism from the seventeenth century.
63. Cf. Chap. 1, above.
64. Cf. Otto Pfleiderer, *Die Ritschl'sche Theologie kritisch beleuchtet* (Braunschweig: Schwetschke, 1891). Pfleiderer's evaluation appeared originally in the *Jahrbücher für protestantische Theologie* but was also published separately. Pfleiderer argued that Ritschl's position in the first edition of the *Justification and Reconciliation*, in which he posited the impossibility of a collision between science and religion because of their different interests, changed in later editions, in which he admitted that an antagonism did exist (see p. 14). For a more sympathetic evaluation of Ritschl's epistemology, see Friedrich Traub, "Ritschls Erkenntnistheorie," *Zeitschrift für Theologie und Kirche* 4 (1894): 91–129. Traub did accuse Ritschl of misunderstanding Kant, and of characterizing Kant's real position as that of Lotze (pp. 96–97). He also noted the difference between the early and late editions of Ritschl's major work regarding the collision between science and religion. But unlike Pfleiderer, Traub argued that the later Ritschl attempted to avoid the collision

by removing metaphysics from theology and religion from philosophy (p. 108). For a more recent assessment of Ritschl's epistemological deficiencies and Herrmann's differences from Ritschl, see Louis Perriraz, *Histoire de la théologie protestante au xixme siècle, surtout en Allemagne,* vol. 1, *Les doctrines: De Kant à Karl Barth* (Neuchâtel: Messeiller, 1949), pp. 168–170, 175.

65. Herrmann, review of Pfleiderer's *Ritschl'sche Theologie* in *Theologische Literaturzeitung* 17 (1892): 383.
66. Ibid., p. 384.
67. O. Ritschl, *Ritschls Leben,* II, 386–387.
68. Ibid., p. 393.
69. Herrmann, *Warum bedarf unser Glaube geschichtliche Thatsachen?* (Halle: Niemeyer, 1884), p. 14.
70. Ibid., p. 16.
71. Ibid., pp. 18–22. The undeniable antagonism toward the Catholic church, evident here and elsewhere in Herrmann's works, is no doubt inherited in part from Ritschl and their common interest in the Reformation and in part from the strong nationalistic sentiment widespread in Germany at the end of the century. German unification, for example, became an illustration of the need to move beyond commitment to a merely intellectual idea of the German spirit. Likewise one must move beyond a purely intellectual idea of religion and insist on its real basis in historical fact (pp. 4–6, 12). Herrmann was even willing to associate Bismarck's accomplishments with the empowerment of his personal faith (p. 30). For Herrmann's anti-Catholicism, see also his essay "Faith as Ritschl Defined It," p. 61; *Die Gewissheit des Glaubens,* pp. 13–14; and esp. "The Moral Law as Understood in Romanism and in Protestantism," pp. 71–193 in *Faith and Morals.*
72. Rade, "Gedächtnisrede," p. xiii.
73. Pfleiderer, *Die Ritschl'sche Theologie,* p. viii. See also Hermelink, *Das Christentum in der Menschheitsgeschichte,* III, 226.
74. Cf. Hermelink, *Das Christentum in der Menschheitsgeschichte,* III, 234–236; Mildenburger, *Geschichte der deutschen Theologie,* p. 249. For Herrmann's observations on the prospects of Ritschl's theology in 1889, see *Die Gewissheit des Glaubens,* pp. 15–17.
75. Both Stählin and Stuckenberg are quoted in Albert Swing, *The Theology of Albrecht Ritschl* (New York: Longmans, Green, and Co., 1901), p. 2.
76. Hasler, *Beherrschte Natur,* p. 302.
77. In 1907 Herrmann openly expressed discontent with his *Religion in*

Relation to the World, noting that he had since come to a better understanding of both religion and science. The earlier work had been constricted and uncertain, but even so Herrmann felt that it had been on the right track. "Die Lage der evangelischen Dogmatik in der Gegenwart," *Zeitschrift für Theologie und Kirche* 16 (1907): 175.

78. "Whoever wants to get rid of facts is turning his back on God, for the omnipotent God dwells in realities." "The Moral Law," p. 132.
79. Quoted in Robert T. Voelkel, *The Shape of the Theological Task* (Philadelphia: Westminster Press, 1968), p. 8, n.
80. Herrmann, *Dogmatik*, pp. 51–53, 83.
81. Herrmann, "Der evangelische Glaube und die Theologie Albrecht Ritschls," pp. 1–25 in *Gesammelte Aufsätze*, pp. 23–24.
82. Herrmann, "Die Erlösung durch Jesus Christus und die Wissenschaft," pp. 336–344 in *Gesammelte Aufsätze*, p. 339.
83. Rade, "Gedächtnisrede," p. xix.
84. Lipsius, "Neue Beiträge," pp. 601–602. Nevertheless Herrmann was convinced that God was "near" to Budda, Socrates, and others, and he "rejoiced in the diversity of the forms of appearance of religion." Cf. F. W. Schmidt, *Wilhelm Herrmann*, p. 7.
85. Cf. Herrmann, *Dogmatik*, pp. 19–21; "Warum bedarf unser Glaube geschichtliche Thatsachen?" pp. 28–29. See also Rade, "Gedächtnisrede," pp. xv-xvi.
86. Herrmann, *Die Gewissheit des Glaubens*, pp. 8–10, 70.
87. A thorough analysis of the mature Herrmann, which examines diverse works from the later years, may be found in Hasler, *Beherrschte Natur*, part 4.
88. Herrmann, "Religion und Sozialdemokratie," pp. 463–489 in *Gesammelte Aufsätze*, p. 472.
89. Ibid., pp. 464, 466–467, 479.
90. Ibid., p. 469.
91. Ibid., p. 473.
92. Ibid., p. 482.
93. Ibid., p. 478.
94. Ibid., p. 483.
95. Voelkel, *Shape of the Theological Task*, p. 19.
96. Herrmann, "Religion und Sozialdemokratie," p. 484.
97. Ibid., p. 485.
98. While this image comes from Martin Heidegger, it was greeted, according to Kohak, by many of Heidegger's contemporaries as

but an acknowledgment of an evident truth. Cf. Kohak, *Embers and the Stars*, pp. 3–8.
99. Herrmann, "Religion und Sozialdemokratie," pp. 486–487.
100. Kohak, *Embers and the Stars*, p. 4.
101. Ibid., p. 22. See also p. 13.

Epilogue

1. Hasler, *Beherrschte Natur*, p. 294.
2. Johannes Iversen, "Zu Wilhelm Herrmanns 70. Geburtstag," *Evangelische Freiheit* 16 (1916): 416.
3. E.g., Rudolf Carnap, "The Elimination of Metaphysics through Logical Analysis of Language," pp. 60–81 in A. J. Ayer, ed., *Logical Positivism* (New York: Free Press, 1959). Carnap's original German article appeared in *Erkenntnis* in 1932.
4. For a brief account of the changing interpretations of science in the twentieth century, see Frederick Gregory, "The Historical Investigation of Science in North America," *Zeitschrift für allgemeine Wissenschaftstheorie* 16 (1985): 151–155.
5. P. Christoph Schönborn, "Schöpfungskatechese und Evolutionstheorie: Vom Burgfrieden zum konstruktiven Konflikt," pp. 91–116 in R. Spaemann, R. Löw, and P. Koslowski, eds., *Evolutionismus und Christentum* (Weinheim: VCH, 1986), pp. 99–106.
6. Ole Jensen, *Theologie zwischen Illusion und Restriktion: Analyse und Kritik der existenz-kritizistischen Theologie bei dem jungen Wilhelm Herrmann und bei Rudolf Bultmann* (Munich: Kaiser, 1975), p. 11. Jensen's claim may be too strong. It might be possible, for example, for Bultmann to define a basis for judgment through the notion of stewardship.
7. Ibid.
8. Thus the title of Jensen's work, *Theology between Illusion and Restriction*. See also p. 209.
9. Ibid., pp. 233–235.
10. Cf. F. Amrine, F. J. Zucker, and H. Wheeler, eds., *Goethe and the Sciences: A Reappraisal* (Dordrecht: Reidel, 1987), pp. 143–145, 169–171, 341–350.
11. Berlin, *Crooked Timber of Humanity*, pp. 10–11.

Index

Agassiz, Louis, 120–122, 124, 137, 144, 145
Age: of human race, 133, 140–147; of earth, 133, 194, 308n87, 309n101
Ahnung, 41, 215
Albert, Prince of Coburg, 168
Altholtz, Joseph, 306n54
Aner, Karl, 25
Anti-realism, 21, 235, 239, 241, 244, 249, 287n45, 326n13
Apelt, Ernst, 20, 62
Apologetics: in German natural theology, 9, 49, 126–127, 187; in mediating theology, 52; Ritschl and, 59; Zöckler and, 114, 115, 119, 126, 127, 130, 147, 159; in English natural theology, 118
Argument from design, 6–7, 15, 95, 305n20
Aristotle, 18, 245

Baader, Franz von, 119
Bad news of science, 3, 5, 215
Baer, Karl Ernst von, 184
Barbarian cultures, 142–143
Barth, Karl, 6, 24, 29, 43, 50, 78, 113, 260, 261, 321n38
Bauer, Bruno, 88
Baur, Ferdinand, 35–36, 45, 49, 55, 69, 71, 72, 80, 86, 88, 106, 163, 299nn52,53
Beck, Tobias, 163, 164
Berlin, Isaiah, 19, 249, 264
Besser, Max, 206
Beweis des Glaubens, 127, 129, 201, 207
Biedermann, Alois, 45, 46, 107, 109, 111, 303nn134,135

Bismarck, Otto von, 48, 56, 57, 106, 233, 330n71
Blumenberg, Hans, 12
Böhme, Jakob, 71
Bopp, Franz, 143
Boyle, Robert, 15
Brahe, Tycho, 135
Brandt, Richard, 52
Braun, Alexander, 184
Brazill, William, 82
Büchner, Ludwig, 46, 130, 137, 152, 153, 181, 182, 303n136, 317n73
Buckland, William, 135
Buffon, Georges-Louis Leclerc, 120
Bultmann, Rudolf, 6, 260, 261, 263
Burdach, Karl, 86
Burnet, Thomas, 135

Camus, Albert, 232
Carus, Gustav, 86, 131
Cavemen, 141, 175
Chance variations, 179–181
Charlotte, Queen of Württemberg, 197
Church union movement, 32–33, 56, 119
Cohen, Hermann, 296n117
Coherence theory of truth: Fries and, 20, 37, 40; Kant as founder, 20–21, 285n38, 286n40, 326n13; Kantian impure form, 21–22, 249, 286n43; Herrmann and, 21, 22, 235, 237–238, 239, 244; Strauss and, 77, 89, 108; Hegelian pure form, 77–78, 89, 108, 286n43; Ritschl and, 207; and pragmatic theory of truth, 287n46
Confessionalist movement, 33–34, 47–48, 56–57, 74, 113, 119, 127

Copernicus, Nicholas, 57, 93, 94, 192, 301n87
Correspondence theory of truth: Kant and, 20, 28; orthodox rationalists and, 26; neologists and, 26; nineteenth century theological rationalism and, 29; nineteenth century conservative theology and, 31; Schleiermacher and, 36; Luther and, 59; Ritschl and, 61, 204; Strauss and, 77–78, 89, 108, 186; Hegel and, 77; Zöckler and, 117, 119, 121, 146, 186; Schmid and, 185–186, 192; Herrmann and, 214, 216, 258, 259; Lipsius and, 222, 247; Pfleiderer and, 228–229; Solms and, 247; Kraus and, 247; twentieth century academic German theologians and, 262; and pragmatic theory of truth, 287n46; in Natorp, 327n30
Cosmology, 207, 208, 231, 250
Cotta, Bernhard von, 142
Creation doctrine: Strauss and, 85, 92–93; Zöckler and, 124, 130, 132, 148–149; Darwin and, 124; in Herbart, 132; restitution theory of, 135–136; in Herrmann, 245
Cremer, Hermann, 128
Cro-Magnon man, 176
Cuvier, Georges, 120
Czolbe, Heinrich, 130

Darwin, Charles: controversies over, 5, 63, 67; and scientific materialists, 46; Hegelian theology and, 47; German theology and, 51, 63; Alexander Schweizer and, 53; impact, 74; in Helmholtz, 103; mediating theologians and, 161. *See also* Hypotheses; Miracles; Schmid, Rudolph; Strauss, David Friedrich; Truth; Zöckler, Otto
Darwin, Erasmus, 151, 311n118
Daub, Karl, 52
Deism, 131–132, 153, 208, 323n55
Delitzsch, Franz, 114
DeMaillet, Benoît, 120
Descartes, René, 69, 99, 155
de Vries, Hugo, 183
DeWette, W. M. L., 42, 62, 212, 322n47
Diestel, Ludwig, 220, 221, 232, 324n77

Dillenberger, John, 7, 25
Dilthey, Wilhelm, 35
Donne, John, 57
Dorner, Isaak, 52, 53, 54, 56, 114, 225, 227
Draper, John, 148
DuBois-Reymond, Emil, 98, 99, 211, 319n106
Duke of Argyll, 168, 181, 185, 187, 197, 314n21
Duke of Sutherland, 169, 314n20
Dupont, Eduard, 143

Ecological crisis, 12, 264
Ehrenfeuchter, Friedrich, 53
Einstein, Albert, 264
Engels, Friedrich, 256, 257
Engis skull, 143
Erlangen program, 47
Essays and Reviews, 120, 129
Evangelical Alliance, 49
Evolution, rate of, 141–142, 306n42

Feeling and willing beings, 237, 238, 240, 327n21
Feuerbach, Ludwig, 46, 84, 86, 88, 89, 96, 103, 154, 226, 299n41
Fichte, Immanuel Hermann, 162
Fichte, Johann Gottlieb, 8, 14, 36, 70, 131, 163, 282n24
Fischer, Kuno, 107, 108
Flood, of Genesis, 135, 140
Florida skull, 309n88
Frankfurt philosophical school, 12
Frederick William II, King of Prussia, 283n24
Frederick William III, King of Prussia, 31, 33, 288n16
Frederick William IV, King of Prussia, 47, 165
French Revolution, 9, 12, 14, 15, 151, 290n32
Fries, Jakob, 20, 37, 38–42, 62, 215, 236, 286n40, 290n39, 291n40, 322n47, 326n17

Galilei, Galileo, 172
Gärtner, Carl von, 167
Gärtner, Emma, 167

Geist, Charlotte, 128
German, Christian, 131
Gesenius, Wilhelm, 30
Giessen University, 113–114, 115, 127, 128, 130, 304n8
Gladstone, William, 169
Glanville, Lord, 169
God: relation to nature, 6, 13, 14; in Schelling, 43; in Hegel, 43; in Strauss, 76, 84, 94, 95–96, 103; in Zöckler, 125, 132, 133, 153, 155, 156; in Schmid, 183, 188; personal, 209, 242, 247
Goethe, Johann Wolfgang von, 42, 100, 101, 106, 162, 192, 264
Goodman, Nelson, 17
Göttingen University, 55, 56, 114, 229
Gottschick, Johannes, 229
Grau, Rudolf, 127
Graue, Georg, 223–225, 229, 324n77
Greg, W. R., 57
Gregory of Nyssa, 206
Greifswald University, 49, 110, 112, 128
Groh, John, 54

Hacking, Ian, 17, 284n29
Haeckel, Ernst, 62, 100, 110, 151, 153, 154, 156, 157, 172, 175–176, 178, 181–182, 183, 184
Halle University, 29, 55, 114, 128, 204, 205, 211, 221, 225, 230, 321n15
Hanne, J. W., 49
Häring, Theodor, 229
Harless, Adolf, 47
Harms, Claus, 32
Harnack, Adolf, 230, 250, 253
Harris, Horton, 86
Hartmann, Eduard von, 227
Harvey, William, 155
Hasler, Ueli, 6, 11–16, 36, 51, 261, 289n32, 327n32
Heer, Oswald, 184
Hefner, Philip, 56, 62
Hegel, Georg Wilhelm Friedrich: public awareness of, 8; critique of Schleiermacher, 37, 43; rise, 38, 42; and Fries, 38–39, 42, 62; Ritschl and, 55; Baur and, 69, 74, 88; identity of philosophy and religion, 72, 73, 78, 80–81, 82, 84, 95, 299n41; on miracles, 73, 77–78; right-wing followers, 83–84; Feuerbach and, 84, 89; Reuschle on, 110; and pantheism, 131, 154; Pfleiderer and, 226. *See also* Coherence theory of truth; Correspondence theory of truth; Herrmann, Wilhelm; Miracles; Schmid, Rudolph; Strauss, David Friedrich
Hegelian theology, 42–47, 72, 80, 96
Heidegger, Martin, 331n98
Heidelberg University, 29–30, 107
Heim, Karl, 12
Helmholtz, Hermann von, 103, 110, 183, 184
Hendry, George, 6
Hengstenberg, Ernst, 30, 48, 55, 80, 113, 114, 116, 119, 130, 204, 289n20
Herbart, Johann, 132
Hermelink, Heinrich, 52
Herrmann, Johann, 204
Herrmann, Wilhelm: Hasler on, 14–15; versus Darwinism, 22; *Metaphysics in Theology*, 62, 201; and Ritschl school, 201; critique of scientific knowledge, 202–203; early career, 203–211; and Ritschl, 205–211, 212, 219, 220, 224, 232, 238, 242, 246, 321n26; and Kant, 205, 218, 234–237, 239, 245, 326nn11,17; and Lipsius, 211, 221–223, 248–250, 324n77, 327n21; and Schleiermacher, 212, 245, 322n38; Graue and, 223–225; Pfleiderer and, 225–229, 251; and pure knowing, 234–236, 325n10; pure natural knowing, 236–237; practical scientific knowing, 236, 237, 238–239, 258; on miracles, 254; on facts, 254, 258, 260; and social democracy, 256–260; on Hegel, 257; Barth and, 260, 321n38; Bultmann and, 260, 263. *See also* Coherence theory of truth; Correspondence theory of truth; Hypotheses; Objectivity; Pantheism
Hieronymous, Saint, 128
Higher criticism, 44, 48, 69, 72, 81, 82, 85, 88, 163, 190, 192, 253, 298n30, 306n54
Hilgenfeld, Adolf, 49

Hinrichs, H. F. W., 37
Hitchcock, Edward, 135
Hodge, Charles, 123, 129, 150–152, 174, 315n39
Hofmann, Johann von, 47
Hooker, Joseph, 119
Humboldt, Alexander von, 106
Humboldt, Wilhelm von, 143
Hume, David, 27
Hütten, Ulrich von, 87, 300n62
Huxley, Thomas, 169
Hypotheses: in Herrmann, 21, 237–238, 240, 244; in Strauss, 85, 99; in Darwin, 153; in Schmid, 174, 176, 191–192, 311n121, 312n129

Incommensurability of science and religion, 222, 223, 243
Incompleteness of scientific knowledge, 216, 235–236, 240, 244, 326nn13,14
Iversen, Johannes, 261

Jäger, Gustav, 188
James, William, 285n37, 323n72
Java man, 176
Jena University, 39, 62, 220, 230
Jensen, Ole, 263, 264
Jerusalem, J. F. W., 27
Jesus: resurrection, 30, 76, 79, 84, 88, 94, 95–96, 103, 193; in Hegelian theology, 44, 72; and Straussian "multitude," 92; birth, 193; in Herrmann, 215, 218, 229, 245, 252, 254, 255; in Pfleiderer, 229, 325n90. *See also* Schleiermacher, Friedrich; Strauss, David Friedrich
Jordan, Hermann, 304n8

Kaftan, Julius, 229, 253, 322n40
Kähler, Martin, 48
Kant, Immanuel: public awareness of, 8, 43; on separation of science and religion, 27–28, 35, 51, 208; on moral foundation of religion, 27–28, 35, 36, 245; Fries and, 39, 40, 41; Ritschl and, 58, 59, 61, 62, 208, 210, 251; and nebular hypothesis, 134; and anti-realism, 249; and historical fact, 252; and denatured theology, 283n24; on intuition, 291n39. *See also* Coherence theory of truth; Correspondence theory of truth; Herrmann, Wilhelm; Schmid, Rudolph; Strauss, David Friedrich
Karl, King of Württemberg, 169, 170, 197
Kattenbusch, Ferdinand, 230
Kerner, Justus, 71
Kierkegaard, Søren, 8
Kingdom of God, 58, 60, 61, 209, 211, 215, 219, 232, 245
Kliefoth, Theodor, 113
Knowing. *See* Herrmann, Wilhelm
Knowledge, societies of, 16, 17
Kohak, Erazim, 331n98
Kohut, Adolph, 67
Kölliker, Albert von, 184
Köstlin, Julius, 167–168, 169, 314n17
Köstlin, Karl, 49, 189, 314n17
Köstlin, Thusnelde, 169
Kraft, Julius, 296n118
Krauss, Alfred, 247–248
Kreibig, Gustav, 250
Kuhn, Thomas, 242, 251

Lakatos, Imre, 17
Lamarck, Jean-Baptiste, 98, 100, 101, 120, 151, 174, 177, 311n118
Lang, Heinrich, 109, 188
Lange, Friedrich, 8, 222, 226, 248
Lange, J. P., 128
Laplace, Pierre-Simon, 98, 134
Leibniz, Gottfried, 69, 135
Lessing, Gotthold, 27, 252
Leuckardt, Rudolph, 119
Liebmann, Otto, 8
Link, Christian, 12
Lipsius, Richard, 211, 220–223, 224–225, 229, 233, 246–247, 248–250, 255, 323n65, 324nn73,77, 327n21
Livingstone, David, 169
Locher, Marie, 302n123
Locke, John, 13, 14
Loofs, Friedrich, 230
Lotze, Hermann, 20, 61, 208, 237, 238, 250, 327n24, 329n64
Lovejoy, Arthur, 10, 43
Löwith, Karl, 46

Lukács, Georg, 12–13
Luthardt, Christoph, 252
Luther, Martin, 58–59, 87, 115, 232
Lutterbeck, Anton, 113–114
Lyell, Charles, 123, 124, 134, 136, 142, 145, 169

Macaulay, Thomas, 169
Mach, Ernst, 234
Marburg University, 231, 246, 247, 260, 328n47
Marheineke, Konrad, 83
Märklin, Christian, 73, 84, 167
Martensen, H. L., 54
Marx, Karl, 88, 89, 256, 257
Matthew, Patrick, 315n39
Mayer, Jules Robert, 170, 191, 192, 318n101
Mechanical explanation, 7, 182, 213, 241, 281n6
Melanchthon, Philipp, 59, 219
Metaphysics: as compulsion to know, 213–214; in relation to natural science, 214, 240, 315n38; separation from theology and religion, 215, 217, 227, 255; dependence on concepts, 216, 217, 240, 245; historical relation to theology, 232, 233; as completion of scientific knowledge, 240–241; as practical solution, 243, 258, 259; in Solms, 246, 328n49
Metaphysics in Theology, 62, 201, 211–219, 221, 227, 229, 323n65
Michelet, Carl, 73
Mimicry, 177
Miracles: in Enlightenment rationalistic philosophy, 25; in orthodox rationalism, 25; in transitional theology, 26; in nineteenth century theological rationalism, 29, 30, 77; in Strauss's *Life of Jesus*, 75, 78; in the second *Life of Jesus*, 88; in *The Old Faith and the New*, 98, 100; in Darwin, 98; antirealism as, 241. *See also* Hegel, Georg Wilhelm Friedrich; Herrmann, Wilhelm; Schleiermacher, Friedrich; Schmid, Rudolph
Mississippi skull, 140–141
Moleschott, Jacob, 19, 46, 303n136

Moltke, Helmuth von, 106
Monism, 154, 182–184, 329n55
Monogenism, 85, 144
Moore, G. E., 286n40
Moral realm, 214, 216, 217–218, 219, 232, 245
Moses, 134, 136
Müller, Max, 143
Murchison, Roderick, 137
Myth, 76–77, 79, 81, 298n21, 299n53

Napoleon Bonaparte, 9, 32, 92
Natorp, Paul, 241, 246, 325n10, 327n30
Natural selection, 101–102, 122–123, 151, 152, 154, 156, 157, 177–181, 183, 191, 196
Natural theology. *See* Apologetics
Nature: middle-class conception of, 12–14, 15, 58; feudal conception of, 13, 14, 51; mastery of, 60, 61; rationality of, 19, 21, 96, 287n46
Naturphilosophie, 36, 70, 76
Neanderthal man, 143, 176
Nebular hypothesis, 98, 134
Nelson, Leonard, 296n118
Nenon, Thomas, 285n38
Neo-Darwinians, 178
Neo-Kantian thought, 8, 15, 42, 113, 185, 201, 221, 226, 227, 228, 239, 241, 245, 248, 295n117, 303n131, 328n47
Neo-Lamarckians, 178
Neologists, on revelation and interpretation, 25, 26–27
New Lutherans, 47, 56
Newton, Isaac, 118, 155, 311n121
Nietzsche, Friedrich, 8, 88, 89–91, 105, 108
Nitzsch, Karl, 52, 55, 114

Objectivity: in Strauss, 108; in Zöckler, 121, 124, 125, 127, 134, 146; in Lipsius, 224, 249, 324n73; in Pfleiderer, 226, 228; Herrmann and, 232–233; Kantians and, 287n44
Oken, Lorenz, 120, 162
Old Lutherans, 34, 47
Olga, Queen of Württemberg, 170
Origin of Species, 46, 51, 63, 68, 74, 104, 112, 119, 122–126, 130, 185

Osiander, Andreas, 56
Ott, Günther, 321n36
Overbye, Dennis, 322n41
Owen, Richard, 169

Paley, William, 123
Palmerston, Henry John, Viscount, 169
Pannenberg, Wolfart, 12
Pantheism: Schleiermacher and, 37; Strauss and, 70, 85, 163–164; Zöckler and, 118, 123, 124, 131, 132; Herbart and, 132; Herrmann and, 323n55
Paulus, H. E. G., 30, 77
Peirce, Charles Sanders, 287n46
Person, as basis for metaphysics, 7, 248, 249, 261
Personal God, 76, 84, 94, 95–96, 103, 183, 188, 209, 242, 247
Pfleiderer, Otto, 225–229, 244, 250, 251, 253, 325n85, 328n43, 329n64
Philippi, F. A., 113
Plato, 43
Platonic viewpoint, 19, 58, 59, 207, 208, 250, 251, 264
Polygenism, 121–122, 144, 145
Popper, Karl, 17, 19, 284n30
Powell, Baden, 120
Pragmatic theory of truth, 287n46
Psychology and religion, 211–212, 222, 224, 323n72
Ptolemaic universe, 134
Ptolemy, 172, 192
Purpose, realm of, 40–41, 62, 103–104, 132, 161, 184, 185, 203, 215, 238, 242, 243
Putnam, Hilary, 19, 327n27

Quantum theory, 281n6
Quenstedt, Friedrich, 163, 285n36

Racial order, 144–145, 309n100
Rade, Martin, 204, 207, 211, 221, 253, 255, 320n15
Rapp, Ernst, 87
Realism, 19, 239, 249, 285n36
Reformation, 208, 232, 326n11
Reforms, Prussian, 9, 31–32
Reimarus, Hermann, 27
Reischle, Max, 229

Renan, Ernst, 92
Restitution theory of creation, 135–136
Resurrection, 30, 76, 79, 84, 88, 94, 95–96, 103, 193, 254, 260, 297n10
Reuschle, Carl, 100, 109–110
Revelation, in Enlightenment theology, 25, 26
Ritschl, Albrecht: as neo-Kantian theologian, 15, 54–63, 329n64; early career, 54–55; and Schleiermacher, 55, 58, 59; *Christian Doctrine*, 57, 220, 250; prophetic role, 57–58; on miracles, 61, 210; and Haeckel, 62; and Strauss, 62; and public, 201; and natural science, 202, 210; and Lipsius, 220–221, 324n77; and Pfleiderer, 329n64. *See also* Coherence theory of truth; Correspondence theory of truth; Hegel, Georg Wilhelm Friedrich; Herrmann, Wilhelm; Kant, Immanuel; Zöckler, Otto
Ritschl, Otto, 230, 325n85
Ritschl school, 22, 54, 57, 58, 62–63, 185, 187, 201, 202–203, 206, 211, 219–230, 234, 251, 253, 320n5, 321n36
Rosenkrantz, Karl, 83
Rothe, Richard, 293n90
Ruben, Peter, 14
Russell, Bertrand, 18
Russell, John, Earl, 169

Saint-Hilaire, Auguste de, 120
Sarton, George, 262
Schebst, Agnese, 87, 302nn119,123
Schelling, Friedrich, 36, 37, 38, 42, 43, 70, 71, 76, 85, 131, 154, 163, 290n39
Schellong, Dieter, 13
Schiller, J. C. Friedrich von, 162
Schleiden, Matthias, 62
Schleiermacher, Friedrich: reputation, 4; on nature, 14–15; as head of school, 30; founder of liberal theology, 34–38, 42; on miracles, 35, 72, 76, 93, 297n10; on religious knowledge, 37, 51, 52; and liberal politics, 37; inspiration of mediation theology, 50–51; Ritschl and, 55, 58, 59, 208; on resurrection, 76, 297n10; Schmid and, 163; Lipsius and, 202; as middle class,

289n32; Johann Herrmann and, 205; on Jesus, 296n10. *See also* Correspondence theory of truth; Hegel, Georg Wilhelm Friedrich; Herrmann, Wilhelm; Pantheism; Strauss, David Friedrich
Schmid, Auguste, 161
Schmid, Karl, 161
Schmid, Leopold, 114, 119
Schmid, Pauline, 167
Schmid, Rudolph: and mediation theology, 53; and Strauss, 162, 167, 182; and Hegel, 163; and Kant, 163, 185; loss of faith, 163–164; early political activities, 164–166, 167; and duke of Argyll, 168; trips to Scotland, 168–169, 197; as mediator, 169–174; on creation, 172, 189–190, 193–194; and Darwin, 173–181, 182, 183, 316n63, 317n82; and "descendance," 174–175; and development, 175–177; and natural selection, 177–181, 183; and monism, 182–184; on miracles, 188, 194; as teacher, 188–197; on resurrection, 193; twentieth century heritage of, 262–263. *See also* Correspondence theory of truth; God; Hypotheses
Schmidt, F. W., 328n47
Schnabel, Franz, 32
Schubert, Gotthilf Heinrich, 119
Schultz, Hermann, 229, 250
Schultze, Victor, 112, 119, 159
Schütz, Werner, 320n15
Schweitzer, Albert, 26, 30, 75, 79
Schweizer, Alexander, 53
Science and religion, separation, 150, 187, 204, 222, 223, 224, 247
Scientific materialism, 46, 63, 89, 99, 104, 130, 243, 303n136
Scientific Revolution (seventeenth century), 10, 20
Semler, J. S., 27
Smith, Norman Kemp, 285n38
Solms, Ludwig zu, 246–247
Spencer, Herbert, 152, 157
Spener, Philipp, 251, 329n62
Spinoza, Baruch, 110, 131, 285n38
Stahl, Georg, 155
Stählin, Leonhard, 253

Steffens, Henrik, 36
Stein, Karl Freiherr vom und zu, 31
Steude, E. G., 110
Stirner, Max, 88
Stowe, Harriet Beecher, 169
Strauss, David Friedrich: reputation, 22; and *The Life of Jesus*, 44–45, 55, 73–82, 83, 84, 106, 107, 162, 255, 299n37; intent of work, 46; and *The Old Faith and the New*, 47, 67, 88–111, 112, 201, 303n131; and Alexander Schweizer, 53; and Ritschl, 62; view of theological schools, 63; and Darwin, 67, 89, 90, 95, 98, 99–100, 101–102, 110; as Hegelian, 67, 68, 86, 88, 89–90, 91, 105; and Hegel, 67, 70, 72–73, 74, 78, 80, 81–82, 83, 84–85, 91, 105, 298n30; formative years, 68–73; and Schleiermacher, 69, 71–72, 73, 76, 79, 90, 93, 95, 300n68, 301n79; and Kant, 70, 89, 90, 97–98; and myth, 76–77, 81, 298n21, 299n53; on resurrection, 76, 88; and theological rationalism, 78, 79, 93; and conservative theology, 78, 79, 84, 85, 100–101; and creation, 85, 92–93; political views, 87, 92; and second *Life of Jesus*, 87–88; Nietzsche and, 89–91; on atonement, 93; on God's existence, 95; on Lamarckian use and disuse, 101; on natural selection, 101–102; on humanity's past, 102–103, 105; on middle-class culture, 105–106, 107; Haeckel on, 110; scientific materialism and, 160, 303n136; Schmid and, 162, 167, 182; twentieth century heritage of, 262–263; and Feuerbach, 299n41. *See also* Coherence theory of truth; Correspondence theory of truth; God; Hypotheses; Kant, Immanuel; Miracles; Objectivity; Pantheism; Worldview
Stuckenberg, John, 253

Teilhard de Chardin, Pierre, 12
Tennyson, Alfred Lord, 169
Tertullian, 135
Theology: twentieth century, 5–6; feudal, 10, 36; normal and extraordinary in nineteenth century, 10–11; Marxist

Theology (*continued*)
history of, 11–16, 36; ecological, 12; de-natured, 12–16, 282n24; Enlightenment, 24–28; orthodox rationalistic, 24, 25–26, 37–38; pietistic, 24, 37–38; critical rationalistic, 25; transitional, 26; nineteenth century rationalistic, 28–30, 38, 78, 79, 93; nineteenth century conservative, 31–34, 38, 47–49, 78, 79, 84, 85, 100–101, 149–150, 215, 225, 254; liberal, 34–38, 49; neo-Kantian, 38–42, 54–63; Hegelian speculative, 42–47, 215, 225, 253; mediation, 50–54, 55, 160–161, 211, 225, 253. *See also* Apologetics
Tholuck, F. A., 205, 206, 320n15, 322n47
Thomasius, Gottfried, 47
Tillich, Paul, 6, 261
Toews, John, 83, 290n3
Traub, Friedrich, 329n64
Truth: interest in, 17; as historiographical tool, 18; correspondence theory of, 18–20, 22, 26, 28, 29, 31, 36, 59, 61, 77–78, 89, 108, 117, 119, 121, 146, 185–186, 192; one not many, 19, 160, 186, 187, 223–224, 228–229, 262, 318n88, 328n9; coherence theory of, 20–21, 27, 37, 40; impure coherence theory of, 21–22; pure coherence theory of, 77–78, 89, 108; revealed, 134; priority for Zöckler, 151, 158–159; Darwin's love of, 181; as utility, 188; in Pfleiderer, 228–229; absolute, 242; pragmatic theory of, 287n46
Tübingen school, 45, 49, 69, 299n52
Tübingen University, 49, 55, 69, 71, 163, 164, 197, 229, 247, 298n21, 325n85
Turner, James, 4
Tuttle, Hudson, 122, 152
Twesten, August, 51, 52
Tyndall, John, 154

Ullmann, Karl, 52
Ulrici, Hermann, 205, 307n68
Unger, Franz, 13

Vaihinger, Hans, 234
Value, role of, 208, 215, 216, 238, 241, 242

Vatke, Wilhelm, 109, 312n133
Vestiges of the Natural History of Creation, 119, 120, 129, 130
Victoria, Queen of England, 168, 169
Vienna Circle, 262
Vilmar, August, 113
Virchow, Rudolf, 99
Vischer, Friedrich, 106, 107, 109, 184, 302n128
Vogt, Karl, 46, 99, 130, 137, 144, 153, 303n136, 306n45
Volta, Alessandro, 304n4
Voltaire, F. M. Arouet de, 92, 106
Voss, Johann, 30

Wagner, Andreas, 135
Wagner, Moritz, 102, 105, 110
Wagner, Rudolph, 119, 144, 306n45
Walker, Ralph, 21, 235, 249, 284n29, 285n38, 287n45, 326n13
Wallace, Alfred, 119, 178, 185
Walsh, W. H., 287n44
Warfare (conflict) historiography, 22, 149, 150, 202
Wegscheider, Julius, 29
Weismann, August, 151, 178
Weisse, C. H., 307n68
Welch, Claude, 25, 29, 42, 48, 50, 58, 328n47
Wellhausen, Julius, 193, 194, 253, 312n133
Wells, William, 315n39
Wendt, Hans, 230
Wettstein, Richard von, 178
White, Andrew, 149, 150
White, Lynn, 12
Wilberforce, Samuel, 169
William, King of Württemberg, 197
William I, King of Prussia, 48, 314n16
Windelband, Wilhelm, 222
Wissenschaft: ideology, 9; vs. belief, 55, 85, 146
Wolfenbüttel Fragments, 27
Wolff, Christian, 28
Worldview: feudal and middle class, 13–14; nineteenth century, 22; naturalistic, 59, 62, 103, 104, 202, 204, 210, 218, 223, 224, 226, 258; religious, 60, 61, 191, 208–209, 210, 213, 216, 255;

of Strauss, 90–91, 94, 97–106, 108, 109, 112, 204, 242; Copernican, 172; teleological, 184; Greek, 209
Wundt, Wilhelm, 222

Young Hegelians, 30, 45–46, 55, 67, 82, 84, 90, 163

Zeller, Eduard, 68, 107, 108–109, 325n85
Ziegler, Theobald, 109, 110, 303nn131,136
Zöckler, Konrad, 113
Zöckler, Otto: reputation, 22; representative of conservative theology, 49; early career, 113–119; relations with Giessen faculty, 114–115, 128; early study of natural science, 113, 115; on theology of nature, 116–119; and Darwin, 119, 120, 122–126, 129, 130, 137, 145, 149, 151–159, 160, 178, 312n133; review of Agassiz's *Essay on Classification*, 120–122; on natural selection, 122–123, 151, 152, 154, 156; and doctrine of creation, 124, 130, 132–133, 148–149, 156; on personal God, 125, 132, 133, 155, 156; and confessionalism, 127; and deism, 131–132, 153; and order of creation, 133–140; concordance hypothesis of, 136–140; and history of natural science and religion, 147–152; and Hodge, 150–152; priority of truth, 151, 158–159; and Ritschl, 204, 207; twentieth century heritage of, 262–263; and Hegelianism, 307n62, 312n133. *See also* Age; Correspondence theory of truth; God; Objectivity; Pantheism
Zöckler, Theodor, 115

BL 245 .G74 1992

Gregory, Frederick, 1942-

Nature lost?